Edoardo Vesentini

Introduction to continuous semigroups

APPUNTI

SCUOLA NORMALE SUPERIORE
2002

ISBN: 978-88-7642-258-4

Edoardo Vesentini
Accademia Nazionale dei Lincei
Via della Lungara, 10
00165 ROMA, Italy
vesentini@lincei.it

Introduction to continuous semigroups
Seconda edizione (2002)

Preface

During the academic year 1994/95 I gave a series of lectures on semi-groups at the Scuola Normale Superiore and at the Politecnico di Torino. The purpose of the lectures was to offer to an audience of graduate students a self-contained introduction to the theory of strongly continuous semigroups of linear operators acting on a Banach space. The main thrust was to present the basic geometrical aspects of the theory as background for applications to concrete problems in the analysis of differential operators. This book is a collection of the lecture notes or, rather, the preparatory notes of the lectures: a sort of a log-book that the author kept during the course. This may help to explain, at least in part, the sloppy style, the repetitions and the generally unorganized character of the notes, that might be considered, with some optimism, the first draft of a book.

Pisa, April 1996   Edoardo Vesentini

Here is an improved and slightly enlarged version of the 1996 lecture notes.

Pisa, March 2002   E. V.

Contents

Chapter 1

Preliminaries

Throughout this book the reader will be assumed to be familiar with the basic geometry of Banach spaces (Hahn-Banach theorem and some of its consequences[1]), and with the main consequences of the Baire theorem (Banach-Steinhaus and open mapping theorems) as they may be found, for example, in [69] and in [81]. He should have also a rudimentary knowledge of the theory of holomorphic functions of one complex variable and of the differential and integral calculus for continuous functions of one real variable, with values in a normed space.

1.1 Closed and bounded linear operators

Let $\mathcal{E}$ and $\mathcal{F}$ be Banach spaces on $\mathbf{C}$ (or on $\mathbf{R}$), and let $A : \mathcal{D}(A) \subset \mathcal{E} \to \mathcal{F}$ be a linear map of a linear subspace $\mathcal{D}(A)$ of $\mathcal{E}$ into $\mathcal{F}$; $\mathcal{D}(A)$ and $A(\mathcal{D}(A)) \subset \mathcal{F}$ are called the *domain* and the *range* of A, the latter linear space being denoted also by $\mathcal{R}(A)$.

Let $\Gamma(A) \subset \mathcal{E} \oplus \mathcal{F}$ be the graph of A,

$$\Gamma(A) = \{(x, Ax) : x \in \mathcal{D}(A)\}.$$

Lemma 1.1.1 *A linear subspace Γ of $\mathcal{E} \oplus \mathcal{F}$ is the graph of a linear map if, and only if, $(0, y) \in \Gamma$ implies $y = 0$.*

---

[1]As, for example, S.Mazur's theorem and its corollary, whereby a closed linear subspace of a Banach space is closed for the weak topology

Proof Let $\mathcal{D}$ be the linear submanifold of $\mathcal{E}$ consisting of those $x \in \mathcal{E}$ for which there is some $y \in \mathcal{F}$ such that $(x, y) \in \Gamma$.

If $(0, y) \in \Gamma$ implies $y = 0$, for every $x \in \mathcal{D}$ there exists one, and only one, $y \in \mathcal{F}$ such that $(x, y) \in \Gamma$. The linear map $x \mapsto y$ has Γ as a graph.

The rest of the lemma is obvious. ∎

Any one of the equivalent norms

$$\|(x, y)\| = \max\{\|x\|, \|y\|\},$$

or

$$\|(x, y)\| = (\|x\|^p + \|y\|^p)^{1/p} \text{ for } 1 \leq p < \infty,$$

introduces in $\mathcal{E} \oplus \mathcal{F}$ the structure of a Banach space.

If $\mathcal{E}$ and $\mathcal{F}$ are Hilbert spaces, the norm

$$\|(x, y)\| = (\|x\|^2 + \|y\|^2)^{1/2}$$

defines on $\mathcal{E} \oplus \mathcal{F}$ a Hilbert space structure.

If $\Gamma(A)$ is closed in $\mathcal{E} \oplus \mathcal{F}$, A is called a *closed operator*.

If A is closed, $\ker A := \{x \in \mathcal{D}(A) : Ax = 0\}$ is closed.

Lemma 1.1.2 *The linear operator A is closed if, and only if, for every sequence $\{x_\nu\}$ in $\mathcal{D}(A)$, converging to some $x \in \mathcal{E}$ and such that $\{Ax_\nu\}$ converges to some $y \in \mathcal{F}$, then $x \in \mathcal{D}(A)$ and $y = Ax$.*

Proof The sequence $\{(x_\nu, Ax_\nu)\}$ converges to (x, y). If A is closed, then $(x, y) \in \Gamma(A)$, and therefore $x \in \mathcal{D}(A)$ and $y = Ax$.

Viceversa, let $\{(x_\nu, y_\nu)\}$ be a sequence in $\Gamma(A)$ converging to some point (x, y) in the closure $\overline{\Gamma(A)}$ of $\Gamma(A)$. Then $y_\nu = Ax_\nu$, and, by the hypothesis, $x \in \mathcal{D}(A)$, $y = Ax$. Hence $(x, y) \in \Gamma(A)$. ∎

Theorem 1.1.3 *(Closed graph theorem) If $\mathcal{D}(A) = \mathcal{E}$, A is continuous if, and only if, A is closed.*

Proof Since $\mathcal{D}(A) = \mathcal{E}$, the linear map $\widetilde{A} : (x, Ax) \mapsto x$ of $\Gamma(A)$ into $\mathcal{E}$ is (surjective, and therefore) bijective and continuous for the norm on $\Gamma(A)$

$$\|(x, y)\| = \|x\| + \|y\|. \tag{1.1}$$

If $\Gamma(A)$ is closed (hence complete), Banach's open mapping theorem implies that $\tilde{A}^{-1}$ is continuous. Thus, there is a positive constant c such that

$$\|x\| + \|Ax\| \leq c\|x\|$$

for all $x \in \mathcal{E}$.

The rest of the proof is obvious. ∎

Let $\mathcal{H}$ be a complex Hilbert space with inner product $(\,|\,)$.

Theorem 1.1.4 *(Theorem of Hellinger and Toeplitz) If $A : \mathcal{H} \to \mathcal{H}$ is a linear map such that $(Ax|y) = (x|Ay)$ for all x and y in $\mathcal{H}$, A is continuous.*

Proof By Theorem 1.1.3 we have only to show that A is closed. Let $\{x_\nu\}$ be a sequence of points $x_\nu \in \mathcal{H}$ converging to some $x \in \mathcal{H}$ and such that the sequence $\{Ax_\nu\}$ converges to some $y \in \mathcal{H}$. For every $z \in \mathcal{H}$,

$$(y|z) = \lim (Ax_\nu|z) = \lim (x_\nu|Az) = (x|Az) = (Ax|z),$$

i.e., $y = Ax$. ∎

If the linear operator $B : \mathcal{D}(B) \subset \mathcal{E} \to \mathcal{F}$ is such that $\mathcal{D}(A) \subset \mathcal{D}(B)$ and $B_{|\mathcal{D}(A)} = A$, i.e., $\Gamma(A) \subset \Gamma(B)$, B is said to be an *extension* of A; in symbols: $A \subset B$.

If $A \subset B$, every linear space Γ such that $\Gamma(A) \subset \Gamma \subset \Gamma(B)$ is the graph of a linear operator which extends A and is extended by B.

If the closure $\overline{\Gamma(A)}$ of $\Gamma(A)$ is a graph, A is said to be *closable*. The operator with graph $\overline{\Gamma(A)}$ is a closed extension of A, that will be denoted by $\overline{A}$, and which is the smallest closed extension of A, in the sense that it is extended by every closed extension of A. Note that, if $A \subset B$ and if B is closed, the inclusion $\Gamma(A) \subset \Gamma(B)$ implies that $\Gamma(A) \subset \overline{\Gamma(A)} \subset \Gamma(B)$, i.e., A is closable.

Lemma 1.1.5 *The linear operator A is closable if, and only if, for every sequence $\{x_\nu\}$ in $\mathcal{D}(A)$ converging to 0 and such that the sequence $\{Ax_\nu\}$ converges to $y \in \mathcal{F}$, then $y = 0$.*

Proof By Lemma 1.1.1, $\overline{\Gamma(A)}$ is a graph if, and only if, $(0,y) \in \overline{\Gamma(A)}$ implies that $y = 0$. The conclusion follows from the fact that $(0,y) \in \overline{\Gamma(A)}$ if, and only if, there is a sequence $\{x_\nu\}$ in $\mathcal{D}(A)$ such that $\{x_\nu\}$ converges to 0 and $\{Ax_\nu\}$ converges to y. ■

Here is a description of the closure of A.

Let A be closable, and let $\{x_\nu\}$ and $\{x'_\nu\}$ be two sequences of points of $\mathcal{D}(A)$ converging to the same point $x \in \mathcal{E}$ and such that $\{Ax_\nu\}$ and $\{Ax'_\nu\}$ converge respectively to y and y' in $\mathcal{F}$. Because $\{x_\nu - x'_\nu\}$ converges to 0 and A is closable, then $y = y'$. Let B be the operator mapping x to y; B is well defined (in the sense that it does not depend on the sequence $\{x_\nu\}$ converging to x), and is linear. Furthermore, it is closed. Indeed, if $\{z_\nu\}$ is a sequence in $\mathcal{D}(B)$ converging to $z \in \mathcal{E}$ and such that $\{Bz_\nu\}$ converges to some point $w \in \mathcal{F}$, there is a sequence $\{x_\nu\}$ in $\mathcal{D}(A)$ with $\|x_\nu - z_\nu\| < 1/\nu$ and $\|Ax_\nu - Bz_\nu\| < 1/\nu$ for $\nu = 1, 2, \ldots$. Then, $\lim_{\nu \to \infty} x_\nu = z$ and $\lim_{\nu \to \infty} Ax_\nu = w$, and thus $z \in \mathcal{D}(B)$ and $Bz = w$.

If a closed linear operator C extends A, then it extends B. Indeed, let $\{x_\nu\}$ be a sequence in $\mathcal{D}(A)$ converging to $x \in \mathcal{E}$ and such that $\{Ax_\nu\}$ converges to $y \in \mathcal{F}$. Since C extends A and is closed, then $x \in \mathcal{D}(C)$ and $Cx = y$. On the other hand, in view of the definition of B, $x \in \mathcal{D}(B)$ and $Bx = y$. Hence $B \subset C$. That implies that B is the closure of A.

Lemma 1.1.6 *If A is continuous in $\mathcal{D}(A)$, A is closable and $\overline{A}$ is continuous in $\overline{\mathcal{D}(A)}$.*

Proof. Since A is uniformly continuous in $\mathcal{D}(A)$, A has a unique continuous extension on $\overline{\mathcal{D}(A)}$, which coincides with $\overline{A}$. ■

If A is closable, $\ker \overline{A}$ is closed. Thus, if A is a closable linear form, $\overline{A}$ is continuous on $\mathcal{D}(\overline{A})$. A *fortiori* A is continuous on $\mathcal{D}(A)$. That proves

Lemma 1.1.7 *If A is closable and if $\mathcal{R}(A)$ has finite dimension, A is continuous on $\mathcal{D}(A)$.*

Set in $\mathcal{E} \oplus \mathcal{F}$ the norm (1.1) and on $\mathcal{D}(A)$ the *graph norm*

$$\|x\| = \|x\| + \|Ax\|.$$

Letting $j(x) = (x, Ax)$, j is a linear map of $\mathcal{D}(A)$ onto $\Gamma(A)$ which is bijective and such that $\|j(x)\| = \|x\|$ for all $x \in \mathcal{D}(A)$. In other words A is a linear isometry of $\mathcal{D}(A)$, endowed with the norm $\|\ \|$, onto $\Gamma(A)$, endowed with the norm $\|\ \|$.

If A is closed, $\Gamma(A)$ is (closed in $\mathcal{E} \oplus \mathcal{F}$, and therefore is) complete, proving thereby

Lemma 1.1.8 *If, and only if, A is closed, $\mathcal{D}(A)$ is complete for the graph norm.*

If A is continuous on $\mathcal{D}(A)$, the norms $\|\ \|$ and $\|\|\ \|\|$ are equivalent. Thus we have

Lemma 1.1.9 *If A is continuous in $\mathcal{D}(A)$, A is closed if, and only if, $\mathcal{D}(A)$ is closed in $\mathcal{E}$.*

The following proposition is a direct consequence of the fact that a closed linear subspace of a Banach space is closed for the weak topology, [81].

Proposition 1.1.10 *If A is closed, A is weakly closed.*

Let $\tau : \mathcal{E} \oplus \mathcal{F} \to \mathcal{F} \oplus \mathcal{E}$ be the surjective isometric isomorphism defined by $\tau : (x, y) \mapsto (y, x)$. If $\tau(\Gamma(A))$ is a graph, A is said to be *invertible*. The linear operator associated to $\tau(\Gamma(A))$ - which has domain $\mathcal{R}(A)$ and range $\mathcal{D}(A)$ - is denoted by A^{-1}. Clearly

$$A^{-1}A(x) = x \text{ for all } x \in \mathcal{D}(A),$$

and

$$A A^{-1}(y) = y \text{ for all } y \in \mathcal{R}(A).$$

The linear operator A is invertible if, and only if, it is injective; the inverse of A, if exists, is unique.

Since $\tau(\Gamma(A))$ is closed if, and only if, $\Gamma(A)$ is closed, then

Lemma 1.1.11 *Let A be invertible. The operator A is closed if, and only if, A^{-1} is closed.*

Let $\mathcal{D}(A)$ be dense in $\mathcal{E}$, and let A be continuous on $\mathcal{D}(A)$. By Lemma 1.1.6, A is closable and $\overline{A}$ is continuous on $\overline{\mathcal{D}(A)} = \mathcal{E}$. Since $\overline{A}$ is the minimal closed extension of A, and - being defined on the entire space $\mathcal{E}$ - cannot be further extended, then $\overline{A}$ is the unique closed extension of A. That proves

Lemma 1.1.12 *If $\mathcal{D}(A)$ is dense in $\mathcal{E}$, and if A is continuous on $\mathcal{D}(A)$, A is closable. The closure has domain $\mathcal{E}$, is continuous and is the unique closed extension of A.*

If A is closed and there exists a linear operator B for which $\overline{B} = A$, the domain $\mathcal{D}(B)$ of B is called a *core* of A.

Let $\mathcal{G}$ be a linear subspace of $\mathcal{D}(A)$. The proof of the following lemma is left as an exercise.

Lemma 1.1.13 *The linear space $\mathcal{G}$ is a core of A if, and only if, $\mathcal{G}$ is dense in $\mathcal{D}(A)$ for the graph norm.*

In other words, $\mathcal{G}$ is a core of A if, and only if, for every $x \in \mathcal{D}(A)$, there is a sequence $\{x_\nu\}$ in $\mathcal{G}$ such that $\lim_{\nu\to\infty} x_\nu = x$, and $\lim_{\nu\to\infty} Ax_\nu = Ax$.

In the following $\mathcal{L}(\mathcal{E}, \mathcal{F})$ will stand for the Banach space of all linear continuous operators $\mathcal{E} \to \mathcal{F}$. When $\mathcal{E} = \mathcal{F}$, the notation $\mathcal{L}(\mathcal{E}, \mathcal{F})$ will be usually replaced by $\mathcal{L}(\mathcal{E})$.

Let $\Xi(\mathcal{E}, \mathcal{F})$ or $\Xi(\mathcal{E})$ be the sets of all closed linear operators mapping linear subspaces of $\mathcal{E}$ into $\mathcal{F}$ or into $\mathcal{E}$.

Exercises 1. If the closed linear operator A is continuous on $\mathcal{D}(A)$, every dense linear subspace of $\mathcal{D}(A)$ is a core of A.

2. If the linear operator $A : \mathcal{D}(A) \subset \mathcal{E} \to \mathcal{F}$ is invertible and such that $A^{-1} \in \mathcal{L}(\mathcal{F}, \mathcal{E})$, a linear subspace $\mathcal{G} \subset \mathcal{D}(\mathcal{A})$ is a core of A if, and only if, $A(\mathcal{G})$ is dense in $\mathcal{F}$.

3. Let $\mathcal{E}$, $\mathcal{F}$, $\mathcal{G}$ be complex Banach spaces. If $A \in \Xi(\mathcal{E}, \mathcal{F})$ and if $B \in \Xi(\mathcal{F}, \mathcal{G})$ is invertible, with $B^{-1} \in \mathcal{L}(\mathcal{G}, \mathcal{F})$, then $BA \in \Xi(\mathcal{E}, \mathcal{G})$.

Here are some examples of linear, unbounded, closed operators.

Let $C[0, 1]$ be the Banach space of all complex-valued continuous functions on the closed interval $[0, 1] = \{t \in \mathbf{R} : 0 \leq t \leq 1\}$, endowed with the norm $\|x\| = \max\{|x(t)| : t \in [0, 1]\}$. Let $C^1[0, 1]$ be the dense linear suspace of $C[0, 1]$ consisting of all $x \in C[0, 1]$ of class C^1 on $[0, 1]$. Let X be the linear operator defined by

$$Xx = \dot{x}, \tag{1.2}$$

where $\dot{x}$ stands for the derivative of x with respect to t.

The domain $\mathcal{D}(X)$ of X must be contained in $C^1[0,1]$. If $\mathcal{D}(X) = C^1[0,1]$, X is *maximal*, and the symbol X will be used here to denote this maximal operator. If $x_n \in C^1[0,1]$ is defined by $x_n(t) = t^n$ for $n = 0, 1, \ldots$, then $\dot{x}_n(t) = n\,t^{n-1}$, and $\|x_n\| = n$. Thus X is unbounded.

Lemma 1.1.14 *The operator X is closed.*

Proof Let $x_\nu \in \mathcal{D}(X)$ be such that the sequences $\{x_\nu\}$ and $\{Xx_\nu\}$ converge to x and to y. Because $\lim_{\nu \to \infty} = x$ uniformly on $[0,1]$, then $x \in C^1[0,1]$ and $\dot{x} = y$. ∎

Since $Xx = 0$ if, and only if, x is constant, the operator X, with domain $C^1[0,1]$ is not invertible.

Let

$$\mathcal{D}_1 = \{x \in C^1[0,1] : x(0) = 0\}, \tag{1.3}$$

$$\mathcal{D}_2 = \{x \in C^1[0,1] : x(1) = 0\}. \tag{1.4}$$

and, for $c \in \mathbf{C} \backslash \{0\}$, let

$$\mathcal{D}_3 = \{x \in C^1[0,1] : x(1) - c\,x(0)\}. \tag{1.5}$$

Finally, let

$$\mathcal{D}_0 = \{x \in C^1[0,1] : x(0) = x(1) = 0\}. \tag{1.6}$$

By Lemma 1.1.14 the restriction, X_α, of X to the linear subspace $\mathcal{D}_\alpha$, is closed for $\alpha = 0, 1, 2, 3$.

Exercises 4. The operator X_1 has range $\mathcal{R}(X_1) = C[0,1]$, is invertible, and $X_1^{-1} \in \mathcal{L}(C[0,1])$ is expressed by the integral

$$(X_1^{-1}(x))(t) = \int_0^t x(s)\,ds.$$

5. A similar conclusion holds for X_2.
   6. The range of X_0 is

$$\mathcal{R}(X_0) = \{x \in C[0,1] : \int_0^1 x(t)\,dt = 0\},$$

and X_0, restriction of X_1 and X_2, is invertible.

7. The operator X_3 is invertible if, and only if, $c \neq 1$. If $c \neq 1$, $\mathcal{R}(X_3) = C[0,1]$ and $X_3^{-1} \in \mathcal{L}(C[0,1])$ is given by

$$(X_3^{-1}(x))(t) = \frac{c}{c-1} \int_0^t x(s)\, ds + \int_t^1 x(s)\, ds.$$

8. Replace now the space $C[0,1]$ by the space $L^p(0,1)$ for $1 \leq p < \infty$ and for the Lebesgue measure on the open interval $(0,1) = \{t \in \mathbf{R} : 0 < t < 1\}$, and let X be still expressed by (1.2) on the domain

$$\mathcal{D}(X) = \{x \in L^p(0,1) : x \text{ absolutely continuous, and } \dot{x} \in L^p(0,1)\}.$$

Extend the preceeding exercises to the case of $L^p(0,1)$.

1.2   Lipschitz functions

Let K be a subset of $\mathcal{E}$. The set $\mathrm{Lip}(K)$ of all maps $f : K \to \mathcal{E}$ such that

$$p_L(f) := \sup \left\{ \frac{\|f(x) - f(y)\|}{\|x - y\|} : x, y \in K, x \neq y \right\} < \infty$$

is a vector space: the space of all *Lipschitz maps* $K \to \mathcal{E}$; p_L is a seminorm on $\mathrm{Lip}(K)$. Chosen any $x_o \in K$, the map

$$f \to \|f(x_o)\| + p_L(f)$$

is a norm on $\mathrm{Lip}(K)$.

Any $f \in \mathrm{Lip}(K)$ is uniformly continuous on K.

Exercises If $A \in \mathcal{L}(\mathcal{E})$, then $A \in \mathrm{Lip}(\mathcal{E})$ and

$$p_L(A) = \|A\| = \sup \left\{ \frac{\|Ax\|}{\|x\|} : x \in \mathcal{E} \backslash \{0\} \right\}. \tag{1.7}$$

If $f \in \mathrm{Lip}(K)$ and $g \in \mathrm{Lip}(f(K))$, then $g \circ f \in \mathrm{Lip}(K)$ and

$$p_L(g \circ f) \leq p_L(f)\, p_L(g). \tag{1.8}$$

Let $\{f_\nu\}$ be a sequence of functions $f_\nu \in \mathrm{Lip}(K)$ such that

$$p_L(f_\nu) \leq M \qquad (1.9)$$

for some $M > 0$ and all ν, and furthermore

$$f(x) := \lim_{\nu \to \infty} f_\nu(x)$$

exists at all $x \in K$.

Let $x \neq y$ in K, and, for any $\epsilon > 0$, let ν be such that

$$\|f(x) - f_\nu(x)\| < \frac{\epsilon}{2}, \quad \|f(y) - f_\nu(y)\| < \frac{\epsilon}{2}.$$

Then,

$$\frac{\|f(x) - f(y)\|}{\|x - y\|} \leq \frac{\|(f(x) - f_\nu(x)) - (f(y) - f_\nu(y))\|}{\|x - y\|} +$$
$$\frac{\|f_\nu(x) - f_\nu(y)\|}{\|x - y\|} < \epsilon + M,$$

proving thereby

Lemma 1.2.1 *If $\{f_\nu\}$ is pointwise convergent on K and (1.9) holds for some $M > 0$ and all ν, then $f \in \mathrm{Lip}(K)$ and $p_L(f) \leq M$.*

Let $K \subset \mathcal{E}$ and let $f : K \to K$. If $f \in \mathrm{Lip}(K)$ and $p_L(f) \leq 1$, f is called a *contraction* of K; a *strict contraction* of K if $p_L(f) < 1$.

Let $n \in \mathbf{N}$ and let $f^n = f \circ \cdots \circ f$ (n times).

The classical Banach fixed point theorem for strict contractions of a complete metric space can be rephrased in terms of Lipschitz conditions, and yields the following statement.

Theorem 1.2.2 *If K is closed in $\mathcal{E}$ and $f : K \to K$ is such that f^n is a strict contraction for some $n \in \mathbf{N}$, there exists a unique $x \in K$ such that $f(x) = x$. For every $z \in K$*

$$x = \lim_{m \to \infty} f^m(z).$$

For $x_o \in \mathcal{E}$ and $r > 0$, let $B(x_o, r)$ be the open ball with center x_o and radius r,

$$B(x_o, r) = \{x \in \mathcal{E} : \|x - x_o\| < r\},$$

and let $\overline{B(x_o, r)}$ be its closure.

Example Let $\mathcal{H}$ be a complex Hilbert space, and let $K \subset \mathcal{H}$ be non-empty, closed and convex. For any $x \in \mathcal{H}$, the set $K(x) := \{x - z : z \in K\}$ is closed and convex. Hence there is a unique point $w \in K(x)$ for which

$$\|w\| = \inf\{\|u\| : u \in K(x)\}.$$

Letting $w = x - y$, y is the closest point of K to x for the distance defined by the norm in $\mathcal{H}$; more exactly, y is the unique point of K such that

$$\|x - y\| = \inf\{\|x - z\| : z \in K\}.$$

Thus

$$\{y\} = \overline{B(x, \|x - y\|)} \cap K. \tag{1.10}$$

The map $\Pi_K : \mathcal{H} \to \mathcal{H}$ defined by $\Pi_K(x) = y$ is called the *projection* or the *projector* of $\mathcal{H}$ onto K.

Since, for $x \in K$, $\Pi_K(x) = x$, then Π_K is an idempotent:

$$\Pi_K \circ \Pi_K = \Pi_K.$$

Furthermore,

$$K = \{x \in \mathcal{H} : \Pi_K(x) = x\}.$$

If K is a closed linear subspace of the Hilbert space $\mathcal{H}$, Π_K is the ortogonal projector onto K (see, *e.g.*, [38]).

We will show now that Π_K is a contraction of $\mathcal{H}$.

First of all, an elementary computation yields the following identity:

$$\|x - y\|^2 - \|x - z\|^2 - \|z - y\|^2 = 2\Re\,(x - z \,|\, z - y) \quad \forall\, x, y, z \in \mathcal{H},$$

which, choosing $z = \Pi_K(x)$, becomes

$$\|x - y\|^2 - \|x - \Pi_K(x)\|^2 - \|\Pi_K(x) - y\|^2 = 2\Re\,(x - \Pi_K(x) \,|\, \Pi_K(x) - y).$$

If $u \in K$ and $y = tu + (1 - t)\Pi_K(x)$, then, for $0 \le t \le 1$, $y \in K$ and, since $\Pi_K(x) - y = t\left(\Pi_K(x) - u\right)$, then

$$\|x-y\|^2 - \|x - \Pi_K(x)\|^2 - t^2 \|\Pi_K(x) - u\|^2 = 2t\Re\left(x - \Pi_K(x)|\Pi_K(x) - u\right).$$

Since

$$\|x - y\| \ge \|x - \Pi_K(x)\|, \tag{1.11}$$

then,

$$t^2 \|\Pi_K(x) - u\|^2 + 2t\Re\left(x - \Pi_K(x)|\Pi_K(x) - u\right) \ge 0 \ \forall\, t \in [0, 1],$$

and therefore

$$t \|\Pi_K(x) - u\|^2 + 2\Re\left(x - \Pi_K(x)|\Pi_K(x) - u\right) \ge 0 \ \forall\, t \in (0, 1],$$

which, letting $t \downarrow 0$, yields

$$\Re\left(x - \Pi_K(x)|\Pi_K(x) - u\right) \ge 0 \ \forall\, x \in \mathcal{H} \ \forall\, u \in K. \tag{1.12}$$

Remark In the real case, this inequality expresses the fact - which can be obtained from (1.11) and from the Carnot theorem - that the angle between the vectors $x - \Pi_K(x)$ and $u - \Pi_K(x)$ is $> \pi/2$.

Choosing $u = \Pi_K(x')$, with $x' \in \mathcal{H}$, (1.12) yields

$$\Re\left(x - \Pi_K(x)|\Pi_K(x) - \Pi_K(x')\right) \ge 0$$

and

$$\Re\left(x' - \Pi_K(x')|\Pi_K(x) - \Pi_K(x')\right) \le 0,$$

whence

$$\Re\left((x - \Pi_K(x)) - (x' - \Pi_K(x'))\,|\Pi_K(x) - \Pi_K(x')\right) \ge 0.$$

Thus,

$$\Re\left(x - x'|\Pi_K(x) - \Pi_K(x')\right) \ge \|\Pi_K(x) - \Pi_K(x')\|^2. \tag{1.13}$$

The Schwarz inequality yields then

$$\|\Pi_K(x) - \Pi_K(x')\| \le \|x - x'\| \ \forall\, x, x' \in \mathcal{H},$$

showing that Π_K is a contraction.

Since

$$\|(x - \Pi_K(x)) - (x' - \Pi_K(x'))\|^2 = \tag{1.14}$$
$$\Re\left((x - \Pi_K(x)) - (x' - \Pi_K(x')) \mid (x - \Pi_K(x)) - (x' - \Pi_K(x'))\right)$$
$$= -\Re\left((x - \Pi_K(x)) - (x' - \Pi_K(x')) \mid \Pi_K(x) - \Pi_K(x')\right) +$$
$$\Re\left((x - \Pi_K(x)) - (x' - \Pi_K(x')) \mid x - x'\right)$$
$$\leq \Re\left((x - \Pi_K(x)) - (x' - \Pi_K(x')) \mid x - x'\right),$$

the Schwarz inequality yields

$$\|(x - \Pi_K(x)) - (x' - \Pi_K(x'))\| \leq \|x - x'\|, \tag{1.15}$$

showing that $I - \Pi_K$ is a contraction (where I is the identity map).

In conclusion, the following lemma holds.

Lemma 1.2.3 *The maps Π_K and $I - \Pi_K$ are contractions of $\mathcal{H}$.*

Let $\mathcal{D}(A)$ be a subset of $\mathcal{H}$, and let $A : \mathcal{D}(A) \to \mathcal{H}$ be a map such that $A(\mathcal{D}(A)) \subset K$ and

$$\Re(x - A(x) \mid u - A(x)) \leq 0 \ \ \forall\, x \in \mathcal{D}(A) \ \forall\, u \in K. \tag{1.16}$$

This latter condition, coupled with (1.12), yields

$$0 \geq \Re\left(x - A(x) \mid \Pi_K(x) - A(x)\right)$$
$$= \Re\left(x - \Pi_K(x) + \Pi_K(x) - A(x) \mid \Pi_K(x) - A(x)\right)$$
$$= \Re\left(x - \Pi_K(x) \mid \Pi_K(x) - A(x)\right) + \|\Pi_K(x) - A(x)\|^2$$
$$\geq \|\Pi_K(x) - A(x)\|^2,$$

whence $A(x) = \Pi_K(x)$, proving thereby

Lemma 1.2.4 *Let K be a closed, convex subset of $\mathcal{H}$. A map $A : \mathcal{D}(A) \to \mathcal{H}$ such that $A(\mathcal{D}(A)) \subset K$ coincides with Π_K on $\mathcal{D}(A)$ if, and only if, (1.16) holds.*

Let now $A : \mathcal{H} \to \mathcal{H}$ be a map such that

$$\Re(x - A(x) \mid A(y) - A(x)) \leq 0 \ \ \forall\, x, y \in \mathcal{H}. \tag{1.17}$$

If $n \geq 1, t_1, \ldots, t_n \in \mathbf{R}_+$ with $\sum_{j=1}^{n} t_j = 1$, and if $u = \sum_{j=1}^{n} t_j A(y_j)$ for $y_1, \ldots, y_n \in \mathcal{H}$, then

$$\Re(x - A(x)|u - A(x)) = \Re(x - A(x)| \sum_{j=1}^{n} t_j A(y_j) - A(x))$$

$$= \sum_{j=1}^{n} t_j \, \Re(x - A(x)|A(y_j) - A(x)) \leq 0.$$

By continuity, $\Re(x - A(x)|u - A(x)) \leq 0$, for all u in the closed, convex hull of $A(\mathcal{H})$. By Lemma 1.2.4, that proves

Proposition 1.2.5 *If the map $A : \mathcal{H} \to \mathcal{H}$ satisfies (1.17), then $A = \Pi_K$, where K is the closed, convex hull of $A(\mathcal{H})$.*

For a more detailed description of projectors, see, *e.g.*, [83], where E.H. Zarantonello develops a theory of spectral projectors defined by closed, convex cones in $\mathcal{H}$.

As at the beginning of this section, let $\mathcal{E}$ be a Banach space, let K be a subset of $\mathcal{E}$, and let $f : K \to \mathcal{E}$.

Proposition 1.2.6 *If $p_L(f) = k < 1$, then:*
1) $I - f$ is injective, and

$$p_L\left((I - f)^{-1}\right) \leq \frac{1}{1 - p_L(f)}; \tag{1.18}$$

2) if $\overline{B(x_o, r)} \subset K$, then

$$(I - f)(\overline{B(x_o, r)}) \supset \overline{B(x_o - f(x_o), r(1 - k))};$$

3) if $K = \mathcal{E}$, $I - f$ is bijective.

Proof 1) For $x_1, x_2 \in K$,

$$\|x_1 - x_2\| \leq \|(x_1 - f(x_1)) - (x_2 - f(x_2))\| + \|f(x_1) - f(x_2)\|$$
$$\leq \|(x_1 - f(x_1)) - (x_2 - f(x_2))\| + k\|x_1 - x_2\|,$$

i.e.,

$$\|x_1 - x_2\| \leq \frac{1}{1 - k}\|(x_1 - f(x_1)) - (x_2 - f(x_2))\|.$$

2) Choose any $y \in \overline{B(x_o - f(x_o), r(1-k))}$, and let $g : \overline{B(x_o,r)} \to \mathcal{E}$ be defined by

$$g(x) = f(x) + y.$$

Then, $g \in \mathrm{Lip}(\overline{B(x_o,r)})$ and $p_L(g) = k$.
Since moreover

$$\|g(x) - x_o\| \le \|f(x) - f(x_o)\| + \|y - x_o + f(x_o)\|$$
$$\le kr + (1-k)r = r,$$

Theorem 1.2.2 yields the conclusion, because

$$g(\overline{B(x_o,r)}) \subset \overline{B(x_o,r)}.$$

3) Since $K = \mathcal{E}$, 1) and 2) yield 3). $\blacksquare$

Let now $K \subset \mathcal{E}$ be closed and convex, and let $f : K \to K$ be such that $f \in \mathrm{Lip}(K)$ with $p_L(f) \le 1$.

Lemma 1.2.7 *If β is a positive constant and $\alpha = \beta + 1$, then: $\alpha - \beta f$ is injective;*

$$(\alpha - \beta f)(K) = K; \tag{1.19}$$

and

$$p_L((\alpha - \beta f)^{-1}) \le 1.$$

Proof Choosing $y \in K$ and setting

$$g(x) = \frac{\beta}{\alpha} f(x) + \frac{1}{\alpha} y$$

for all $x \in K$, then $(\alpha - \beta g) \in \mathrm{Lip}(K)$. Since

$$\frac{\beta}{\alpha} + \frac{1}{\alpha} = 1$$

and $y \in K$, then $g(K) \subset K$. Furthermore,

$$p_L(g) = \frac{\beta}{\alpha} p_L(f) < 1. \tag{1.20}$$

Theorem 1.2.2 and (1.20) imply that, for any $y \in K$, there exists a unique point $x \in K$ such that $g(x) = x$, that is to say,

$$\alpha x - \beta f(x) = y,$$

proving thereby that (1.19) holds (and that $\alpha - \beta f$ is a bijective map of K onto K).

Since

$$p_L\left(\frac{\beta}{\alpha}f\right) \leq \frac{\beta}{\alpha},$$

and therefore

$$1 - p_L\left(\frac{\beta}{\alpha}f\right) \geq 1 - \frac{\beta}{\alpha} = \frac{1}{\alpha},$$

(1.18) yields

$$p_L\left((\alpha I - \beta f)^{-1}\right) = \frac{1}{\alpha}p_L\left(\left(I - \frac{\beta}{\alpha}f\right)^{-1}\right)$$

$$\leq \frac{1}{\alpha}\frac{1}{1 - p_L\left(\frac{\beta}{\alpha}f\right)} \leq \frac{1}{\alpha}\alpha = 1;$$

which completes the proof of the lemma. ∎

1.3   Fixed point theorems for contractions

Following [7], we will investigate fixed points of contractions of closed subsets of the Banach space $\mathcal{E}$.

Lemma 1.3.1 *Let K be a closed, convex subset of $\mathcal{E}$ and let $A : K \rightarrow K$ be a contraction. Then, for every $\alpha > 0$,*

$$(I + \alpha(I - A))(K) \supset K. \tag{1.21}$$

Furthermore, the map $(I + \alpha(I - A))^{-1}$ is a contraction $K \rightarrow K$.

Proof In order to prove that (1.21) holds, one has to show that, for every $y \in K$, there is $x \in K$ such that

$$x + \alpha(I - A))(x) = y, \tag{1.22}$$

i.e.,

$$x = \frac{1}{1+\alpha}(y + \alpha A(x)).$$

Since

$$\left\|\frac{1}{1+\alpha}(y + \alpha A(u_1)) - \frac{1}{1+\alpha}(y + \alpha A(u_2))\right\| =$$
$$= \frac{\alpha}{1+\alpha}\|A(u_1) - A(u_2)\| \le \frac{\alpha}{1+\alpha}\|u_1 - u_2\|,$$

the map

$$K \ni u \mapsto \frac{1}{1+\alpha}(y + \alpha Au)$$

is a strict contraction of K. By Theorem 1.2.2 it fixes a unique point $x \in K$. Hence there is (a unique point) $x \in K$ satisfying (1.22).

Furthermore, if x_1, x_2 are the unique solutions of the equation (1.22) for $y = y_1$, $y = y_2$, then

$$(1+\alpha)\|x_1 - x_2\| \le \alpha\|A(x_1) - A(x_2)\| + \|y_1 - y_2\|$$
$$\le \alpha\|x_1 - x_2\| + \|y_1 - y_2\|,$$

and therefore $\|x_1 - x_2\| \le \|y_1 - y_2\|$. $\blacksquare$

If $K \subset \mathcal{E}$ and $A : K \to \mathcal{E}$, Fix(A) will indicate the set of all fixed points of A.

Lemma 1.3.2 *Let $\mathcal{E}$ be uniformly convex. Let $K \subset \mathcal{E}$ be closed, bounded and convex, and let $A : K \to \mathcal{E}$ be a contraction. If $\{x_\nu\}$ is a sequence of points $x_\nu \in K$ converging weakly to some z and such that $\{x_\nu - A(x_\nu)\}$ converges strongly to y, then ($z \in K$ and) $z - A(z) = y$.*

Proof in the case in which $\mathcal{E}$ is a Hilbert space[2]. Assume $y = 0$ and let Π_K be the projection $\mathcal{E} \to K$. By Lemma 1.2.3 and by (1.8), $A \circ \Pi_K : \mathcal{E} \to \mathcal{E}$ is a contraction of $\mathcal{E}$.

Hence, for any $v \in \mathcal{E}$,

$$(A(x_\nu) - A(\Pi_K(v))\,|x_\nu - v) = (A(\Pi_K(x_\nu)) - A(\Pi_K(v))\,|x_\nu - v)$$
$$\le \|A(\Pi_K(x_\nu)) - A(\Pi_K(v))\|\,\|x_\nu - v\|$$
$$\le \|x_\nu - v)\|^2.$$

---

[2]For the proof in the general case of a uniformly convex Banach space, see [7].

As a consequence,

$$(x_\nu - v - A(x_\nu) + A(\Pi_K(v)) | x_\nu - v) =$$
$$\|x_\nu - v)\|^2 - (A(x_\nu) - A(\Pi_K(v)) | x_\nu - v)$$
$$\geq \|x_\nu - v)\|^2 - \|x_\nu - v)\|^2 = 0$$

for all $v \in \mathcal{E}$.

Hence, letting $\nu \to \infty$,

$$(-v + A(\Pi_K(v)) | z - v) \geq 0 \ \forall v \in \mathcal{E}.$$

Replacing now v by $z + tv$ with $t > 0$ and $v \in \mathcal{E}$, this latter inequality reads

$$(-z - tv + A(\Pi_K(z + tv)) | v) \leq 0 \ \forall v \in \mathcal{E} \ \forall t > 0,$$

whence

$$(-z + A(\Pi_K(z)) | v) = \lim_{t \downarrow 0} (-z - tv + A(\Pi_K(z + tv)) | v) \leq 0,$$

i.e.,

$$(-z + A(\Pi_K(z)) | v) = 0 \ \forall v \in \mathcal{E},$$

whence: $z - A(z) = 0$. ∎

Theorem 1.3.3 *Let $\mathcal{E}$ be uniformly convex. Let $K \subset \mathcal{E}$ be closed, bounded and convex, and let $A : K \to K$ be a contraction. Then Fix(A) is non-empty and convex.*

Proof To show that Fix(A) is convex we need recall that $\mathcal{E}$, being uniformly convex, is strictly convex[3].

---

[3]**Proof** Let B be the open unit ball of $\mathcal{E}$. If $\mathcal{E}$ is not strictly convex, there is some $x \in \partial B$ which is not an extreme of $\overline{B}$, that is, there is $u \in \mathcal{E} \backslash \{0\}$ such that

$$\|x + tu\| = 1 \ \forall t \in (-1, 1).$$

Let $\epsilon > 0$ and let

$$0 \neq |t| < \min \left(\frac{\epsilon}{\|u\|}, 1 \right).$$

Let $x_0, x_1 \in \text{Fix}(A)$, with $x_0 \neq x_1$, and let

$$x_t = tx_0 + (1-t)x_1 \text{ for } t \in (0,1).$$

Being

$$\|A(x_t) - x_0\| = \|A(x_t) - A(x_0)\| \leq p_L(A)\|x_t - x_0\|$$
$$\leq \|x_t - x_0\| \leq (1-t)\|x_1 - x_0\|,$$

and

$$\|A(x_t) - x_1\| = \|A(x_t) - A(x_1)\| \leq p_L(A)\|x_t - x_1\|$$
$$\leq \|x_t - x_1\| \leq t\|x_0 - x_1\|,$$

then

$$\|x_1 - x_0\| \leq \|A(x_t) - x_0\| + \|A(x_t) - x_1\|$$
$$\leq (1-t)\|x_1 - x_0\| + t\|x_0 - x_1\| = \|x_1 - x_0\|,$$

and therefore
$$\|A(x_t) - x_0\| = (1-t)\|x_1 - x_0\|,$$
$$\|A(x_t) - x_1\| = t\|x_0 - x_1\|,$$

whence

$$\|x_0 - A(x_t)\| + \|A(x_t) - x_1\| = \|x_0 - x_1\| = \|x_0 - A(x_t) + A(x_t) - x_1\|.$$

Since $\mathcal{E}$ is strictly convex, then

$$A(x_t) = tx_0 + (1-t)x_1 = x_t,$$

i.e., $x_t \in \text{Fix}(A)$ for all $t \in (0,1)$.

Then
$$\|x + tu - x\| = |t|\,\|u\| < \epsilon,$$

$$\|x + tu + x\| = \|2x + tu\| = 2\|2x + \frac{t}{2}u\| = 2,$$

showing that $\mathcal{E}$ is not uniformly convex. ∎

To prove that $\text{Fix}(A) \neq \emptyset$, let $x_o \in K$ and, for $s > 0$, let

$$x_s = (I + \frac{1}{s}(I - A))^{-1}(x_o),$$

which is contained in K by Lemma 1.3.1. Choose next a sequence $\{s_\nu\}$ such that $s_\nu \downarrow 0$ and that $\{x_{s_\nu}\}$ converges weakly to some $x \in K$.

Letting $\nu \to \infty$ in the equation

$$s_\nu(x_{s_\nu} - x_o) + x_{s_\nu} - A(x_{s_\nu}) = 0,$$

then

$$\lim_{\nu \to \infty}(x_{s_\nu} - A(x_{s_\nu})) = 0$$

(because the sequence $\{\|x_{s_\nu}\|\}$ is bounded). Lemma 1.3.2 implies then that $x - A(x) = 0$. $\blacksquare$

Example The following example, taken from [7], shows that some restricting condition on $\mathcal{E}$ is essential for Theorem 1.3.3 to hold.

Let $\mathcal{E}$ be the Banach space of all numerical sequences $x = \{x_0, x_1, x_2, \ldots\}$ converging to 0, endowed with the norm $\|x\| = \sup\{|x_n| : n = 0, 1, 2, \ldots\}$. Let $K = \overline{B(0,1)}$ and

$$A : \{x_0, x_1, x_2, \ldots\} \mapsto \{1, x_0, x_1, x_2, \ldots\}.$$

Since $A(x) = x$ would imply $x_n = 1$ for all $n = 1, 2, \ldots$, then $\text{Fix}(A) = \emptyset$.

Theorem 1.3.4 *Let $\mathcal{E}$ be uniformly convex, let K be a closed convex subset of $\mathcal{E}$ and let $\mathcal{A}$ be a family of contractions $K \to K$ such that, if $A_1, A_2 \in \mathcal{A}$, then $A_1 A_2 \in \mathcal{A}$, $A_2 A_1 \in \mathcal{A}$ and $A_1 A_2 = A_2 A_1$.*

If there is some $u \in K$ for which the set $\{A(u) : A \in \mathcal{A}\}$ is bounded, then there exists $x_o \in K$ such that

$$A(x_o) = x_o \ \forall A \in \mathcal{A}. \tag{1.23}$$

Proof Let

$$k := \sup\{\|S(u)\| : S \in \mathcal{A}\},$$
$$R := \|u\| + k.$$

For $A \in \mathcal{A}$, let

$$K_A = \{x \in K : \|x - AS(u)\| \leq R \ \forall \, S \in \mathcal{A}\}.$$

Then:

$$u \in K_A$$

because

$$\|u - AS(u)\| \leq \|u\| + \|AS(u)\|$$
$$\leq \|u\| + \sup\{\|S(u)\| : S \in \mathcal{A}\} = R;$$

K_A *is convex* because, if $x, y \in K_A$ and $0 \leq t \leq 1$,

$$\|tx + (1-t)y - AS(u)\| = \|tx + (1-t)y - (tAS(u) + (1-t)AS(u))\|$$
$$\leq t\,\|x - AS(u)\| + (1-t)\,\|y - AS(u)\|$$
$$\leq tR + (1-t)R = R;$$

$$x \in K_A \Longrightarrow x \in K_{AS} = K_{SA} \ \forall \, S \in \mathcal{A}$$

because, for $S' \in \mathcal{A}$,

$$\|x - SAS'(u)\| = \|x - S\,(AS'(u))\| \leq R;$$

K_A *is uniformly bounded* because, if $x \in K_A$ and $S \in \mathcal{A}$, then

$$\|x\| = \|x - AS(u) + AS(u)\|$$
$$\leq \|x - AS(u)\| + \|AS(u)\| \leq R + k.$$

As a consequence of this latter observation, the set

$$\tilde{K} := \cup\{K_A : A \in \mathcal{A}\}$$

is bounded.

Furthermore:

$u \in K_A$ for all $A \in \mathcal{A}$, and therefore $\tilde{K} \neq \emptyset$;

$\tilde{K}$ *is convex* because, if $x_1, x_2 \in \tilde{K}$, there are $A_1, A_2 \in \mathcal{A}$ for which $x_1 \in K_{A_1}, x_2 \in K_{A_2}$. As a consequence,

$$x_1 \in K_{AA_1} \ \forall \, A \in \mathcal{A},$$

and, in particular,
$$x_1 \in K_{A_2 A_1}.$$

Similarly,
$$x_2 \in K_{A_1 A_2} = K_{A_2 A_1}.$$

Hence, if $0 \le t \le 1$,
$$t x_1 + (1 - t) x_2 \in K_{A_2 A_1} \subset \tilde{K}.$$

Furthermore,
$$A(\tilde{K}) \subset \tilde{K} \quad \forall A \in \mathcal{A}.$$

Since $\mathcal{E}$ is uniformly convex, by Theorem 1.3.3 Fix(A) is closed, convex and non-empty.

If $x \in$ Fix(A) and $S \in \mathcal{A}$, then $SA(x) = S(x)$, and therefore $AS(x) = S(x)$, i.e., $S(x) \in$ Fix(A). Hence,
$$S(\text{Fix}(A)) \subset \text{Fix}(A) \quad \forall \ A, S, \in \mathcal{A}.$$

Since K is bounded, and therefore also the closed convex set Fix$(A) \subset K$ is bounded, there is $y \in$ Fix(A) such that $S(y) = y$. Hence,
$$\text{Fix}(A) \cap \text{Fix}(S) \ne \emptyset \quad \forall \ A, S \in \mathcal{A}.$$

Iterating this argument we see that
$$\text{Fix}(A_1) \cap \text{Fix}(A_2) \cap \cdots \cap \text{Fix}(A_n) \ne \emptyset \qquad (1.24)$$

for $n = 1, 2, \ldots$ and all $A_1, A_2, \ldots, A_n$ in $\mathcal{A}$.

The convex set Fix(A), being closed, is also weakly closed by Mazur's theorem. Since $\mathcal{E}$ is uniformly convex, and therefore reflexive, $\tilde{K}$ is weakly compact.

By (1.24), the collection $\{\text{Fix}(A) : A \in \mathcal{A}\}$ has the finite intersection property in $\tilde{K}$. Thus,
$$\cap \{\text{Fix}(A) : A \in \mathcal{A}\} \ne \emptyset,$$

and there exists some $x_o \in K$ for which (1.23) holds. ■

Corollary 1.3.5 *If A is a contraction of K into K, Fix$(A) \ne \emptyset$ if, and only if there is some $u \in K$ such that the sequence $\{A^n(x) : n = 1, 2, \ldots\}$ is bounded.*

1.4 The minimax theorem

Let $\mathcal{E}$ and $\mathcal{F}$ be two finite-dimensional, real vector spaces, and let $H \subset \mathcal{E}, K \subset \mathcal{F}$ be compact and convex.

Following [7], we shall establish the following theorem.

Theorem 1.4.1 *Let $\Phi : H \times K \to \mathbf{R}$ be such that $\Phi(\bullet, y)$ is convex and lower semicontinuous on H for all $y \in K$, and $\Phi(x, \bullet)$ is concave and upper semicontinuous on K for all $x \in H$. Then there exists $(x_o, y_o) \in H \times K$ for which*

$$\Phi(x_o, y) \leq \Phi(x_o, y_o) \leq \Phi(x, y_o) \ \forall \, x \in H, \forall \, y \in K. \tag{1.25}$$

We begin by recalling a few elementary facts on upper semicontinuous and lower semicontinuous functions.

Lemma 1.4.2 *If K is compact in $\mathcal{F}$, and $g : K \to \mathbf{R}$ is upper semicontinuous, then g is bounded and assumes its maximum.*

Proof For any $n = 1, 2, \ldots$ the sets

$$A_n := \{y \in K : g(y) < n\}$$

are open in K, and $\cup A_n = K$. Since K is compact,

$$K = A_1 \cup A_2 \cup \cdots \cup A_N = A_N$$

for some positive integer N, showing that $g(y) < N$ for all $y \in K$.

Let $k = \sup\{g(y) : y \in K\}$. The sets

$$K_n := \left\{y \in K : g(y) \geq k - \frac{1}{n}\right\}$$

are closed and non empty.

Since

$$K_1 \cap K_2 \cap \cdots \cap K_n = K_n,$$

the family $\{K_n\}$ has the finite intersection property and K is compact, then

$$\cap\{K_n : n = 1, 2, \ldots\} \neq \emptyset.$$

Thus, there is some $y_0 \in K$ such that

$$g(y_o) = \sup\{g(y) : y \in K\}.$$

∎

As a consequence,

Lemma 1.4.3 *If H is compact in $\mathcal{E}$, and $f : H \to \mathbf{R}$ is lower semi-continuous, then g is bounded from below and assumes its minimum.*

Lemma 1.4.4 *Under the hypotheses of Theorem 1.4.1, (1.25) is equivalent to*

$$\min_{x\in H} \max_{y\in K} \Phi(x,y) = \max_{y\in K} \min_{x\in H} \Phi(x,y). \tag{1.26}$$

Proof If (1.25) holds, then

$$\Phi(x_o, y_o) \leq \min_{x\in H} \Phi(x, y_o)$$
$$\leq \max_{y\in K} \min_{x\in H} \Phi(x,y),$$

$$\Phi(x_o, y_o) \geq \max_{y\in K} \Phi(x_o, y)$$
$$\geq \min_{x\in H} \max_{y\in K} \Phi(x,y).$$

Hence

$$\max_{y\in K} \min_{x\in H} \Phi(x,y) \geq \min_{x\in H} \max_{y\in K} \Phi(x,y).$$

Since, on the other hand,

$$\max_{y\in K} \min_{x\in H} \Phi(x,y) \leq \max_{y\in K} \Phi(x,y) \ \ \forall x \in H,$$

and therefore

$$\max_{y\in K} \min_{x\in H} \Phi(x,y) \leq \min_{x\in H} \max_{y\in K} \Phi(x,y), \tag{1.27}$$

(1.26) follows from (1.25).

To prove the converse, we begin by showing that the function

$$y \mapsto \min_{x \in H} \Phi(x, y) \tag{1.28}$$

is upper semicontinuous on K.

Let $\bar{y} \in K$, and let $a \in \mathbf{R}$ be such that

$$\min_{x \in H} \Phi(x, \bar{y}) < a.$$

Since $\Phi(\bullet, \bar{y})$ reaches its minimum on H, there is $\bar{x} \in H$ for which

$$\Phi(\bar{x}, \bar{y}) < a.$$

Because $\Phi(\bar{x}, \bullet)$ is upper semicontinuous on K, there is a neighbourhood U of $\bar{y}$ in K such that

$$\Phi(\bar{x}, y) < a \ \forall \, y \in U.$$

A fortiori,

$$\min_{x \in H} \Phi(x, y) < a \ \forall \, y \in U.$$

That proves that the function (1.28) is upper semicontinuous on K. Similarly, the function

$$x \mapsto \max_{y \in K} \Phi(x, y)$$

is lower semicontnuous on H.

As a consequence, there is $x_o \in H$ such that

$$\max_{y \in K} \Phi(x_o, y) = \min_{x \in H} \max_{y \in K} \Phi(x, y),$$

and there is $y_o \in K$ such that

$$\min_{x \in H} \Phi(x, y_o) = \max_{y \in K} \min_{x \in H} \Phi(x, y).$$

If (1.26) holds, denoting by c the common value of its two terms, we have

$$\max_{y \in K} \Phi(x_o, y) = c = \min_{x \in H} \Phi(x, y_o),$$

and therefore

$$\Phi(x_o, y) \leq c \leq \Phi(x, y_o) \quad \forall x \in H, \forall y \in K,$$

whence $\Phi(x_o, y_o) = c$, proving thereby that (1.25) holds. ∎

Remark The proof of the lemma shows that (1.27) holds whenever the two terms of the inequality do exist. Hence the following slightly more general statement has been established.

Lemma 1.4.5 *Under the hypotheses of Theorem 1.4.1, (1.25) is equivalent to*

$$\min_{x \in H} \max_{y \in K} \Phi(x, y) \leq \max_{y \in K} \min_{x \in H} \Phi(x, y).$$

Let $\| \ \|$ be a Hilbert space norm in $\mathcal{E}$.

Lemma 1.4.6 *Let $H \subset \mathcal{E}$ be compact and convex, and let the function $f : H \to \mathbf{R}$ be lower semicontinuous and convex. For every $\epsilon > 0$, the function*

$$f_\epsilon : x \mapsto f(x) + \epsilon \|x\|^2$$

reaches its minimum at a unique point of H.

Proof Since f_ϵ is lower semicontinuous, there exists at least one minimum point of f_ϵ in H.

We shall see now that the epigraph of f_ϵ,

$$\mathrm{epi}(f_\epsilon) = \{(x, y) : x \in H, y \geq f_\epsilon(x)\},$$

is strictly convex, *i.e.*, every point in the boundary, $\partial \, \mathrm{epi}(f_\epsilon)$ of $\mathrm{epi}(f_\epsilon)$ is an extreme point of $\overline{\mathrm{epi}(f_\epsilon)}$.

If $x \in H$ and $u \in \mathcal{E} \backslash \{0\}$ are such that $x + u \in H$, then

$$f(tx + (1 - t)(x + u)) \leq tf(x) + (1 - t)f(x + u),$$

i.e.,

$$f(x + (1 - t)u) \leq tf(x) + (1 - t)f(x + u)$$

for all $t \in [0, 1]$.

If epi(f_ϵ) is not strictly convex, then

$$f_\epsilon(tx + (1-t)(x+u)) \leq tf_\epsilon(x) + (1-t)f_\epsilon(x+u)$$

for some x and $x + u$ as above, and therefore

$$tf(x) + (1-t)f(x+u) + \epsilon\left(t\|x\|^2 + (1-t)\|x+u\|^2\right) =$$
$$f(tx + (1-t)(x+u)) + \epsilon\|tx + (1-t)(x+u)\|^2$$
$$\leq tf(x) + (1-t)f(x+u) + \epsilon\|x + (1-t)u\|^2.$$

Then, since $\epsilon > 0$,

$$t\|x\|^2 + (1-t)\|x+u\|^2 \leq \|x + (1-t)u\|^2,$$

that is,

$$t\|x\|^2 + (1-t)\left(\|x\|^2 + 2(x|u) + \|u\|^2\right)$$
$$\leq \|x\|^2 + 2(1-t)(x|u) + (1-t)^2\|u\|^2,$$

and, in conclusion,

$$\|u\|^2 \leq (1-t)\|u\|^2 \ \forall\, t \in [0,1),$$

which is absurd. That proves that, if $\epsilon > 0$, epi(f_ϵ) is strictly convex.

If x_1 and x_2 are two distinct minimum points of f_ϵ in H, then $f_\epsilon(x_1) = f_\epsilon(x_2)$, and therefore, if $t \in [0,1]$,

$$f_\epsilon(tx_1 + (1-t)(x_2)) \leq tf_\epsilon(x_1) + (1-t)f_\epsilon(x_2) = f_\epsilon(x_1).$$

Hence,

$$f_\epsilon(tx_1 + (1-t)(x_2)) = f_\epsilon(x_1) \ \forall\, t \in [0,1],$$

contradicting the strict convexity of H. ∎

We come now to the proof of Theorem 1.4.1.
For $\epsilon > 0$, let

$$\Phi_\epsilon(x,y) = \Phi(x,y) + \epsilon\|x\|^2,$$

where, as before, $\|\ \|$ is a Hilbert space norm in the finite dimensional vector space $\mathcal{E}$.

The function $\Phi_\epsilon(\bullet, y)$ is lower semicontinuos on H. Since $\Phi(\bullet, y)$ is convex on H, Lemma 1.4.6 implies that, for every $y \in K$, there is a unique point $\omega(y) \in H$ such that

$$\min_{x \in H} \Phi_\epsilon(x, y) = \Phi_\epsilon(\omega(x), y);$$

that is to say, setting

$$f_\epsilon(y) = \min_{x \in H} \Phi_\epsilon(x, y),$$

then

$$f_\epsilon(y) = \Phi_\epsilon(\omega(x), y).$$

As in the proof of Lemma 1.4.4, one shows that f_ϵ is lower semicontinuous. Furthermore, for $y_1, y_2 \in K$, $t \in [0, 1]$,

$$f_\epsilon(ty_1 + (1 - t)y_2) = \min_{x \in H} \Phi_\epsilon(x, ty_1 + (1 - t)y_2)$$
$$\geq t \min_{x \in H} \Phi_\epsilon(x, y_1) + (1 - t) \min_{x \in H} \Phi_\epsilon(x, y_2)$$
$$= tf_\epsilon(y_1) + (1 - t)f_\epsilon(y_2),$$

showing that f_ϵ is concave on K.

The function f_ϵ, being lower semicontinuous, reaches its maximum value on K. If $y_o \in K$ is one of its maximum points, then

$$f_\epsilon(y_o) = \max_{y \in K} f_\epsilon(y) = \max_{y \in K} \min_{x \in H} \Phi_\epsilon(x, y) = \min_{x \in H} \Phi_\epsilon(x, y_o).$$

If $x \in H$, $y \in K$ and $t \in [0, 1]$,

$$\Phi_\epsilon(x, (1 - t)y_o + ty) \geq (1 - t)\Phi_\epsilon(x, y_o) + t\Phi_\epsilon(x, y)$$
$$\geq (1 - t)f_\epsilon(y_o) + t\Phi_\epsilon(x, y).$$

As a consequence, for $x = \omega(ty + (1 - t)y_o) \in H$,

$$f_\epsilon(y_o) \geq f_\epsilon(ty + (1 - t)y_o)$$
$$\geq \Phi_\epsilon(\omega(ty + (1 - t)y_o), y) + (1 - t)f_\epsilon(y_o)$$

for all $t \in [0, 1]$. Hence,

$$tf_\epsilon(y_o) \geq t\Phi_\epsilon(\omega(ty + (1 - t)y_o), y),$$

and therefore

$$f_\epsilon(y_o) \geq \Phi_\epsilon(\omega(ty + (1-t)y_o), y) \; \forall \, t \in (0,1), \; \forall \, y \in K. \qquad (1.29)$$

Let $y_1, y_2 \in K$, $t \in (0,1)$, and let $\xi_t = \omega((1-t)y_1 + ty_2)$. We show now that

$$\lim_{t\downarrow 0} \xi_t = \omega(y_1).$$

For any $x \in H$,

$$\Phi_\epsilon(\xi_t, (1-t)y_1 + ty_2) \leq \Phi_\epsilon(x, (1-t)y_1 + ty_2),$$

and therefore

$$(1-t)\Phi_\epsilon(\xi_t, y_1) + t\Phi_\epsilon(\xi_t, y_2) \leq \Phi_\epsilon(\xi_t, (1-t)y_1 + ty_2) \; \forall \, x \in H \quad (1.30)$$

because $\Phi_\epsilon(\xi_t, \bullet)$ is concave.

Let $t_\nu \downarrow 0$ as $\nu \to \infty$, and let $\xi_{t_\nu} = \omega((1-t_\nu)y_1 + t_\nu y_2)$ converge to some $\xi \in H$.

Since $\Phi_\epsilon(\bullet, y_2)$ is lower semicontinuous in H, and therefore is bounded from below, then

$$\lim_{\nu \to \infty} t_\nu \Phi_\epsilon(\xi_{t_\nu}, y_2) = 0.$$

Furthermore, since $\Phi_\epsilon(x, \bullet)$ is upper semicontinuous on K, then,

$$\limsup_{\nu \to \infty} \Phi_\epsilon(x, (1-t_\nu)y_1 + t_\nu y_2) \leq \Phi_\epsilon(x, y_1)$$

(Proof: exercise, or see, e.g., [68], pp. 48-49.).

Since $\Phi_\epsilon(\bullet, y_1)$ is lower semicontinuous on H, then

$$\Phi_\epsilon(\xi, y_1) \leq \liminf_{\nu \to \infty} \Phi_\epsilon(\xi_{t_\nu}, y_1).$$

Hence, (1.29) yields

$$\Phi_\epsilon(\xi, y_1) \leq \Phi_\epsilon(x, y_1) \; \forall \, x \in H.$$

Thus, $\xi = \omega(y_1)$.

Setting in (1.29) $t = t_\nu$, and letting $\nu \to \infty$, we have

$$\Phi_\epsilon(\omega(y_o), y_o) = f_\epsilon(y_o) \geq \liminf_{\nu \to \infty} \Phi_\epsilon(\omega(t_\nu y + (1-t_\nu)y_o), y)$$
$$\geq \Phi_\epsilon(\omega(y_o), y)$$

for all $y \in K$.

Since, on the other hand,

$$f_\epsilon(y_o) = \min_{x \in H} \Phi_\epsilon(x, y_o) \le \Phi_\epsilon(x, y_o) \ \forall \ x \in H,$$

then

$$\Phi_\epsilon(\omega(y_o), y) \le \Phi_\epsilon(\omega(y_o), y_o) \le \Phi_\epsilon(x, y_o)$$

for all $x \in H$, $y \in K$.

Being

$$\Phi(\omega(y_o), y) \le \Phi_\epsilon(\omega(y_o), y) = \min_{x \in H} \Phi_\epsilon(x, y)$$

and

$$\Phi(\omega(y_o), y) \le \max_{y \in K} \min_{x \in H} \Phi_\epsilon(x, y)$$
$$\le \max_{y \in K} \min_{x \in H} \Phi(x, y) + \epsilon \max_{x \in H} \|x\|^2,$$

then, for any $y \in K$,

$$\min_{x \in H} \Phi(x, y) \le \max_{y \in K} \min_{x \in H} \Phi(x, y) + \epsilon \max_{x \in H} \|x\|^2,$$

whence

$$\min_{x \in H} \max_{y \in K} \Phi(x, y) \le \max_{y \in K} \min_{x \in H} \Phi(x, y) + \epsilon \max_{x \in H} \|x\|^2$$

for all $\epsilon > 0$.

Lemma 1.4.5 completes the proof of Theorem 1.4.1.

1.5 Subadditive functions

If Γ is an (additive) subgroup of $\mathbf{R}$, then the intersection $\Gamma_+ = \Gamma \cap \mathbf{R}_+$ is a semigroup. However, not all semigroups $\Gamma_+ \subset \mathbf{R}_+$ are intersections of $\mathbf{R}_+$ with subgroups of $\mathbf{R}$. Consider, for example, the semigroup $\Gamma_+ = \{p + q\sqrt{2} : p, q \in \mathbf{N}\}$, which is a discrete semigroup in $\mathbf{R}_+$. Any group $\Gamma \subset \mathbf{R}$ containing Γ_+ as a semigroup, contains also the group $\{p + q\sqrt{2} : p, q \in \mathbf{Z}\}$, which is dense in $\mathbf{R}$.

A semigroup $\Gamma_+ \subset \mathbf{R}_+$ will be said to be *complete* if there is a subgroup Γ of $\mathbf{R}$ such that $\Gamma_+ = \Gamma \cap \mathbf{R}_+$.

If $\alpha, \beta \in \Gamma_+$ with $\alpha < \beta$ implies that $\beta - \alpha \in \Gamma_+$, then $\Gamma := \Gamma_+ \cup (-\Gamma_+)$ is a group, and therefore the semigroup Γ_+ is complete.

Since every subgroup of $\mathbf{R}$ is either dense in $\mathbf{R}$ or discrete - in which case it consists of all integral multiples of one of its elements - this dichotomy extends to all complete semigroups in Γ_+. If 0 is not isolated in Γ_+, no point of Γ_+ is isolated, and - in particular - Γ_+ is not discrete.

From now on, Γ_+ will denote a complete semigroup in $\mathbf{R}_+$.

A function $\phi \to [-\infty, +\infty)$ is said to be *subadditive* if $\phi(\gamma_1 + \gamma_2) \le \phi(\gamma_1) + \phi(\gamma_2)$ for all $\gamma_1, \gamma_2 \in \Gamma_+$.

If $\phi(\gamma) > -\infty$ for some $\gamma \in \Gamma_+$, then $\phi(0) \ge 0$.

Lemma 1.5.1 *If the subadditive function ϕ is bounded from above on a neighbourhood of 0 in Γ_+, then it is bounded from above on $[0, t] \cap \Gamma_+$ for every $t > 0$.*

Proof The conclusion holds trivially if Γ_+ is discrete. Thus, let Γ_+ be dense in $\mathbf{R}_+$, and let $k > 0$ and $U = [0, a]$, with $a > 0$, be such that, if $\gamma \in U \cap \Gamma_+$, then $\phi(\gamma) \le k$. For any $t > 0$, let n be an integer larger than the integral part of $\frac{t}{a}$. For every $\gamma \in [0, t] \cap \Gamma_+$ there is $\delta \in U \cap \Gamma_+$ such that

$$0 \le \frac{1}{n}\gamma - \delta \in \frac{1}{n}U.$$

Hence $0 \le \gamma - n\delta \in U$, and therefore

$$\begin{aligned}
\phi(\gamma) &= \phi(\gamma - n\delta + n\delta) \\
&\le \phi(\gamma - n\delta) + \phi(n\delta) \\
&\le \phi(\gamma - n\delta) + n\phi(\delta) \\
&\le (n+1)k.
\end{aligned}$$

∎

Theorem 1.5.2 *If the subadditive function ϕ is bounded from above in a neighbourhood of 0 in Γ_+, the limit $\lim_{\gamma \to +\infty} \frac{\phi(\gamma)}{\gamma}$ exists, and*

$$-\infty \le \inf \left\{ \frac{\phi(\gamma)}{\gamma} : \gamma \in \Gamma_+ \backslash \{0\} \right\}$$

$$= \lim_{\gamma \to +\infty} \frac{\phi(\gamma)}{\gamma} < +\infty.$$

Proof If $\phi(\gamma) = -\infty$ for some $\gamma \in \Gamma_+$, then $\phi(\delta) = -\infty$ for all $\delta \in \Gamma_+$ with $\delta \geq \gamma$, and the conclusion follows. Assume that $\phi(\gamma) > -\infty$ for all $\gamma \in \Gamma_+$. For any $\gamma \in \Gamma_+ \backslash \{0\}$, let $\delta \in \Gamma_+$ with $\delta \geq \gamma$, and let n be a positive integer such that $n\gamma \leq \delta < (n+1)\gamma$. Then

$$
\begin{aligned}
\frac{1}{\delta}\phi(\delta) &= \frac{1}{\delta}\phi(n\gamma + \delta - n\gamma) \\
&\leq \frac{1}{\delta}(\phi(n\gamma) + \phi(\delta - n\gamma)) \\
&\leq \frac{n}{\delta}\phi(\gamma) + \frac{1}{\delta}\phi(\delta - n\gamma) \\
&\leq \frac{n}{\delta}\phi(\gamma) + \frac{1}{\delta}\sup\{\phi(\eta) : \eta \in [0,\gamma] \cap \Gamma_+\}.
\end{aligned}
$$

Since

$$
\frac{n}{n+1}\frac{1}{\gamma} < \frac{n}{\delta} \leq \frac{1}{\gamma},
$$

then

$$
\limsup_{\delta \to +\infty} \frac{\phi(\delta)}{\delta} \leq \inf\{\frac{\phi(\gamma)}{\gamma} : \gamma \in \Gamma_+ \backslash \{0\}\} \leq \liminf_{\delta \to +\infty} \frac{\phi(\delta)}{\delta},
$$

and the conclusion follows.

As was noticed before, the assumption on the boundedness of ψ in a neighbourhood of 0 in Γ_+ is void when Γ_+ is discrete. If $\Gamma_+ = \mathbf{N}$, the above theorem yields

Corollary 1.5.3 *If $\phi : \mathbf{N} \to [-\infty, +\infty)$ is subadditive, then*

$$
-\infty \leq \inf\left\{\frac{\phi(n)}{n} : n \in \mathbf{N}^*\right\} = \lim_{n \to +\infty} \frac{\phi(n)}{n} < +\infty.
$$

Besides the case $\Gamma_+ = \mathbf{N}$, in the following we shall only deal with the semigroup $\Gamma_+ = \mathbf{R}_+$. Actually, we will also consider the slightly more general situation in which the function ϕ, with values in $[-\infty, +\infty)$ is subadditive on the half-line $[a, +\infty)$, for some $a \in \mathbf{R}_+$, in the sense that

$$
\phi(t_1 + t_2) \leq \phi(t_1) + \phi(t_2)
$$

for all $t_1 \geq a$ and $t_2 \geq a$. The same kind of argument as the one proving the above theorem yields

Theorem 1.5.4 *If the subadditive function $\phi : [a, +\infty) \to [-\infty, +\infty)$ is bounded from above on any compact set of $[a, +\infty)$ then*

$$-\infty \leq \inf \left\{ \frac{\phi(\gamma)}{\gamma} : \gamma \in (a, +\infty) \right\} = \lim_{\gamma \to +\infty} \frac{\phi(\gamma)}{\gamma} < +\infty.$$

The following two results establish sufficient conditions for the local boundedness of subadditive functions.

Theorem 1.5.5 *If the function $\phi : \mathbf{R}_+ \to [-\infty, +\infty)$ is subadditive and measurable, ϕ is bounded from above on every compact set in $\mathbf{R}_+^*$. If, moreover, $\phi \neq -\infty$ at every point of $\mathbf{R}_+^*$, then ϕ is bounded on all compact sets of $\mathbf{R}_+^*$.*

Proof I. Let $a > 0$ and let $b = \phi(a)$. If $t_1 > 0$ and $t_2 > 0$ are such that $t_1 + t_2 = a$, then

$$b = \phi(t_1 + t_2) \leq \phi(t_1) + \phi(t_2),$$

and therefore, for every $t \in (0, a)$, either $\phi(t) \geq \frac{b}{2}$ or $\phi(a-t) \geq \frac{b}{2}$. Thus, setting $E = \{t : 0 < t < a, \phi(t) \geq \frac{b}{2}\}$, one has $(0, a) = E \cup (a - E)$, and therefore the Lebesgue measure $m(E)$ of E is $m(E) \geq \frac{a}{2}$. Assume now that ϕ is not bounded from above on some interval (α, β), with $0 < \alpha < \beta < +\infty$. There is a sequence $\{t_n\}$ in (α, β), converging to some $t_o \geq \alpha$, such that $\phi(t_n) \geq 2n$. If $E_n := \{t : 0 < t < \beta, \phi(t) \geq n\}$, then $m(E_n) \geq \frac{\alpha}{2}$, and $\phi(t) = +\infty$ on a set whose measure is $\geq \frac{\alpha}{2}$. Contradiction.

II. Assume now that, furthermore, $\phi \neq -\infty$ at all points of $\mathbf{R}_+^*$. If ϕ is not bounded from below on (α, β), with $0 < \alpha < \beta < +\infty$, there is a sequence $\{t_n\}$ in (α, β), converging to some $t_o \geq \alpha$, such that $\phi(t_n) \leq -n$. Let $M = \{\phi(t) : 2 < t < 5\}$. For every $t' \in (2, 5)$,

$$\phi(t' + t_n) \leq \phi(t') + \phi(t_n) \leq M - n. \tag{1.31}$$

If $n \gg 0$, then $(t_o + 3, t_o + 4) \subset (t_n + 2, t_n + 5)$. Hence, for any $t \in (t_o + 3, t_o + 4)$, there exist $t' \in (2, 5)$ and t_n with $n \gg 0$, such that $t = t' + t_n$. It follows from (1.31) that $\phi(t) = -\infty$ for all $t \in (t_o + 3, t_o + 4)$. Contradiction. ∎

Theorem 1.5.6 *If $\phi : \mathbf{R}_+^* \to [-\infty, +\infty)$ is subadditive and measurable, and if is finite the integral*

$$\int_0^{+\infty} e^{\phi(t)}\, dt = M,$$

then

$$\phi(t) \le 2 \log \left(\frac{M}{t} \right) \qquad (1.32)$$

if $t > 0$, and

$$\phi(t) \le 2 - \frac{2t}{Me} \qquad (1.33)$$

if $t > Me$.

Proof The positive, measurable function ψ defined by $\psi(t) = e^{\phi(t)}$, is submultiplicative, in the sense that

$$\psi(t_1 + t_2) \le \psi(t_1)\, \psi(t_2) \qquad (1.34)$$

for all $t_1 > 0$, $t_2 > 0$. For any fixed $t > 0$ and for all $s \in (0, t)$,

$$\psi(t) \le \psi(s)\, \psi(t - s) \le \frac{1}{2}(\psi(s)^2 + \psi(t - s)^2)$$

$$= \frac{1}{2}[(\psi(s) + \psi(t - s))^2 - 2\,\psi(s)\,\psi(t - s)],$$

whence

$$2(\psi(t) + \psi(s)\,\psi(t - s)) \le (\psi(s) + \psi(t - s))^2,$$

and therefore

$$2\,\psi(t)^{\frac{1}{2}} \le \psi(s) + \psi(t - s)$$

for all $s \in (0, t)$. Thus

$$t\,\psi(t)^{\frac{1}{2}} = 2 \int_0^{\frac{t}{2}} \psi(t)^{\frac{1}{2}}\, ds$$

$$\le \int_0^{\frac{t}{2}} \psi(s)\, ds + \int_0^{\frac{t}{2}} \psi(t - s)\, ds$$

$$= \int_0^{\frac{t}{2}} \psi(s)\, ds + \int_{\frac{t}{2}}^t \psi(s)\, ds$$

$$= \int_0^t \psi(s)\, ds \le M,$$

i.e.

$$\psi(t) \leq \left(\frac{M}{t}\right)^2, \tag{1.35}$$

which proves (1.32).

Let now $t > Me$, and let $n \in \mathbf{N}^*$ be such that

$$nMe \leq t < (n+1)Me. \tag{1.36}$$

Then (1.34) and (1.35) imply that

$$\psi(t) = \psi\left(n\frac{t}{n}\right) \leq \left(\frac{Mn}{t}\right)^{2n}. \tag{1.37}$$

Since, by (1.36), $t \geq nMe$ and $n > \frac{t}{Me} - 1$, (1.37) yields

$$\psi(t) \leq \left(\frac{Mn}{nMe}\right)^{2\left(\frac{t}{Me}-1\right)} = e^{2-\frac{2t}{Me}},$$

whence (1.33) follows. ∎

The following two statements shed some light on the relationship between subadditivity and convexity.

Lemma 1.5.7 *If the function $t \mapsto \frac{\phi(t)}{t}$ is decreasing on $\mathbf{R}_+^*$, ϕ is subadditive.*

Proof If $t_1 > 0$ and $t_2 > 0$, then

$$\phi(t_1 + t_2) = t_1 \frac{\phi(t_1 + t_2)}{t_1 + t_2} + t_2 \frac{\phi(t_1 + t_2)}{t_1 + t_2}$$
$$\leq t_1 \frac{\phi(t_1)}{t_1} + t_2 \frac{\phi(t_2)}{t_2} = \phi(t_1) + \phi(t_2).$$

∎

Lemma 1.5.8 *If the function $\phi : \mathbf{R}_+^* \to \mathbf{R}$ is convex and subadditive, the function $t \mapsto \frac{\phi(t)}{t}$ is decreasing.*

Proof If $t_1 > 0$, $t_2 > 0$ and $0 \leq s \leq 1$, then

$$\phi(s\,t_1 + (1-s)\,t_2) \leq s\phi(t_1) + (1-s)\phi(t_2).$$

Setting $t_1 = a$, $t_2 = a + b$, with $0 < a < b$, and $s = \frac{a}{b}$, then

$$s\,t_1 + (1-s)t_2 = \frac{a}{b}\,a + \left(1 - \frac{a}{b}\right)(a+b)$$
$$= \frac{1}{b}(a^2 + b^2 - a^2) = b,$$

and therefore

$$\phi(b) \leq \frac{a}{b}\phi(a) + \left(1 - \frac{a}{b}\right)\phi(a+b)$$
$$\leq \frac{a}{b}\phi(a) + \left(1 - \frac{a}{b}\right)(\phi(a) + \phi(b))$$
$$= \phi(a) - \frac{a}{b}\phi(b) + \phi(b),$$

whence $a\,\phi(b) \leq b\,\phi(a)$. ∎

Remark. The function $t \mapsto \frac{1}{t} + \sqrt{t}$, defined on $\mathbf{R}_+^*$, satisfies the hypothesis of Lemma 1.5.7, but is not convex.

Note. The material in this section is taken, essentially, from [40].

1.6 Spectral radius

Theorem 1.5.2 has a first important application in the theory of Banach algebras. A complex normed space $\mathcal{A}$ which carries a structure of an associative algebra a, $b \mapsto ab$ for a, $b \in \mathcal{A}$, such that $\|ab\| \leq \|a\|\,\|b\|$, is called a (complex) *normed algebra*. Such an algebra will be said to be *unital* if it contains an identity element, which henceforth will be denoted by 1, whose norm is $\|1\| = 1$. If the normed space $\mathcal{A}$ is complete, $\mathcal{A}$ will be called a *Banach algebra*. In the following, we shall be mainly concerned with Banach algebras.

If $\mathcal{A}$ is any complex normed space, carrying the structure of an (associative) algebra endowed with an identity element 1 and such that the product $\mathcal{A} \times \mathcal{A} \to \mathcal{A}$ is continuous, there exists an equivalent norm in $\mathcal{A}$ with respect to which $\mathcal{A}$ is a unital normed algebra.

For example, if $\mathcal{E}$ is a complex Banach space, the complex Banach space $\mathcal{L}(\mathcal{E})$ considered at the beginning of this chapter, with the norm $\|A\| = \sup\{\|Ax\| : x \in \mathcal{E}\}$ is a Banach algebra for the operation of composition of operators.

Let $\mathcal{A}$ be a normed algebra and let $\tilde{\mathcal{A}}$ be the Banach space which is the completion of $\mathcal{A}$. The product $\mathcal{A} \times \mathcal{A} \to \mathcal{A}$ is uniformly continuous, and thus extends (uniquely) to a uniformly continuous map $\tilde{\mathcal{A}} \times \tilde{\mathcal{A}} \to \tilde{\mathcal{A}}$, that is a product with respect to which $\tilde{\mathcal{A}}$ is a complex Banach algebra (unital if, and only if, $\mathcal{A}$ is unital). The imbedding $\mathcal{A} \subset \tilde{\mathcal{A}}$ is an isometric algebra homomorphism.

If $\mathcal{A}$ is a normed algebra, for any $a \in \mathcal{A}$ let $L_a \in \mathcal{L}(\mathcal{A})$ be the bounded linear operator $L_a : x \mapsto ax$. Then $\|L_a\| \leq \|a\|$, and the map $a \mapsto L_a$ is a continuous algebra-homomorphism of $\mathcal{A}$ into $\mathcal{L}(\mathcal{A})$, which is injective if, and only if, $\mathcal{A}$ has no (non-trivial) left zero divisor. If $\mathcal{A}$ is unital, $\|L_a\| = \|a\|$, and the homomorphism $a \mapsto L_a$ is an isometry.

Let $\mathcal{A}$ be a Banach algebra. For $x \in \mathcal{A}\backslash\{0\}$, the function $\phi : \mathbf{N} \to [-\infty, +\infty)$ defined by $\phi(n) = \log\|x^n\|$, is a subadditive function. Corollary 1.5.3 yields

Theorem 1.6.1 *For every $x \in \mathcal{A}$ the limit $\lim_{n\to+\infty} \|x^n\|^{\frac{1}{n}}$ exists, and*

$$\lim_{n\to+\infty} \|x^n\|^{\frac{1}{n}} = \inf\{\|x^n\|^{\frac{1}{n}} : n = 1, 2, \ldots\} \in [0, +\infty).$$

The limit on the left is called the *spectral radius* of x, and will be denoted by $\rho(x)$.

Exercise 1. Show that the sequence $\{\|x^{2^n}\|^{\frac{1}{2^n}}\}$ converges *decreasingly* to $\rho(x)$ as $n \to +\infty$.

2. In the Hilbert space $\mathbf{C}^2$, endowed with the scalar product

$$((x_1, y_1)|(x_2, y_2)) = x_1\overline{x_2} + y_1\overline{y_2},$$

consider the linear operator

$$A = \begin{pmatrix} 0 & a^2 \\ b^2 & 0 \end{pmatrix}$$

with $a > b > 0$. Show that $\rho(A) = ab$, but that $\{\|A^n\|^{\frac{1}{n}}\}$ does not converge decreasingly to $\rho(A)$.

Theorem 1.6.1 yields

$$\rho(x) \leq \|x\|. \tag{1.38}$$

Lemma 1.6.2 *One has $\rho(x) = \|x\|$ for all $x \in \mathcal{A}$ if, and only if, $\|x^2\| = \|x\|^2$ identically on $\mathcal{A}$.*

Proof If $\|x^2\| = \|x\|^2$ for all $x \in \mathcal{A}$, then

$$\|x^4\| = \|(x^2)^2\| = \|x^2\|^2 = \|x\|^4,$$

and, for any $n \in \mathbf{N}$,

$$\|x^{2^n}\| = \|x\|^{2^n}.$$

Thus

$$\rho(x) = \lim_{n \to +\infty} \|x^{2^n}\|^{\frac{1}{2^n}} = \|x\|.$$

Viceversa, if $\|x\| = \rho(x)$, then

$$\|x\|^2 = \rho(x)^2 = \rho(x^2) \leq \|x^2\| \leq \|x\|^2,$$

whence $\|x^2\| = \|x\|^2$. ∎

If $\zeta \in \mathbf{C}$, then $\rho(\zeta x) = |\zeta|\, \rho(x)$.

Lemma 1.6.3 *If $xy = yx$, then*

$$\rho(x + y) \leq \rho(x) + \rho(y) \text{ and } \rho(xy) \leq \rho(x)\rho(y).$$

Proof First of all

$$\rho(xy) = \lim_{n \to +\infty} \|(xy)^n\|^{\frac{1}{n}} = \lim_{n \to +\infty} \|x^n y^n\|^{\frac{1}{n}}$$
$$\leq \lim_{n \to +\infty} (\|x^n\|^{\frac{1}{n}} \|y^n\|^{\frac{1}{n}}) = \rho(x)\rho(y).$$

In order to establish the first inequality, choose $\alpha > \rho(x)$, $\beta > \rho(y)$ and let $u = \frac{1}{\alpha}x$, $v = \frac{1}{\beta}y$. For every positive integer n

$$\|(x+y)^n\|^{\frac{1}{n}} = \|x^n + nx^{n-1}y + \cdots + y^n\|^{\frac{1}{n}}$$
$$\leq (\alpha^n\|u^n\| + n\alpha^{n-1}\beta\|u^{n-1}\|\,\|v\| + \cdots + \beta^n\|v\|^n)^{\frac{1}{n}}$$
$$\leq (\alpha + \beta)\|u^{n'}\|^{\frac{1}{n}}\|v^{n''}\|^{\frac{1}{n}},$$

where n' and n'' are non-negative integers such that $n' + n'' = n$ and $\|u^{n'}\| \|v^{n''}\| \geq \|u^k\| \|v^{n-k}\|$ for $k = 0, 1, \ldots, n$.

Let $\{n_j\}$ be an increasing sequence of positive integers such that the sequences $\{\frac{n'_j}{n_j}\}$, $\{\frac{n''_j}{n_j}\}$ converge. Let $\delta = \lim_{j \to +\infty} \frac{n'_j}{n_j}$. Then $\delta \geq 0$. If $\delta > 0$,

$$\lim_{j \to +\infty} \|u^{n'_j}\|^{\frac{1}{n_j}} = \lim_{j \to +\infty} (\|u^{n'_j}\|^{\frac{1}{n'_j}})^{\frac{n'_j}{n_j}}$$

$$= \rho(u)^\delta \leq (\frac{1}{a}\rho(x))^\delta < 1.$$

If $\delta = 0$, then

$$\limsup_{j \to +\infty} \|u^{n'_j}\|^{\frac{1}{n_j}} \leq \limsup_{j \to +\infty} \|u\|^{\frac{n'_j}{n_j}} \leq 1.$$

Hence we have always

$$\limsup_{j \to +\infty} \|u^{n'_j}\|^{\frac{1}{n_j}} \leq 1,$$

and, similarly,

$$\limsup_{j \to +\infty} \|u^{n''_j}\|^{\frac{1}{n_j}} \leq 1.$$

Thus, $\rho(x + y) \leq \alpha + \beta$ whenever $\alpha > \rho(x)$, $\beta > \rho(y)$, and also the first inequality in the lemma has been established. ∎

Corollary 1.6.4 *If the Banach algebra $\mathcal{A}$ is abelian, the spectral radius is a seminorm.*

As we hall see later on, as a consequence of the relationship between spectrum and spectral radius, $\rho(xy) = \rho(yx)$.

Example Let $\mathcal{A}$ be the Banach algebra of all 2×2 complex matrices and let

$$x = \begin{pmatrix} 0 & 0 \\ 1 & 0 \end{pmatrix}$$

and

$$y = \begin{pmatrix} 0 & 1 \\ 0 & 0 \end{pmatrix}$$

for which $x^2 = y^2 = 0$, and therefore $\rho(x) = \rho(y) = 0$. On the other hand

$$xy = \begin{pmatrix} 0 & 0 \\ 0 & 1 \end{pmatrix},$$

$$yx = \begin{pmatrix} 1 & 0 \\ 0 & 0 \end{pmatrix},$$

$$x + y = \begin{pmatrix} 0 & 1 \\ 1 & 0 \end{pmatrix},$$

and therefore $\rho(x + y) = 1$, $\rho(xy) = \rho(yx) = 1$.

This example shows that Corollary 1.6.4 fails if the Banach algebra $\mathcal{A}$ is not abelian.

The elements $x \in \mathcal{A} \backslash \{0\}$ for which $x^n = 0$ for some positive integer n are said to be *nilpotent*. More in general, if $x \neq 0$ but $\rho(x) = 0$, x is said to *topologically nilpotent* or *quasinilpotent*.

Lemma 1.6.5 *If $\rho(x) = 0$, then $\lim_{n \to +\infty} \zeta^n x^n = 0$ for every $\zeta \in \mathbb{C}$.*

Proof Setting $\alpha_n = \|x^n\|$, then

$$\lim_{n \to +\infty} (t^n \alpha_n)^{\frac{1}{n}} = t \lim_{n \to +\infty} \alpha_n^{\frac{1}{n}} = 0,$$

and therefore the sequence $\{t^n \alpha_n\}$ is bounded for every $t > 0$. Let s be a limit value, and let $\{t^{n_j} \alpha_{n_j}\}$ be a subsequence converging to s. If $s > 0$ there exist some $\epsilon > 0$ and some index j_o such that $t^{n_j} \alpha_{n_j} > \epsilon$ for every $j > j_o$. Hence $t \alpha_{n_j}^{\frac{1}{n_j}} > \epsilon^{\frac{1}{n_j}}$, and $\lim_{j \to +\infty} \epsilon^{\frac{1}{n_j}} = 1$. Thus $t\rho(x) \geq 1$. Contradiction. ∎

Lemma 1.6.6 *If the sequence $\{\|\zeta^n x^n\|\}$ is bounded when $|\zeta| \gg 0$, then $\rho(x) = 0$.*

More exactly, if there is some $k > 0$ such that, whenever $|\zeta| > k$ the sequence $\{\|\zeta^n x^n\|\}$ is bounded, then $\rho(x) = 0$.

Proof With the same notations as before, if $\rho(x) > 0$, there exist $\epsilon > 0$ and n_o such that $\alpha_n^{\frac{1}{n}} > \epsilon$, i.e. $\alpha_n > \epsilon^n$, for all $n > n_o$. Thus, $t^n \alpha_n > (t\epsilon)^n$ for all $t > 0$ and $n > n_o$. If $t \leq \frac{2}{\epsilon}$, then $t^n \alpha_n > 2^n$, and the sequence $\{\|t^n x^n\|\}$ is not bounded. ∎

Example Going back to the differential operators considered in the first section, let X be the maximal operator defined by $Xx = \dot{x}$ in the space $C^1[0,1] \subset C[0,1]$, and let $Y \in \mathcal{L}[0,1]$ be the multiplication operator defined by $(Yx)(t) = t\,x(t)$. Obviously,

$$XY - YX = I$$

on $\mathcal{D}(X) = C^1[0,1]$, where I is the identity operator. The fact that one of the two operators appearing in this anticommutation relation is not bounded, is not accidental, as the following proposition shows[4].

Proposition 1.6.7 *If $\mathcal{A}$ is a unital Banach algebra, for all a and b in $\mathcal{A}$,*

$$ab - ba \neq 1.$$

Proof We begin by showing that, if $ab - ba = 1$, and if

$$a^n b - b a^n = n a^{n-1} \neq 0 \tag{1.39}$$

for some positive integer n, then $(a^n \neq 0$, and$)$

$$a^{n+1} b - b a^{n+1} = (n+1) a^n \neq 0.$$

It follows from (1.39) that

$$a^{n+1} b - b a^{n+1} = a^n (ab - ba) + (a^n b - b a^n) a$$
$$= a^n + n a^n = (n+1) a^n.$$

---

[4]See [70] and references therein also for the connections of this proposition with the commutation relations in quantum mechanics.

An inductive argument shows then that, if $ab - ba = 1$, (1.39) holds for every integer $n \geq 1$. As a consequence,

$$n \left\| a^{n-1} \right\| \leq 2 \left\| a^n \right\| \left\| b \right\| \leq 2 \left\| a^{n-1} \right\| \left\| a \right\| \left\| b \right\|,$$

and, being $\left\| a^{n-1} \right\| > 0$, then $n \leq 2 \left\| a \right\| \left\| b \right\|$ for all $n \geq 1$. ∎

1.7 Differentiable functions

This and the following section will collect a few facts on operator valued functions that will be useful throughout the following[5]. One of the main tools will be the Banach-Steinhaus theorem.

Theorem 1.7.1 *(Theorem of Banach-Steinhaus) Let $\mathcal{F}_j$ be a family of complex normed spaces, and, for each j, let $A_j : \mathcal{E} \to \mathcal{F}_j$ be a linear continuous map. Either there is a finite constant $a > 0$ such that*

$$\sup\{\left\| A_j \right\| : j\} \leq a$$

or

$$\sup\{\left\| A_j x \right\| : j\} = \infty$$

for all x in a dense subset of $\mathcal{E}$.

For a proof of the theorem, see, *e.g.*, [69], [40].

Let $\mathcal{E}'$ be the topological dual of the complex Banach space $\mathcal{E}$. For $\lambda \in \mathcal{E}'$, $< x, \lambda > \in \mathbf{C}$ will denote the value of λ on $x \in \mathcal{E}$.

A closed linear suspace $\mathcal{G}'$ of $\mathcal{E}'$ is called a *determining manifold* of $\mathcal{E}$ if, for every $x \in \mathcal{E}$,

$$\left\| x \right\| = \sup\{| < x, \lambda > | : \lambda \in \mathcal{G}', \left\| \lambda \right\| \leq 1\}.$$

For example, let $\mathcal{E}_o$ be a dense linear submanifold of $\mathcal{E}$ and let $\mathcal{H}'$ be a determining manifold of a Banach space $\mathcal{F}$. The set of continuous linear forms on $\mathcal{L}(\mathcal{E}, \mathcal{F})$ defined by $A \mapsto < Ax, \lambda >$ for $x \in \mathcal{E}_o$ and $\lambda \in \mathcal{H}'$, is a determining manifold of $\mathcal{L}(\mathcal{E}, \mathcal{F})$.

---

[5]For a more systematic account of the theory of vector-valued or operator valued functions, see, *e. g.*, [40], [19], [6], [26], [76]

Theorem 1.7.2 *Let $\mathcal{G}'$ be a determining manifold of $\mathcal{E}$, and let $\{x_j\}$ be a family of vectors in $\mathcal{E}$. Then, either*

$$\sup\{|<x_j,\lambda>| : j\} < \infty$$

for all $\lambda \in \mathcal{G}'$, and

$$\sup\{\|x_j\| : j\} < \infty,$$

or

$$\sup\{|<x_j,\lambda>| : j\} = \infty$$

for all λ in a dense subset of $\mathcal{G}'$.

The proof of the theorem consist in applying Theorem 1.7.1 to the continuous linear maps $A_j \in \mathcal{L}(\mathcal{G}', \mathbf{C})$ defined on the Banach space $\mathcal{G}'$ by $A_j : \lambda \mapsto <x_j,\lambda>$.

We shall now establish some elementary properties of differentiable functions $f : [a,b] \to \mathcal{E}$ and $\phi : [a,b] \to \mathcal{E}'$, defined on an interval $[a,b] \subset \mathbf{R}$.

Let $\mathcal{G}'$ be a determining manifold of $\mathcal{E}$.

Lemma 1.7.3 *If $t \mapsto <f(t),\lambda>$ is of class C^1 on $[a,b]$ for all $\lambda \in \mathcal{G}'$, then f is continuous on $[a,b]$.*

Proof By Theorem 1.7.1 there is a finite constant $k > 0$ such that $\|f(t)\| < k$ for all $t \in [a,b]$. For $t \in [a,b]$,

$$\lim_{h\to 0} \frac{<f(t+h)-f(t),\lambda>}{h} = \frac{d}{dt}<f(t),\lambda>.$$

Again by the Banach-Steinhaus theorem, and by the boundedness of $t \mapsto \|f(t)\|$, there is a finite constant $M > 0$ such that

$$\|f(t+h)-f(t)\| \le M|h|$$

whenever $|h|$ is sufficiently small. ∎

A similar argument shows that, if $t \mapsto <x,\phi(t)>$ is of class C^1 on $[a,b]$ for all $x \in \mathcal{E}$, then ϕ is continuous on $[a,b]$.

Lemma 1.7.4 *If* $t \mapsto < f(t), \lambda >$ *is of class* C^2 *for all* $\lambda \in \mathcal{E}'$, *then* f *is of class* C^1 *on* $[a, b]$.

Proof By Theorem 1.7.1, for every $t \in [a, b]$, $\lambda \mapsto \frac{d}{dt} < f(t), \lambda >$ is a continuous linear form on $\mathcal{E}'$. Hence, there is a unique $\psi(t) \in \mathcal{E}''$ (the topological dual of $\mathcal{E}'$) such that

$$\frac{d}{dt} < f(t), \lambda >=< \psi(t), \lambda > . \tag{1.40}$$

By Lemma 1.7.3, the function $\psi : [a, b] \to \mathcal{E}''$ is continuous. Thus, the integral

$$\int_a^t \psi(s)ds \in \mathcal{E}'',$$

exists and is a continuous function of $t \in [a, b]$. By (1.40),

$$\frac{d}{dt} \left(< f(t) - f(a), \lambda > - < \int_a^t \psi(s)ds, \lambda > \right) =$$
$$\frac{d}{dt} < f(t), \lambda > - < \psi(t), \lambda >= 0.$$

Hence $\int_a^t \psi(s)ds$ is the image of $f(t) - f(a)$ by the canonical isometric imbedding $\mathcal{E} \to \mathcal{E}''$:

$$f(t) - f(a) = \int_a^t \psi(s)ds.$$

Hence

$$\lim_{h \to 0} \frac{1}{h}(f(t + h) - f(t)) = \psi(t)$$

for all $t \in [a, b]$. ∎

By a similar argument one shows that, if $t \mapsto < x, \phi(t) >$ is of class C^2 for all $x \in \mathcal{E}$, then ϕ is of class C^1 on $[a, b]$.

Iteration of the above considerations yields

Theorem 1.7.5 *1) If* $t \mapsto < f(t), \lambda >$ *is of class* C^∞ *on* $[a, b]$ *for all* $\lambda \in \mathcal{E}'$, *the function* $f : [a, b] \to \mathcal{E}$ *is of class* C^∞.

2) If $t \mapsto < x, \phi(t) >$ *is of class* C^∞ *on* $[a, b]$ *for all* $x \in \mathcal{E}$, *the function* $\phi \to \mathcal{E}'$ *is of class* C^∞.

1.8 Holomorphic functions

We shall now develop the foundations of the theory of vector-valued or operator-valued holomorphic functions of one complex variable.

For $\zeta_o \in \mathbf{C}$ and $r > 0$, let $\Delta(\zeta_o, r) = \{\zeta \in \mathbf{C} : |\zeta - \zeta_o| < r\}$, and let $\Delta = \Delta(0, 1)$.

Given $\zeta \in \mathbf{C}$ and a sequence $\{a_n : n = 0, 1, \dots\}$ of vectors $a_n \in \mathcal{E}$, the sum

$$\sum_{n=0}^{+\infty} \zeta^n a_n \tag{1.41}$$

is a (formal) *power series*.

Theorem 1.8.1 *(Abel lemma) If there exists $\zeta_o \neq 0$ such that*

$$\sup\{\|a_n\| |\zeta_o|^n : n = 0, 1, \dots\} < \infty,$$

the power series (1.41) is absolutely and uniformly convergent on all compact sets contained in $\Delta(0, |\zeta_o|)$.

The proof of this theorem follows closely the classical argument holding for scalar-valued power series, and is left as an exercise.

By definition, *radius of convergence* of the power series is the supremum of those r such that the power series is uniformly convergent on the closure $\overline{\Delta(0, r)}$ of $\Delta(0, r)$. If the radius of convergence, R, is positive, the power series is said to be convergent. The sum of the series (1.41) is a continuous function of $\zeta \in \Delta(0, R)$.

Theorem 1.8.2 *(Theorem of Cauchy-Hadamard) The radius of convergence, R, of the power series (1.41) is given by*

$$\frac{1}{R} = \limsup_{n \to +\infty} \|a_n\|^{\frac{1}{n}}.$$

Proof Let R be the supremum of all r such that the power series (1.41) converges uniformly on all compact subsets of $\Delta(0, r)$.

1) We show first that

$$\frac{1}{R} \geq \limsup_{n \to +\infty} \|a_n\|^{\frac{1}{n}}. \tag{1.42}$$

If the power series (1.41) converges uniformly on $\overline{\Delta(0,r)}$, setting

$$f(\zeta) = \sum_{n=0}^{+\infty} \zeta^n a_n,$$

there exists a positive integer N_0 such that

$$\left\| f(\zeta) - \sum_{n=0}^{N} \zeta^n a_n \right\| \leq 1$$

for all $\zeta \in \overline{\Delta(0,r)}$ and all integers $N \geq N_0$. Since

$$\left| \left\| f(\zeta) - \sum_{n=0}^{N-1} \zeta^n a_n \right\| - \left\| \zeta^N a_N \right\| \right| \leq$$

$$\left\| f(\zeta) - \sum_{n=0}^{N} \zeta^n a_n \right\| \leq 1$$

for every $\zeta \in \overline{\Delta(0,r)}$, then $\left\| \zeta^N a_N \right\| \leq 2$, i.e.

$$\|a_N\| \leq \frac{2}{|\zeta|^N}$$

for every $N > N_0$ and all $\zeta \in \overline{\Delta(0,r)}$. Hence

$$\frac{1}{|\zeta|} \geq \limsup_{N \to +\infty} \|a_N\|^{\frac{1}{N}},$$

and therefore

$$\frac{1}{r} \geq \limsup_{N \to +\infty} \|a_N\|^{\frac{1}{N}}.$$

That proves (1.42).

2) The opposite inequality holds obviously when $\limsup_{n \to +\infty} \|a_n\|^{\frac{1}{n}} = +\infty$. To establish it when $\limsup_{n \to +\infty} \|a_n\|^{\frac{1}{n}} < +\infty$, choose r such that

$$0 \leq r < \frac{1}{\limsup_{n \to +\infty} \|a_n\|^{\frac{1}{n}}}. \tag{1.43}$$

For every

$$s \in \left(r, \frac{1}{\limsup_{n \to +\infty} \|a_n\|^{\frac{1}{n}}} \right)$$

there is some integer $N > 0$ such that

$$s < \frac{1}{\|a_n\|^{\frac{1}{n}}}$$

i.e.

$$\|a_n\| < \frac{1}{s^n}$$

whenever $n \geq N$. If $|\zeta| \leq r$, then

$$\|\zeta^n a_n\| \leq \left(\frac{r}{s}\right)^n$$

for all $n \geq N$. Since $\frac{r}{s} < 1$, then

$$\sum_{n=N}^{+\infty} \|\zeta^n a_n\| \leq \left(\frac{r}{s}\right)^N \left(1 + \frac{r}{s} + \left(\frac{r}{s}\right)^2 + \cdots\right)$$

$$= \left(\frac{r}{s}\right)^N \frac{1}{1 - \frac{r}{s}},$$

and therefore the power series (1.41) converges uniformly on $\overline{\Delta(0, r)}$ for all r satisfying (1.43). Thus $r \leq R$, and in conclusion

$$R \geq \frac{1}{\limsup_{n \to +\infty} \|a_n\|^{\frac{1}{n}}}.$$

∎

Let U be an open set of $\mathbf{C}$. A function $f : U \to \mathcal{E}$ is said to be *holomorphic* on U, if, for every $\zeta_o \in U$. there is a convergent power series (1.41) with radius of convergence $R > 0$ such that, for all $\zeta \in U \cap \Delta(\zeta_o, R)$,

$$f(\zeta) = \sum_{n=0}^{+\infty} (\zeta - \zeta_o)^n a_n.$$

The holomorphic function f is continuous on U.

The set of all holomorphic functions $U \to \mathcal{E}$ will be denoted by $\mathrm{Hol}(U, \mathcal{E})$. Let ϖ be a closed rectifiable path, which is contained and contractible in U.

Theorem 1.8.3 *(Cauchy's integral theorem) If $f \in \mathrm{Hol}(U, \mathcal{E})$, then*

$$\int_{\varpi} f(\zeta)d\zeta = 0.$$

Theorem 1.8.4 *(Cauchy's integral formula) For all $\zeta_o \notin \varpi$*

$$I(\zeta_o, l)f(\zeta_o) = \frac{1}{2\pi i} \int_{\varpi} \frac{1}{\zeta - \zeta_o} f(\zeta)d\zeta.$$

Here $I(\zeta_o, l)$ is the index of ζ_o with respect to l, [69].

Proof Exercise. (Hint: Using the Hahn-Banach theorem, reduce the proofs to the case of scalar valued functions)

Theorem 1.8.5 *(Liouville's theorem) If $f \in \mathrm{Hol}(\mathbf{C}, \mathcal{E})$ is bounded in norm on $\mathbf{C}$, then f is constant.*

Proof For any $\lambda \in \mathcal{E}'$ the function $\zeta \mapsto < f(\zeta), \lambda >$ is an entire function which is bounded on $\mathbf{C}$. By the classical Liouville theorem, $< f(\bullet), \lambda >$ is constant. The Hahn-Banach theorem yields the conclusion. ∎

As was noticed already, if $f \in \mathrm{Hol}(U, \mathcal{E})$ and $\lambda \in \mathcal{E}$, then $< f(\bullet), \lambda >$ is a scalar-valued holomorphic function on U.

Does the converse hold?

An answer - actually, a positive answer - to this question is provided by the following theorem, where, as before, U is a non-empty open set in $\mathbf{C}$, $\mathcal{E}$ and $\mathcal{F}$ are complex Banach spaces, $\mathcal{H}'$ is a determining manifold of $\mathcal{F}$.

Theorem 1.8.6 *If $f : U \to \mathcal{L}(\mathcal{E}, \mathcal{F})$ is such that, for all $x \in \mathcal{E}$ and all $\lambda \in \mathcal{H}'$, the scalar-valued function $\zeta \mapsto < f(\zeta)x, \lambda >$ is holomorphic on U, then $f \in \mathrm{Hol}(U, \mathcal{L}(\mathcal{E}, \mathcal{F}))$.*

We begin by establishing a few technical facts.

Lemma 1.8.7 *For every $\zeta_o \in U$, the limit*

$$\lim_{\zeta \to 0} \frac{1}{\zeta}(f(\zeta_o + \zeta) - f(\zeta_o))$$

exists for the norm topology in $\mathcal{L}(\mathcal{E}, \mathcal{F})$.

Proof Let $D \subset\subset U$ be a domain bounded by a closed, rectifiable, Jordan curve ϖ, for which $\zeta_o \in D$. Let D_o be a domain such that $\zeta_o \in D_o \subset\subset D$. If ζ_1, ζ_2 are such that $0 \neq \zeta_1 \neq \zeta_2 \neq 0$ and that $\zeta_o + \zeta_1 \in D_o$, $\zeta_o + \zeta_2 \in D_o$, the Cauchy integral formula yields

$$< (f(\zeta_o + \zeta_j) - f(\zeta_o))x, \lambda > =$$
$$= \frac{\zeta_j}{2\pi i} \int_\varpi \frac{< f(\zeta)x, \lambda >}{(\zeta - \zeta_o)(\zeta - \zeta_o - \zeta_j)} d\zeta$$

for $j = 1, 2$, whence

$$\frac{1}{\zeta_1 - \zeta_2} \left(\frac{< (f(\zeta_o + \zeta_1) - f(\zeta_o))x, \lambda >}{\zeta_1} - \right.$$
$$\left. - \frac{< (f(\zeta_o + \zeta_2) - f(\zeta_o))x, \lambda >}{\zeta_2} \right)$$
$$= \frac{1}{2\pi i} \int_\varpi \frac{< f(\zeta)x, \lambda >}{(\zeta - \zeta_o)(\zeta - \zeta_o - \zeta_1)(\zeta - \zeta_o - \zeta_2)} d\zeta.$$

Let $r := \text{dist}(D_o, \varpi) > 0$. Since $|\zeta - \zeta_o| > r$, $|\zeta - \zeta_o - \zeta_1| > r$, $|\zeta - \zeta_o - \zeta_2| > r$ for every $\zeta \in \varpi$, then, for all $x \in \mathcal{E}$ and all $\lambda \in \mathcal{H}'$,

$$\sup \left\{ \frac{1}{|\zeta_1 - \zeta_2|} \left| \frac{< (f(\zeta_o + \zeta_1) - f(\zeta_o))x, \lambda >}{\zeta_1} - \right. \right.$$
$$\left. \left. \frac{< (f(\zeta_o + \zeta_2) - f(\zeta_o))x, \lambda >}{\zeta_2} \right| : \right.$$
$$\zeta_o, \zeta_o + \zeta_1, \zeta_o + \zeta_2 \in D_o, 0 \neq \zeta_1 \neq \zeta_2 \neq 0 \}$$
$$\leq \frac{1}{2\pi r^3} \sup\{| < f(\zeta)x, \lambda > | : \zeta \in \varpi \} \int_\varpi |d\zeta| < \infty.$$

By the theorem of Banach-Steinhaus there exists a finite constant $k > 0$ such that

$$\sup \left\{ \left\| \frac{1}{\zeta_1}(f(\zeta_o + \zeta_1) - f(\zeta_o)) - \frac{1}{\zeta_2}(f(\zeta_o + \zeta_2) - f(\zeta_o)) \right\| : \right.$$
$$\left. \zeta_o, \zeta_o + \zeta_1, \zeta_o + \zeta_2 \in D_o, 0 \neq \zeta_1 \neq \zeta_2 \neq 0 \right\} \leq k|\zeta_1 - \zeta_2|.$$

Letting ζ_1 and ζ_2 tend to 0, one sees that the limit

$$\lim_{\zeta \to 0} \frac{1}{\zeta}(f(\zeta_o + \zeta) - f(\zeta_o))$$

exists for the norm topology in $\mathcal{L}(\mathcal{E}, \mathcal{F})$. In other words, there exists $f'(\zeta_o) \in \mathcal{L}(\mathcal{E}, \mathcal{F})$ for which

$$\lim_{\zeta \to 0} \left\| \frac{1}{\zeta}(f(\zeta_o + \zeta) - f(\zeta_o)) - f'(\zeta_o) \right\| = 0.$$

■

Exercise Show that the limit is uniform when ζ_o varies on compact subsets of U.

Corollary 1.8.8 *The function f is continuous.*

As a consequence, f can be integrated along ϖ.

Lemma 1.8.9 *For every $\zeta_o \in U$,*

$$f(\zeta_o) = \frac{1}{2\pi i} \int_\varpi \frac{1}{\zeta - \zeta_o} f(\zeta) d\zeta. \tag{1.44}$$

Proof The Cauchy integral formula yields

$$< f(\zeta_o)x, \lambda > = \frac{1}{2\pi i} \int_\varpi \frac{< f(\zeta)x, \lambda >}{\zeta - \zeta_o} d\zeta$$
$$= < \frac{1}{2\pi i} \int_\varpi \frac{1}{\zeta - \zeta_o} f(\zeta)x d\zeta, \lambda >,$$

because the linear form λ is continuous. By the Hahn-Banach theorem,

$$f(\zeta_o)x = \frac{1}{2\pi i} \int_\varpi \frac{1}{\zeta - \zeta_o} f(\zeta)x d\zeta$$

for all $x \in \mathcal{E}$, and (1.44) follows. ∎

We come now to the proof of Theorem 1.8.6.

For all $\zeta \in \Delta(\zeta_o, \mathrm{dist}(\zeta_o, \mathbf{C} \backslash D))$ and all $\tau \in \varpi$,

$$\frac{1}{\tau - \zeta} = \frac{1}{\tau - \zeta_o - (\zeta - \zeta_o)} = \frac{1}{\tau - \zeta_o} \sum_{n=0}^{+\infty} \left(\frac{\zeta - \zeta_o}{\tau - \zeta_o} \right)^n,$$

the convergence being absolute and uniform for $\tau \in \varpi$. Integrating term by term, (1.44) yields the Taylor expansion

$$f(\zeta) = \sum_{n=0}^{+\infty} \frac{(\zeta - \zeta_o)^n}{n!} d^n f(\zeta_o),$$

whose coefficients are

$$d^n f(\zeta_o) = \frac{n!}{2\pi i} \int_{\varpi} \frac{1}{(\tau - \zeta_o)^{n+1}} f(\tau) d\tau.$$

The convergence of the Taylor series is uniform on all compact subsets of $\Delta(\zeta_o, \mathrm{dist}(\zeta_o, \mathbf{C} \backslash D))$. ∎

The proofs of the following statements are left as exercises.

Lemma 1.8.10 *If the limit*

$$\lim_{\zeta \to 0} \frac{1}{\zeta} (f(\zeta_o + \zeta) - f(\zeta_o))$$

exists for all $\zeta_o \in U$ and for the strong operator topology, or for the norm-topology, on $\mathcal{L}(\mathcal{E}, \mathcal{F})$, then $f \in \mathrm{Hol}(U, \mathcal{L}(\mathcal{E}, \mathcal{F}))$.

Proposition 1.8.11 *The map $\zeta \mapsto d^n f(\zeta)$ from U to $\mathcal{L}(\mathcal{E}, \mathcal{F})$ is holomorphic for $n = 1, 2, \ldots$.*

1.9 The spectrum of a linear operator

Let $\mathcal{A}^{-1}$ be the set of all elements of a unital complex normed algebra $\mathcal{A}$ (whose identity element will be denoted by 1), which are invertible.

For any $a \in \mathcal{A}$, the *resolvent set* of a is, by definition,

$$r(a) = \{\zeta \in \mathbf{C} : \zeta 1 - a \in \mathcal{A}^{-1}\}.$$

The *spectrum* of a is, by definition,

$$\sigma(a) = \mathbf{C} \backslash r(a).$$

In the following, $\mathcal{A}$ will be assumed to be complete, *i.e.*, $\mathcal{A}$ will be a complex Banach algebra, although some of the results that will be established hold for any unital complex normed algebra. The norm in $\mathcal{A}$ will be chosen in such a way that $\|ab\| \le \|a\| \, \|b\|$ for all $a, b \in \mathcal{A}$.

All these conditions are satisfied when $\mathcal{A} = \mathcal{L}(\mathcal{E})$. Thus, for a linear operator $X \in \mathcal{L}(\mathcal{E})$, the resolvent set $r(X)$ is defined as the set of all complex numbers ζ for which $\zeta I - X$ is invertible, and its inverse is contained in $\mathcal{L}(\mathcal{E})$;

$$\sigma(X) = \mathbf{C} \backslash r(X) \tag{1.45}$$

is the set of all $\zeta \in \mathbf{C}$ such that either $\zeta I - X$ is not invertible, or its inverse is not continuous.

Lemma 1.9.1 *For all $a, b \in \mathcal{A}$, $\sigma(ab) \backslash \{0\} = \sigma(ba) \backslash \{0\}$.*

Proof. Let $\zeta \in r(ab)$, $\zeta \ne 0$, and let $u \in \mathcal{A}$ be such that $u(ab - \zeta 1) = (ab - \zeta 1)u = 1$, i.e.

$$uab = abu = \zeta u + 1.$$

Then

$$(ba - \zeta 1)(bua - 1) = \zeta 1$$
$$(bua - 1)(ba - \zeta 1) = \zeta 1,$$

and therefore $\zeta \in r(ba)$. ∎

Given now a linear operator X with domain $\mathcal{D}(X) \subset \mathcal{E}$ and range $\mathcal{R}(X) \subset \mathcal{E}$, we shall define the spectrum $\sigma(X)$ of X, which, when

$X \in \mathcal{L}(\mathcal{E})$, will be shown, later on, to coincide with the one given above.

The resolvent set $r(X)$ of X will be assumed, by definition, to be the set of all $\zeta \in \mathbf{C}$ such that:

$\mathcal{R}(\zeta I - X)$ is dense in $\mathcal{E}$;

$\zeta I - X$ is a bijection of $\mathcal{D}(X)$ onto $\mathcal{R}(\zeta I - X)$;

$(\zeta I - X)^{-1}$ is continuous on $\mathcal{R}(\zeta I - X)$.

The spectrum of X is given by (1.45).

Lemma 1.9.2 *If X is closed, $r(X)$ consists of all $\zeta \in \mathbf{C}$ such that: $\mathcal{R}(\zeta I - X) = \mathcal{E}$, $\zeta I - X$ is injective, and $(\zeta I - X)^{-1} \in \mathcal{L}(\mathcal{E})$.*

Proof It suffices to show that, if $\zeta \in \mathbf{C}$ is such that: $\overline{\mathcal{R}(\zeta I - X)} = \mathcal{E}$, $\zeta I - X$ is injective, and

$$(\zeta I - X)^{-1} \in \mathcal{L}(\mathcal{R}(\zeta I - X)), \tag{1.46}$$

then $\mathcal{R}(\zeta I - X) = \mathcal{E}$ and $\zeta I - X$ is injective.

In view of (1.46), there is a constant $c > 0$ such that

$$\|(\zeta I - X)(x)\| \geq c \, \|x\| \tag{1.47}$$

for all $x \in \mathcal{D}(X)$. For any $v \in \mathcal{E}$, let $\{x_\nu\}$ be a sequence in $\mathcal{D}(X)$ such that the sequence $\{(\zeta I - X)x_\nu\}$ converges to v. By (1.47), also the sequence $\{x_\nu\}$ converges. Let $u = \lim x_\nu$. Since X is closed, $u \in \mathcal{D}(X)$ and $v = (\zeta I - X)u$. Hence $\mathcal{R}(\zeta I - X) = \mathcal{E}$, and (1.47) implies that $\|(\zeta I - X)u\| \geq c \, \|u\|$, showing that $\zeta I - X$ maps $\mathcal{D}(X)$ bijectively onto $\mathcal{E}$. ∎

Lemma 1.9.3 *If X is closed, and if $\zeta I - X$ maps $\mathcal{D}(X)$ bijectively onto $\mathcal{E}$, then $\zeta \in r(X)$.*

Proof Since $\zeta I - X$ is closed, by Lemma 1.1.11 $(\zeta I - X)^{-1}$ is closed. The closed graph theorem yields the conclusion. ∎

Summing up, the following result has been established, which, incidentally, shows that, when $X \in \mathcal{L}(\mathcal{E})$, the definition of $\sigma(X)$ given now coincides with the one introduced at the beginning.

Theorem 1.9.4 *If X is closed, $\zeta \in r(X)$ if, and only if, $\zeta I - X$ maps $\mathcal{D}(X)$ bijectively onto $\mathcal{E}$.*

The spectrum $\sigma(X)$ is the union of the three mutually disjoint subsets:

  the *point spectrum*, $p\sigma(X)$, consisting of all points $\zeta \in \mathbf{C}$ for which $\zeta I - X$ is not injective. Thus, $\zeta \in p\sigma(X)$ if, and only if, ζ is an eigenvalue of X;

  the *continuous spectrum*, $c\sigma(X)$, consisting of all points $\zeta \in \mathbf{C}$ for which $\zeta I - X$ is injective, has a dense range, but $(\zeta I - X)^{-1} : \mathcal{R}(\zeta I - X) \to \mathcal{D}(X)$ is not continuous;

  the *residual spectrum*, $r\sigma(X)$, consisting of all points $\zeta \in \mathbf{C}$ for which $\zeta I - X$ is invertible, but its range is not dense.

If $\mathcal{E}$ is a real Banach space, the spectrum of X is defined for X acting on the complexification $\mathcal{E}_{\mathbf{C}}$ of $\mathcal{E}$, endowed with the norm defined by

$$\|x + iy\| = \sup\{\|(\cos\theta)x + (\sin\theta)y\| : 0 \leq \theta \leq 2\pi\} \quad x, y \in \mathcal{H},$$

for which the embedding of $\mathcal{E}$ into $\mathcal{E}_{\mathbf{C}}$ is an isometry.

Theorem 1.9.5 *If $\zeta \in c\sigma(X)$. there exists a sequence $\{x_\nu\}$ of vectors $x_\nu \in \mathcal{D}(X)$, with $\|x_\nu\| = 1$, such that $\lim_{\nu \to +\infty}(\zeta I - X)x_\nu = 0$.*

Proof There is a sequence $\{y_\nu\}$ in $\mathcal{D}(X)$, with $\|y_\nu\| = 1$, such that

$$\|(\zeta I - X)^{-1} y_\nu\| \geq \nu.$$

If $z_\nu := (\zeta I - X)^{-1} y_\nu$, then $\|z_\nu\| \geq \nu$, and, setting $x_\nu = \frac{1}{\|z_\nu\|} z_\nu$, then

$$\|(\zeta I - X)x_\nu\| \leq \frac{1}{\nu}.$$

∎

We will investigate now the behaviour of the spectrum under the extension of the operator.

  If $X \subset X'$, then $\zeta I - X \subset \zeta I - X'$ for all $\zeta \in \mathbf{C}$. If $\zeta I - X'$ is invertible, also $\zeta I - X$ is invertible, and $(\zeta I - X)^{-1} \subset (\zeta I - X')^{-1}$.

Theorem 1.9.6 *If $X \subset X'$, then*

$$p\sigma(X) \subset p\sigma(X'),$$
$$c\sigma(X) \subset c\sigma(X') \cup p\sigma(X'),$$
$$r\sigma(X') \subset r\sigma(X),$$
$$r(X') \subset r(X) \cup r\sigma(X).$$

Proof The first inclusion is obvious.

To establish the second one, let $\zeta \in c\sigma(X)$. Since $\mathcal{R}(\zeta I - X) \subset \mathcal{R}(\zeta I - X')$, then $\overline{\mathcal{R}(\zeta I - X')} = \mathcal{E}$, and therefore $\zeta \notin r\sigma(X')$. If $\zeta \in r(X')$, then $(\zeta I - X')^{-1}$ (exists and) is continuous on $\mathcal{R}(\zeta I - X')$. Therefore, also $(\zeta I - X)^{-1}$ (exists and) is continuous on $\mathcal{R}(\zeta I - X)$, contradicting the hypothesis whereby $\zeta \in c\sigma(X)$. Thus, either $\zeta \in c\sigma(X')$ or $\zeta \in p\sigma(X')$.

Since $\overline{\mathcal{R}(\zeta I - X)} \subset \overline{\mathcal{R}(\zeta I - X')}$, if $\zeta \in r\sigma(X')$, then $\overline{\mathcal{R}(\zeta I - X)} \neq \mathcal{E}$. Since, moreover, $\zeta I - X'$ is invertible, and therefore also $\zeta I - X$ is invertible, then $\zeta \in r\sigma(X)$, and the third inclusion is established.

As for the last one, let $\zeta \in r(X')$. Then $\mathcal{R}(\zeta I - X')$ is dense in $\mathcal{E}$, and $\zeta I - X'$ has a continuous inverse on $\mathcal{R}(\zeta I - X')$. Thus, either $\overline{\mathcal{R}(\zeta I - X)} = \mathcal{E}$, in which case $\zeta \in r(X)$, or $\overline{\mathcal{R}(\zeta I - X)} \neq \mathcal{E}$, and therefore $\zeta \in r\sigma(X)$. ∎

Theorem 1.9.7 *If X is closed, anf if X' is an effective extension of X, then*

$$r(X) \subset p\sigma(X'),$$
$$r(X') \subset r\sigma(X).$$

Proof If $\zeta \in r(X)$, $\zeta I - X$ maps $\mathcal{D}(X)$ bijectively onto $\mathcal{E}$. Since $\mathcal{D}(X)$ is properly contained in $\mathcal{D}(X')$, $\zeta I - X'$ cannot be injective on $\mathcal{D}(X')$. Therefore $\zeta \in p\sigma(X')$.

In view of this fact and of the fourth inclusion in Theorem 1.9.6,

$$r(X') \subset r(X) \cup r\sigma(X) \subset p\sigma(X') \cup r\sigma(X).$$

Since $r(X') \cap p\sigma(X') = \emptyset$, the conclusion follows. ∎

Theorem 1.9.8 *If X is closable, then*

$$\sigma(\overline{X}) = \sigma(X)$$

and

$$co(\overline{X}) \subset co(X).$$

Proof I. If $\zeta \in co(\overline{X})$, then $\overline{\mathcal{R}(\zeta I - \overline{X})} = \mathcal{E}$, $\zeta I - \overline{X}$ is invertible on $\mathcal{R}(\zeta I - X)$, and $(\zeta I - \overline{X})^{-1}$ is not bounded. Hence $\zeta I - X$ is invertible on $\mathcal{R}(\zeta I - X)$, which is dense in $(\mathcal{R}(\zeta I - \overline{X})$ and therefore also in$)$ $\mathcal{E}$. If $(\zeta I - X)^{-1}$ were continuous, also $(\zeta I - \overline{X})^{-1}$ should be continuous, by Lemma 1.1.12. Thus $(\zeta I - X)^{-1}$ is not continuous, and therefore $\zeta \in co(X)$.

II. If $\zeta \in r(\overline{X})$, $\zeta I - \overline{X}$ maps $\mathcal{D}(\overline{X})$ bijectively onto $\mathcal{E}$, and $(\zeta I - \overline{X})^{-1} \in \mathcal{L}(\mathcal{E})$. As a consequence, $\zeta I - X$ maps $\mathcal{D}(X)$ bijectively onto $\mathcal{R}(\zeta I - X)$, which is dense in $\mathcal{E}$, and $(\zeta I - X)^{-1}$ is continuous. Thus $\zeta \in r(X)$, i.e.,

$$r(\overline{X}) \subset r(X).$$

.

III. We shall establish the opposite inclusion. If $\zeta \in r(X)$, $\zeta I - X$ is injective, $\mathcal{R}(\zeta I - X)$ is dense in $\mathcal{E}$ and $(\zeta I - X)^{-1} \in \mathcal{L}(\mathcal{R}(\zeta I - X))$. Hence, there is a positive constant k such that

$$\|(\zeta I - X)^{-1}x\| \leq k\|x\|$$

for all $x \in \mathcal{R}(\zeta I - X)$. We show now that $\zeta I - \overline{X}$ is injective. Let $y \in \mathcal{D}(\overline{X})$ be such that $(\zeta I - \overline{X})y = 0$. Since $\Gamma(\zeta I - \overline{X}) = \overline{\Gamma(\zeta I - X)}$, for every $\epsilon > 0$ there is $x \in \mathcal{D}(X)$ for which $\|x - y\| < \epsilon$ and

$$\|(\zeta I - X)x\| = \|(\zeta I - X)x - (\zeta I - \overline{X})y\| < \epsilon.$$

Being

$$\begin{aligned}
\|y\| &\leq \|y - x\| + \|x\| < \epsilon + \|x\| \\
&= \epsilon + \|(\zeta I - X)^{-1}(\zeta I - X)x\| \\
&\leq \epsilon + k\|(\zeta I - X)x\| < (1 + k)\epsilon,
\end{aligned}$$

letting $\epsilon \to 0$ yields $y = 0$, showing that $\zeta I - \overline{X}$ is injective. Hence $(\zeta I - \overline{X})^{-1}$ exists and is a closed extension of $(\zeta I - X)^{-1}$. Since

$\mathcal{D}((\zeta\,I{-}X)^{-1}) = \mathcal{R}(\zeta\,I{-}X)$ is dense in $\mathcal{E}$ and $(\zeta\,I{-}X)^{-1}$ is continuous, $(\zeta\,I - \overline{X})^{-1}$ is continuous. By Lemma 1.1.12, $\mathcal{R}(\zeta\,I - \overline{X})^{-1} = \mathcal{E}$ and $(\zeta\,I - \overline{X})^{-1} \in \mathcal{L}(\mathcal{E})$. Thus $\zeta \in r(\overline{X})$. ∎

Examples Let $\mathcal{E} = C[0,1]$ and let X be the linear operator defined by (1.2), with domain $\mathcal{D}(X) = C^1[0,1]$. For every $\zeta \in \mathbf{C}$, the function $x \in C^1[0,1]$ defined by $x(t) = e^{t\zeta}$, is such that

$$Xx = \zeta x.$$

Hence $p\sigma(X) = \mathbf{C}$.

Let now $X_1 = X_{|\mathcal{D}_1}$, where $\mathcal{D}_1$ is given by (1.3), and let $\zeta \in \mathbf{C}$. For every $y \in C[0,1]$, the differential equation

$$\dot{x} - \zeta\,x = y$$

has a unique solution $x \in \mathcal{D}_1$, which is given by the integral

$$x(t) = e^{\zeta t} \int_0^t e^{-\zeta s} y(s)\,ds.$$

As a consequence, $r(X_1) = \mathbf{C}$, and

$$((\zeta\,I - X_1)^{-1}(y))(t) = -e^{\zeta t} \int_0^t e^{-\zeta s} y(s)\,ds.$$

In a similar way one sees that, $r(X_2) = \mathbf{C}$, where $X_2 = X_{|\mathcal{D}_2}$ and $\mathcal{D}_2$ is given by (1.4).

Exercises 1. For $c \in \mathbf{C}\backslash\{0\}$, let $X_3 = X_{|\mathcal{D}_3}$, where $\mathcal{D}_3$ is given by (1.5). Show that $\sigma(X_3) = p\sigma(X_3) = \{\log c + 2n\pi i : n \in \mathbf{Z}\}$, and compute $(\zeta\,I - X_3)^{-1}y$ for $\zeta \notin p\sigma(X_3)$ and $y \in C[0,1]$.

2. Le $X_0 = X_{|\mathcal{D}_0}$, where $\mathcal{D}_0$ is given by (1.6). Show that $\sigma(X_0) = r\sigma(X_0) = \mathbf{C}$.

1.10 The spectrum of a closed linear operator

Let X be a closed linear operator with domain $\mathcal{D}(X) \subset \mathcal{E}$ and range $\mathcal{R}(X) \subset \mathcal{E}$, for which $r(X) \neq \emptyset$ (i.e. $\sigma(X) \neq \mathbf{C}$).

Theorem 1.10.1 *The resolvent set $r(X)$ is open, and the function $\zeta \mapsto (\zeta I - X)^{-1}$ is holomorphic on every connected component of $r(X)$.*

Proof For $\tau \in r(X)$ let $\|(\tau I - X)^{-1}\| = \frac{1}{r}$. The series

$$\sum_{n=0}^{+\infty} (\tau - \zeta)^n (\tau I - X)^{-(n+1)} \tag{1.48}$$

is norm-convergent when ζ varies on all compact subsets of the open disc $\Delta(\tau, r) = \{\zeta \in \mathbf{C} : |\zeta - \tau| < r\}$, and thus defines an operator $Y(\zeta) \in \mathcal{L}(\mathcal{E})$.

Writing $\zeta I - X = \tau I - X + (\zeta - \tau)I$, for $x \in \mathcal{D}(X)$ and $\zeta \in \Delta(\tau, r)$ one has

$$\begin{aligned}
Y(\zeta)(\zeta I - X)x &= x + \sum_{n=1}^{+\infty} (\tau - \zeta)^n (\tau I - X)^{-n} x \\
&\quad + (\zeta - \tau) Y(\zeta) x \\
&= x + (\tau - \zeta) \sum_{n=0}^{+\infty} (\tau - \zeta)^n (\tau I - X)^{-(n+1)} x \\
&\quad + (\zeta - \tau) Y(\zeta) x \\
&= x + (\tau - \zeta) Y(\zeta) x + (\zeta - \tau) Y(\zeta) x,
\end{aligned}$$

i.e.

$$Y(\zeta)(\zeta I - X) = I$$

on $\mathcal{D}(X)$ for all $\zeta \in \Delta(\tau, r)$.

Now, let $x \in \mathcal{E}$, and, for $m = 1, 2, \ldots$, and $\zeta \in \Delta(\tau, r)$, let

$$x_m = \sum_{n=0}^{m} (\tau - \zeta)^n (\tau I - X)^{-(n+1)}(x).$$

Then $x_m \in \mathcal{D}(X)$ and $x = \lim_{m \to +\infty} x_m = Y(\zeta)x$. Since $(\tau I - X)^{-1}(\mathcal{E}) \subset \mathcal{D}(X)$ and

$$X(\tau I - X)^{-1} = -I + \tau(\tau I - X)^{-1}, \tag{1.49}$$

then

$$X x_m = \sum_{n=0}^{m} (\tau - \zeta)^n X (\tau I - X)^{-(n+1)} x$$

$$= \sum_{n=0}^{m} (\tau - \zeta)^n X (\tau I - X)^{-1} (\tau I - X)^{-n} x$$

$$= -\sum_{n=0}^{m} (\tau - \zeta)^n (\tau I - X)^{-n} x$$

$$+ \tau \sum_{n=0}^{m} (\tau - \zeta)^n (\tau I - X)^{-(n+1)} x$$

$$= -x - (\tau - \zeta) \sum_{n=0}^{m-1} (\tau - \zeta)^n (\tau I - X)^{-(n+1)} x$$

$$+ \tau \sum_{n=0}^{m} (\tau - \zeta)^n (\tau I - X)^{-(n+1)} x$$

$$= -x - (\tau - \zeta) x_{m-1} + \tau x_m,$$

and therefore

$$\lim_{m \to +\infty} X x_m = -x - (\tau - \zeta) Y(\zeta) x + \tau Y(\zeta) x$$

$$= -x + \zeta Y(\zeta) x.$$

Since X is closed, then $Y(\zeta) x \in \mathcal{D}(X)$ and $X Y(\zeta) x = -x + \zeta Y(\zeta) x$, i.e.

$$(\zeta I - X) Y(\zeta) = I$$

on $\mathcal{E}$.

In conclusion

$$(\zeta I - X)^{-1} = Y(\zeta) \tag{1.50}$$

for any $\zeta \in \Delta(\tau, r)$. Thus $\Delta(\tau, r) \subset r(X)$ - and therefore $r(X)$ is open in $\mathbf{C}$ - and moreover the function $\zeta \mapsto (\zeta I - X)^{-1}$, with values in $\mathcal{L}(\mathcal{E})$, is holomorphic on each connected component of the open set $r(X)$. ∎

For $\tau \in r(X)$, let $d(\tau)$ be the distance of τ from $\sigma(X)$: $d(\tau) = \inf\{|\tau - \zeta| : \zeta \in \sigma(X)\}$.

Then $\Delta(\tau, d(\tau)) \subset r(X)$ and the series (1.48) converges uniformly in $\mathcal{L}(\mathcal{E})$, when ζ varies on all compact subsets of $\Delta(\tau, d(\tau))$.

Proposition 1.10.2 *For all $\tau \in r(X)$,*

$$\|(\tau I - X)^{-1}\| \geq \frac{1}{d(\tau)}.$$

Proof If $d(\tau)\,\|(\tau I - X)^{-1}\| < 1$. there exists some $\epsilon > 0$ such that

$$\|(\tau I - X)^{-1}\| \leq \frac{1}{d(\tau) + \epsilon}.$$

Thus, the series (1.48) converges in $\mathcal{L}(\mathcal{E})$ uniformly on all compact subsets of $\Delta(\tau, d(\tau) + \epsilon)$, defining an operator $Y(\zeta) \in \mathcal{L}(\mathcal{E})$ satisfying (1.50) and contradicting the fact that $\Delta(\tau, d(\tau) + \epsilon) \cap \sigma(X) \neq \emptyset$. ∎

Now, let $X \in \mathcal{L}(\mathcal{E})$. If $\zeta \in \mathbf{C}$ is such that $|\zeta| > \|X\|$, the numerical series

$$\sum_{n=0}^{+\infty} \frac{1}{\zeta^n} \|X^n\|$$

converges uniformly on all compact subsets of the open set $\mathbf{C}\backslash\overline{\Delta(0, \|X\|)}$. Being

$$(\zeta I - X)^{-1} = \frac{1}{\zeta}(I - \frac{1}{\zeta}X)^{-1} = \frac{1}{\zeta}\sum_{n=0}^{+\infty} \frac{1}{\zeta^n} X^n,$$

then $(\zeta I - X)^{-1} \in \mathcal{L}(\mathcal{E})$, and therefore

$$\sigma(X) \subset \overline{\Delta(0, \|X\|)}. \tag{1.51}$$

Since

$$\|(\zeta I - X)^{-1}\| \leq \frac{1}{|\zeta|} \sum_{n=0}^{+\infty} \frac{1}{|\zeta|^n} \|X^n\|$$

$$\leq \frac{1}{|\zeta|} \sum_{n=0}^{+\infty} \left(\frac{\|X\|}{|\zeta|}\right)^n = \frac{1}{|\zeta| - \|X\|},$$

the holomorphic function $\zeta \mapsto (\zeta I - X)^{-1}$ vanishes at infinity. If $\sigma(X) = \emptyset$, the function would be holomorphic on $\mathbf{C}$, and therefore, by the Liouville theorem, would vanish identically. This contradiction, together with (1.51) proves

Lemma 1.10.3 *If $X \in \mathcal{L}(\mathcal{E})$, then $\sigma(X)$ is a non-empty compact subset of* **C**.

By the Cauchy-Hadamard theorem,

$$
\begin{aligned}
\max\{|\zeta| : \zeta \in \sigma(X)\} &= \min\{t > 0 : (\zeta I - X)^{-1} \in \mathcal{L}(\mathcal{E}) \\
&\qquad \text{for all } |\zeta| > t\} \\
&= (\max\{t > 0 : (I - \zeta X)^{-1} \in \mathcal{L}(\mathcal{E}) \\
&\qquad \text{whenever } |\zeta| < t\})^{-1} \\
&= (\text{radius of convergence of } \sum_{n=0}^{+\infty} \zeta^n X^n)^{-1} \\
&= \limsup_{n \to +\infty} \|X^n\|^{\frac{1}{n}}.
\end{aligned}
$$

Theorem 1.6.1 implies then

Theorem 1.10.4 *The spectral radius $\rho(X)$ of any $X \in \mathcal{L}(\mathcal{E})$ is given by*

$$\rho(X) = \max\{|\zeta| : \zeta \in \sigma(X)\}.$$

Let $X_o \in \mathcal{L}(\mathcal{E})$ be invertible, with $X_o^{-1} \in \mathcal{L}(\mathcal{E})$. For $X \in \mathcal{L}(\mathcal{E})$, let $Z = X - X_o$. If $\|X_o^{-1}\| \, \|Z\| < 1$, i.e., if

$$\|X - X_o\| = \|Z\| < \frac{1}{\|X_o^{-1}\|},$$

the numerical series $\sum_{n=0}^{+\infty} \|(X_o^{-1} Z)^n\|$ converges. Therefore

$$X^{-1} = (I + X_o^{-1} Z)^{-1} X_o^{-1} = (\sum_{n=0}^{+\infty} (-1)^n (X_o^{-1} Z)^n) X_o^{-1}$$

exists and $X^{-1} \in \mathcal{L}(\mathcal{E})$. That proves

Lemma 1.10.5 *The set*

$$\{X \in \mathcal{L}(\mathcal{E}) : X^{-1} \text{ exists, and } X^{-1} \in \mathcal{L}(\mathcal{E})\} \tag{1.52}$$

is open in the Banach space $\mathcal{L}(\mathcal{E})$.

1.11   Spectra in complex Banach algebras

Let $\mathcal{A}$ be a unital Banach algebra, and, for $a \in \mathcal{A}$, let $L_a \in \mathcal{L}(\mathcal{A})$, be the operator $L_a : x \mapsto ax$. Since, if a is invertible,

$$L_a \, L_{a^{-1}} = L_{a^{-1}} \, L_a = L_1 = I,$$

then L_a is invertible, and $L_a^{-1} \in \mathcal{L}(\mathcal{A})$, if, and only if, a is invertible in $\mathcal{A}$. As a consequence, for any $a \in \mathcal{A}$,

$$\sigma(a) = \sigma(L_a), \tag{1.53}$$

and therefore $\rho(a) = \rho(L_a)$. Lemma 1.10.3 and (1.53) imply then

Lemma 1.11.1 *If $\mathcal{A}$ is a unital complex Banach algebra and $a \in \mathcal{A}$, then $\sigma(a)$ is a non-empty compact subset of $\mathbf{C}$.*

The previous results established for $\mathcal{L}(\mathcal{E})$ hold, more in general, for unital complex Banach algebras.

Consider the function $a \mapsto \sigma(a)$ from $\mathcal{A}$ to the compact sets of $\mathbf{C}$. We shall prove

Theorem 1.11.2 *The function $a \mapsto \sigma(a)$ is upper semicontinuous.*

That means that, for any $a_o \in \mathcal{A}$ and for any open neighbourhood U of $\sigma(a_o)$ in $\mathbf{C}$, there is a neighbourhood V of a_o in $\mathcal{A}$ such that

$$a \in V \ \Rightarrow \ \sigma(a) \subset U.$$

Proof Let $\{a_\nu\}$ be a sequence in $\mathcal{A}$ converging to a_o and such that $\sigma(a_\nu) \not\subset U$. Replacing, if necessary, $\{a_\nu\}$ by a subsequence, we can assume

$$\|a_o - a_\nu\| \le \frac{1}{\nu} \text{ for } \nu = 1, 2, \dots .$$

Let $\zeta_\nu \in \sigma(a_\nu) \cap (\mathbf{C} \backslash U)$. Since

$$\|a_\nu\| \le \|a_o\| + \frac{1}{\nu},$$

and therefore

$$\rho(a_\nu) \le \|a_\nu\| \le \|a_o\| + \frac{1}{\nu} \le \|a_o\| + 1,$$

the sequence $\{\zeta_\nu\}$ is bounded. Let ζ_o be a limit value of $\{\zeta_\nu\}$, and let $\{\zeta_{\nu_j}\}$ be a subsequence of $\{\zeta_\nu\}$ converging to ζ_o. Since $\zeta_o \notin \mathbf{C}\backslash U$, then $\zeta_o \notin \sigma(a_o)$.

If $\zeta_o = 0$, a_o is invertible. Since $\mathcal{A}^{-1}$ is open, and since

$$\|a_{\nu_j} - \zeta_{\nu_j}1 - a_o\| \le \|a_{\nu_j} - a_o\| + |\zeta_{\nu_j}| \to 0$$

as $j \to +\infty$, then $a_{\nu_j} - \zeta_{\nu_j}1 \in \mathcal{A}^{-1}$ when $j \gg 0$; which is absurd.

If $\zeta_o \ne 0$, then $\zeta_{\nu_j} \ne 0$ when $j \gg 0$. Since

$$\left\| 1 - \frac{1}{\zeta_{\nu_j}}a_{\nu_j} - \left(1 - \frac{1}{\zeta_o}a_o\right) \right\| = \left\| \frac{1}{\zeta_o}a_o - \frac{1}{\zeta_{\nu_j}}a_{\nu_j} \right\| \to 0$$

as $j \to +\infty$, then

$$\lim_{j \to +\infty} \left(\zeta_{\nu_j}1 - a_{\nu_j}\right) = \zeta_o 1 - a_o.$$

Therefore $\zeta_{\nu_j}1 - a_{\nu_j} \in \mathcal{A}^{-1}$ when $j \gg 0$; which is absurd. This contradiction completes the proof. ∎

Corollary 1.11.3 *The function $a \mapsto \rho(a)$ is upper semicontinuous.*

1.12 Resolvent functions of closed operators

We have seen before that, for any $X \in \mathcal{L}(\mathcal{E})$, $\sigma(X)$ is compact. What happens if $\sigma(X)$ is compact, and X is closed but not continuous? The answer to this question is given by the following

Theorem 1.12.1 *Let the linear operator X be closed. Then either the holomorphic function $\zeta \mapsto (\zeta I - X)^{-1}$, from $r(X)$ to $\mathcal{L}(\mathcal{E})$, vanishes at infinity and $X \in \mathcal{L}(\mathcal{E})$, or the function has an essential singularity at infinity,*

Proof Assume that $(\bullet I - X)^{-1}$ has not an essential singularity at infinity. Since $(\bullet I - X)^{-1} \neq 0$, there exist:

$R > 0$; a polynomial P of degree $p \geq 0$, given by

$$P(\zeta) = A_o + \zeta A_1 + \cdots + \zeta^p A_p,$$

with coefficients $A_j \in \mathcal{L}(\mathcal{E})$ for $j = 0, 1, \ldots, p$; a power series

$$Q(\zeta) = \zeta B_1 + \zeta^2 B_2 + \cdots$$

with coefficients $B_j \in \mathcal{L}(\mathcal{E})$ for $j = 1, 2, \ldots$, and positive radius of convergence,

such that

$$(\zeta I - X)^{-1} = P(\zeta) + Q\left(\frac{1}{\zeta}\right)$$

for all $\zeta \in \mathbf{C}$ with $|\zeta| > R$.

I. We show now that $P = 0$. For all $x \in \mathcal{E}$,

$$\lim_{\zeta \to \infty} \frac{1}{\zeta^{p+1}} (\zeta I - X)^{-1} x = 0.$$

Since, by (1.49), $X(\zeta I - X)^{-1} \in \mathcal{L}(\mathcal{E})$, then, again by (1.49),

$$\lim_{\zeta \to \infty} \frac{1}{\zeta^{p+1}} X(\zeta I - X)^{-1} x = \lim_{\zeta \to \infty} \frac{1}{\zeta^p} (\zeta I - X)^{-1} x = A_p x.$$

The operator X being closed implies then that $A_p x = 0$ for all $x \in \mathcal{E}$, and therefore $A_p = 0$. Iteration of this argument yields: $P = 0$.

II. As a consequence,

$$(\zeta I - X)^{-1} = Q\left(\frac{1}{\zeta}\right) = \frac{1}{\zeta} B_1 + \frac{1}{\zeta^2} B_2 + \cdots \qquad (1.54)$$

whenever $|\zeta| > R$. Thus,

$$\lim_{\zeta \to \infty} (\zeta I - X)^{-1} x = 0 \qquad (1.55)$$

and, by (1.49),

$$\lim_{\zeta \to \infty} X(\zeta I - X)^{-1} x = (-I + B_1) x \qquad (1.56)$$

for all $x \in \mathcal{E}$. On the other hand, since X is closed, (1.49) implies that

$$\lim_{\zeta \to \infty} X(\zeta I - X)^{-1} x = 0$$

for all $x \in \mathcal{E}$. Thus,

$$B_1 = I, \tag{1.57}$$

and, by (1.54),

$$\lim_{\zeta \to \infty} \zeta(\zeta I - X)^{-1} x = x. \tag{1.58}$$

Furthermore, in view of (1.54) and (1.57), (1.49) yields

$$\lim_{\zeta \to \infty} \zeta X(\zeta I - X)^{-1} x = \lim_{\zeta \to \infty} (-\zeta I + \zeta I + B_2$$
$$+ \frac{1}{\zeta} B_3 + \cdots) x = B_2 x.$$

The operator X being closed, (1.58) yields $X x = B_2 x$ for all $x \in \mathcal{E}$, i.e.

$$X = B_2 \in \mathcal{L}(\mathcal{E}).$$

∎

Remark The proof of the theorem uses only the fact that $(\zeta I - X)^{-1}$ is the right inverse of $\zeta I - X$, Hence the following proposition holds.

Proposition 1.12.2 *Let X be a closed linear operator. If there are a compact set $K \subset \mathbf{C}$ and a holomorphic function from $\mathbf{C} \backslash K$ to $\mathcal{L}(\mathcal{E})$ such that*

$$(\zeta I - X) F(\zeta) = I$$

whenever $\zeta \notin K$, and if F does not have an essential singularity at infinity, then $X \in \mathcal{L}(\mathcal{E})$.

If X is a closed linear operator, the *extended spectrum* and the *extended resolvent set* of X are, by definition, the subsets $\tilde{\sigma}(X)$ and $\tilde{r}(X)$ of the Riemann sphere S^2, defined by: $\tilde{\sigma}(X) = \sigma(X)$ if X is a continuous operator, and by $\tilde{\sigma}(X) = \sigma(X) \cup \{\infty\}$ if X is unbounded; $\tilde{r}(X) = S^2 \backslash \tilde{\sigma}(X)$.

Theorem 1.12.3 *Let X be closed and invertible. If $\zeta \in r(X)$, with $\zeta \neq 0$, then $\frac{1}{\zeta} \in r(X^{-1})$. If, moreover, $\mathcal{R}(X) = \mathcal{E}$, i.e., if $0 \in r(X)$, then $\tilde{r}(X^{-1})$ is the image of $\tilde{r}(X)$ by the holomorphic automorphism $\tau \mapsto \frac{1}{\tau}$ of the Riemann sphere.*

Proof. If $\zeta \in r(X)$, the linear operator $Z(\zeta)$ defined by

$$Z(\zeta) = X\,(\zeta\,I - X)^{-1} = -I + \zeta(\zeta\,I - X)^{-1}$$

is bounded on $\mathcal{E}$. For any $x \in \mathcal{E}$,

$$Z(\zeta)x \in \mathcal{R}(X) = \mathcal{D}(X^{-1}),$$

and therefore, if $\zeta \neq 0$,

$$X^{-1}\,Z(\zeta)x = (\zeta\,I - X)^{-1}x = \frac{1}{\zeta}(I + Z(\zeta))x.$$

Thus

$$\left(\frac{1}{\zeta}I - X^{-1}\right) Z(\zeta)x = -\frac{1}{\zeta}x,$$

i.e.

$$x = -\zeta\left(\frac{1}{\zeta}I - X^{-1}\right) Z(\zeta)x$$

for all $x \in \mathcal{E}$, showing that

$$\mathcal{R}\left(\frac{1}{\zeta}I - X^{-1}\right) = \mathcal{E}.$$

In order to prove that $\frac{1}{\zeta}I - X^{-1}$ is injective, let $y \in \mathcal{D}(X^{-1}) = \mathcal{R}(X)$ be such that

$$\left(\frac{1}{\zeta}I - X^{-1}\right) y = 0,$$

i.e. $\zeta X^{-1}y = y$. Then $y \in \mathcal{R}(X^{-1}) = \mathcal{D}(X)$, and

$$Xy = \zeta y.$$

Thus $y = 0$ (because $\zeta \in r(X)$), and $\frac{1}{\zeta}I - X^{-1}$ maps $\mathcal{D}(X^{-1})$ bijectively onto $\mathcal{E}$. Since X is closed - and therefore also X^{-1} is closed - then $\frac{1}{\zeta} \in r(X^{-1})$.

If $0 \in r(X)$, then $X^{-1} \in \mathcal{L}(\mathcal{E})$, and therefore $\infty \in \tilde{r}(X^{-1})$.

If $\infty \in \tilde{r}(X)$, then $X \in \mathcal{L}(\mathcal{E})$. Since $\mathcal{R}(X) = \mathcal{E}$ and X is invertible, then $0 \in r(X^{-1})$. ∎

Corollary 1.12.4 *Let $\zeta \in \mathbf{C}$ be such that $\zeta I - X$ be invertible. If X is closed. the extended spectrum of $(\zeta I - X)^{-1}$ contains the image of the spectrum of X by the map $\tau \mapsto \frac{1}{\tau - \zeta}$.*

Lemma 1.12.5 *Let X be a closed linear operator. If $\emptyset \neq \sigma(X) \neq \mathbf{C}$ and if $\zeta \notin \sigma(X)$, then*

$$\mathrm{dist}(\zeta, \sigma(X)) = (\rho((\zeta I - X)^{-1}))^{-1}.$$

Proof By Corollary 1.12.4,

$$\sigma((\zeta I - X)^{-1}) \cup \{0\} = \{0\} \cup \left\{ \frac{1}{\zeta - \tau} : \tau \in \sigma(X) \right\},$$

and therefore

$$\rho((\zeta I - X)^{-1}) = \sup \left\{ \frac{1}{\zeta - \tau} : \tau \in \sigma(X) \right\}$$

$$= \frac{1}{\inf\{|\zeta - \tau| : \tau \in \sigma(X)\}} = \frac{1}{\mathrm{dist}(\zeta, \sigma(X))}.$$

∎

Thus, both $\rho((\zeta I - X)^{-1})$ and $\|(\zeta I - X)^{-1}\|$ ($\geq \rho((\zeta I - X)^{-1})$) tend to infinity when ζ tends to $\sigma(X)$.

We shall now investigate the structure of the boundary of the spectrum of a bounded linear operator. We begin by establishing a technical lemma.

An element a of a Banach algebra $\mathcal{A} \backslash \{0\}$ is called a *generalized left zero divisor* if there is a sequence of elements b_n of $\mathcal{A}$ with $\|b_n\| = 1$, such that $\lim_{n \to +\infty} ab_n = 0$. In a similar way one defines a *generalized right zero divisor*.

Lemma 1.12.6 *Let $\{a_n\}$ be a sequence of elements of A converging to some $a \in A$. If every a_n is invertible, then either*

$$\limsup_{n \to +\infty} \|a_n^{-1}\| < \infty \qquad (1.59)$$

and a is invertible, or ($\limsup_{n \to +\infty} \|a_n^{-1}\| = \infty$ and) a is zero or a generalized left zero divisor.

Proof. I. If (1.59) holds, there is a constant $k > 0$ such that $\|a_n^{-1}\| \leq k$ for all $n \in \mathbf{N}$. Since

$$a_m^{-1} - a_n^{-1} = a_m^{-1}(a_n - a_m)a_n^{-1},$$

and therefore

$$\|a_m^{-1} - a_n^{-1}\| \leq k^2 \|a_n - a_m\|,$$

then $\{a_n^{-1}\}$ is a Cauchy sequence. Denoting by b its limit, then

$$ab = (a - a_n)b + a_n(b - a_n^{-1}) + 1 \to 1$$

and

$$ba = (b - a_n^{-1})a + a_n^{-1}(a - a_n) + 1 \to 1$$

as $n \to +\infty$, showing that a is invertible.

II. If

$$\limsup_{n \to +\infty} \|a_n^{-1}\| = \infty,$$

there is a subsequence $\{c_n\}$ of $\{a_n\}$ such that $\lim_{n \to +\infty} \|c_n^{-1}\| = \infty$. Setting $d_n = \frac{1}{\|c_n^{-1}\|} c_n^{-1}$ for $n \gg 0$, then

$$ad_n = (a - c_n)d_n + c_n d_n = (a - c_n)d_n + \frac{1}{\|c_n^{-1}\|} 1,$$

whence

$$\|ad_n\| \leq \|a - c_n\| + \frac{1}{\|c_n^{-1}\|} \to 0$$

as $n \to +\infty$, showing that a is either zero or a generalized left zero divisor. ∎

Theorem 1.12.7 *Let $X \in \mathcal{L}(\mathcal{E})$. If ζ is contained in the boundary of $\sigma(X)$, then $\zeta I - X$ is either zero or a generalized left zero divisor.*

Proof. Since ζ belongs to the boundary of $\sigma(X)$, there is a sequence $\{\zeta_n\}$ of complex numbers $\zeta_n \notin \sigma(X)$ converging to ζ, and therefore such that $\lim_{n \to +\infty}(\zeta_n I - X) = \zeta I - X$.

Since $\zeta_n I - X$ is invertible in $\mathcal{L}(\mathcal{E})$ for all $n \in \mathbf{N}$, Lemma 1.12.6 yields the conclusion. ∎

Letting

$$A_n = \frac{1}{\|(\zeta_n I - X)^{-1}\|}(\zeta_n I - X)^{-1} \in \mathcal{L}(\mathcal{E}),$$

then

$$\lim_{n \to +\infty} (\zeta I - X) A_n = 0.$$

Since $\|A_n\| = 1$, for every $\epsilon \in (0, 2)$ and every $n = 1, 2, \ldots$, there is some $y_n \in \mathcal{E}$ for which $1 \le \|y_n\| \le 1 + \epsilon$ and $\|A_n y_n\| = 1$. For any $\delta > 0$, there is some $n_o \in \mathbf{N}$ such that, whenever $n \ge n_o$, $\|(\zeta I - X) A_n\| < \delta$, and therefore

$$\|(\zeta I - X) A_n y_n\| \le \|(\zeta I - X) A_n\| \, \|y_n\|$$
$$< (1 + \epsilon)\delta < 3\delta,$$

whence

$$\lim_{n \to +\infty} (\zeta I - X) A_n y_n = 0.$$

Letting $x_n = A_n y_n$, that proves

Lemma 1.12.8 *If $X \in \mathcal{L}(\mathcal{E})$, for any ζ in the boundary of $\sigma(X)$, there is a sequence $\{x_n\}$ of vectors $x_n \in \mathcal{E}$, with $\|x_n\| = 1$ for all $n \in \mathbf{N}$, such that*

$$\lim_{n \to +\infty} (\zeta I - X)x_n = 0. \tag{1.60}$$

As an application of this lemma, we will investigate the spectral structure of the linear isometries of $\mathcal{E}$.

Lemma 1.12.9 *If $X \in \mathcal{L}(\mathcal{E})$ is an isometry and if ζ is a boundary point of $\sigma(X)$, then $|\zeta| = 1$.*

Proof. With the same notations of Lemma 1.12.8, since

$$| \,\|X x_n\| - |\zeta| \,| \leq \|(X - \zeta I)x_n\|,$$

(1.60) yields

$$\lim_{n \to +\infty} \|X x_n\| = |\zeta|.$$

On the other hand, $\|X x_n\| = \|x_n\| = 1$. ∎

Let $\Delta = \Delta(0, 1) = \{\zeta \in \mathbf{C} : |\zeta| < 1 \}$, and let $\partial \Delta$ be the boundary of Δ.

Corollary 1.12.10 *If $X \in \mathcal{L}(\mathcal{E})$ is an isometry, either $\sigma(X) = \overline{\Delta}$ or $\sigma(X) \subset \partial \Delta$ and X is surjective.*

Lemma 1.12.11 *If $X \in \mathcal{L}(\mathcal{E})$ is a non-surjective isometry, then $\Delta \subset r\sigma(X)$.*

Proof. Clearly, every $\zeta \in \Delta$ is not an eigenvalue of X. Assume that $\zeta \in \Delta$ is a point of $c\sigma(X)$. Given any $y \in \mathcal{E}$, there is a sequence $\{x_n\}$ in $\mathcal{E}$ such that

$$\lim_{n \to +\infty} (\zeta I - X)x_n = y.$$

Since

$$\|(\zeta I - X)(x_m - x_n)\| \geq |\,\|X(x_m - x_n)\| - |\zeta|\|x_m - x_n\| \,|$$
$$= (1 - |\zeta|)\|x_m - x_n\|,$$

the fact that $\{(\zeta I - X)x_n\}$ is a Cauchy sequence implies that $\{x_n\}$ is a Cauchy sequence, and therefore converges to some $x \in \mathcal{E}$. But then $(\zeta I - X)x = \lim_{n \to +\infty}(\zeta I - X)x_n = y$, and so $(\zeta I - X)\mathcal{E} = \mathcal{E}$. Thus $\zeta \in r(X)$, contradicting Corollary 1.12.10. ∎

1.13 Pseudoresolvents

Let $X : \mathcal{D}(X) \subset \mathcal{E} \to \mathcal{E}$ be a closed linear operator for which $r(X) \neq \emptyset$. If $\zeta, \tau \in r(X)$, then

$$
\begin{aligned}
(\zeta I - X)^{-1} - (\tau I - X)^{-1} &= (\zeta I - X)^{-1}(\tau I - X - \\
&\quad (\zeta I - X))(\tau I - X)^{-1} \\
&= (\tau - \zeta)(\zeta I - X)^{-1}(\tau I - X)^{-1}.
\end{aligned}
$$

This identity, known also as the *resolvent equation*, characterizes the *resolvent function* $(\bullet I - X)^{-1}$, as will be shown now.

Let U be a non-empty subset of $\mathbf{C}$, and, for every $\zeta \in U$, let $R(\zeta) : \mathcal{F} \to \mathcal{F}$ be a linear operator defined on a complex vector space $\mathcal{F}$, with values in $\mathcal{F}$, such that

$$
R(\zeta) - R(\tau) = (\tau - \zeta)R(\zeta)\, R(\tau) \tag{1.61}
$$

for $\zeta, \tau \in U$. The function $R : \zeta \mapsto R(\zeta)$ is called a *pseudoresolvent*.

Clearly, the resolvent function is a pseudoresolvent.

Note the $R(\zeta)$ and $R(\tau)$ commute.

Let R be a pseudoresolvent.

Lemma 1.13.1 $\ker R(\zeta)$ *and* $\mathcal{R}(R(\zeta))$ *are independent of* $\zeta \in U$.

Proof If $R(\zeta)x = 0$ for some $\zeta \in U$ and some $x \in \mathcal{F}$, then, for all $\tau \in U$,

$$
\begin{aligned}
R(\tau)x &= (R(\tau) - R(\zeta))x \\
(\zeta - \tau)R(\tau)\, R(\zeta)x &= 0.
\end{aligned}
$$

For any $y \in \mathcal{F}$,

$$
\begin{aligned}
R(\zeta)y &= R(\tau)y + (R(\zeta) - R(\tau))y \\
&= R(\tau)y + (\tau - \zeta)R(\tau)\, R(\zeta)y \\
&= R(\tau)(y + (\tau - \zeta)R(\zeta)y) \in \mathcal{R}(R(\tau)).
\end{aligned}
$$

∎

If, and only if, $\ker R(\zeta) = \{0\}$, $R(\zeta)$ is invertible (for some $\zeta \in U$, and therefore for all $\zeta \in U$), and

$$R(\zeta)^{-1} : \mathcal{D}(R(\zeta)^{-1}) = \mathcal{R}(R(\zeta)) \rightarrow$$
$$\rightarrow \mathcal{R}(R(\zeta)^{-1}) = \mathcal{D}(R(\zeta)) = \mathcal{F}.$$

Multiplying both terms of (1.61) by $R(\zeta)^{-1}$ on the left, and by $R(\tau)^{-1}$ on the right, we obtain

$$\zeta I - R(\zeta)^{-1} = \tau - R(\tau)^{-1}.$$

That proves

Lemma 1.13.2 *If* $\ker R(\zeta) = \{0\}$ *for some* $\zeta \in U$, *there exists a unique linear operator* X *such that* $R(\tau)^{-1} = \tau I - X$ *for all* $\tau \in U$. *The domain of* X *is* $\mathcal{D}(X) = \mathcal{R}(R(\zeta))$ *and its range is* $\mathcal{R}(X) = \mathcal{D}(R(\zeta)) = \mathcal{F}$.

From now on, the complex vector space $\mathcal{F}$ will be replaced by the complex Banach space $\mathcal{E}$.

Since $\mathcal{D}(R(\zeta)) = \mathcal{E}$, the linear operator $R(\zeta)$ is bounded if, and only if, it is closed. Hence, the operator X in Lemma 1.13.2 is closed if, and only if, $R(\zeta)$ is bounded. Furthermore, the condition $\ker R(\zeta) = \{0\}$ is equivalent to $\sigma(X) \cap U = \emptyset$.

Suppose now that $U = (a, +\infty)$ for some $a \in \mathbf{R}$.

Theorem 1.13.3 *Let* $R(\zeta) \in \mathcal{L}(\mathcal{E})$ *and suppose that there is a finite constant* $M > 0$ *such that*

$$\|R(\zeta)\| \le \frac{M}{\zeta - a} \quad \forall \zeta \in (a, +\infty). \tag{1.62}$$

If $\mathcal{R}(R(\zeta))$ *is dense in* $\mathcal{E}$, *there exists a (unique) linear operator* X *for which*

$$R(\zeta) = (\zeta I - X)^{-1}. \tag{1.63}$$

The operator X *is closed, densely defined and* $\mathcal{D}(X) = \mathcal{R}(R(\zeta))$. *Vice-versa, if there exists a densely defined, closed, linear operator* X *for which (1.63) holds, then* $\mathcal{R}(R(\zeta))$ *is dense in* $\mathcal{E}$.

Proof I. Let

$$\overline{\mathcal{R}(R(\zeta))} = \mathcal{E} \tag{1.64}$$

For $y \in \mathcal{E}$ and $\tau \in U$, let $x = R(\tau)y$. If $\tau \neq \zeta$,

$$\begin{aligned}
\zeta R(\zeta)x &= \zeta R(\zeta)\,R(\tau)y \\
&= \frac{\zeta}{\tau - \zeta}(R(\zeta) - R(\tau))y \\
&= \frac{\zeta}{\tau - \zeta}(R(\zeta)y - x).
\end{aligned}$$

Since, by (1.62),

$$\lim_{\zeta \to +\infty} R(\zeta)y = 0,$$

then

$$\lim_{\zeta \to +\infty} \zeta R(\zeta)x = x$$

for all $x \in \mathcal{R}(\zeta)$, and - by (1.64) and the continuity of $R(\zeta)$ - for all $x \in \mathcal{E}$. As a consequence, $R(\zeta)$ is injective.

II. The second part of the theorem follows from the fact that, if X exists, then $\mathcal{R}(R(\zeta)) = \mathcal{D}(X)$. ∎

1.14 Isolated points of the spectrum

Let X be a closed linear operator on $\mathcal{E}$, and let ζ_o be an isolated point of $\sigma(X)$[6]. As before, let $\Delta(\zeta_o, r_o)$ be the open disc in $\mathbf{C}$ with center ζ_o and radius $r_o > 0$, and choose $r_o > 0$ in such a way that $\overline{\Delta(\zeta_o, r_o)} \cap \sigma(X) = \{\zeta_o\}$. The resolvent function is holomorphic on $\Delta(\zeta_o, r_o) \setminus \{\zeta_o\}$ and, in this domain, has a Laurent expansion:

$$(\zeta I - X)^{-1} = \sum_{\nu=-\infty}^{+\infty} (\zeta - \zeta_o)^\nu A_\nu, \tag{1.65}$$

---

[6]For further information on isolated points of the spectrum, see also [81], [40], [19], [76].

whose coefficients $A_\nu \in \mathcal{L}(\mathcal{E})$ are given by the integrals

$$A_\nu = \frac{1}{2\pi i} \int_\gamma (\zeta - \zeta_o)^{-(\nu+1)}(\zeta I - X)^{-1} d\zeta, \qquad (1.66)$$

where γ is the circle with center ζ_o and radius $r \in (0, r_o]$, oriented counterclockwise. By the Cauchy integral theorem, the integral does not depend on r.

Writing $\zeta_o I - X = \zeta I - X - (\zeta - \zeta_o)I$, equation (1.66) yields, on $\mathcal{D}(X)$,

$$A_\nu(\zeta_o I - X) = \left(\frac{1}{2\pi i} \int_\gamma (\zeta - \zeta_o)^{-(\nu+1)} d\zeta \right) I$$
$$- \frac{1}{2\pi i} \int_\gamma (\zeta - \zeta_o)^{-\nu}(\zeta I - X)^{-1} d\zeta,$$

i.e.

$$A_\nu(\zeta_o I - X) = \delta_{\nu,0} I - A_{\nu-1} \qquad (1.67)$$

on $\mathcal{D}(X)$, where $\delta_{\nu,0}$ is the Kronecker delta.

We prove now that

$$\mathcal{R}(A_\nu) \subset \mathcal{D}(X). \qquad (1.68)$$

For every $x \in \mathcal{E}$, $A_\nu x$ is expressed by the integral

$$A_\nu x = \frac{r^{-\nu}}{2\pi} \int_0^{2\pi} e^{-i\nu\theta}((\zeta_o + re^{i\theta})I - X)^{-1} x \, d\theta.$$

We approximate now the integrand uniformly on $[0, 2\pi]$ by the elementary functions

$$\varphi_n(\theta) = \sum_\mu \chi_{[\theta_\mu^n, \theta_{\mu+1}^n]}(\theta) e^{-i\nu\eta_\mu^n}((\zeta_o + re^{i\eta_\mu^n})I - X)^{-1} x,$$

where: $0 = \theta_0^n < \theta_1^n < \cdots < \theta_n^n = 2\pi$, $\mu = 0, 1, \ldots, n-1$, $\chi_{[\theta_\mu^n, \theta_{\mu+1}^n]}$ is the characteristic function of the closed interval $[\theta_\mu^n, \theta_{\mu+1}^n]$, and $\theta_\mu^n \leq \eta_\mu^n \leq \theta_{\mu+1}^n$.

Then $\int_0^{2\pi} \varphi_n(\theta)\, d\theta \in \mathcal{D}(X)$ and

$$\lim_{n \to +\infty} \frac{r^{-\nu}}{2\pi} \int_0^{2\pi} \varphi_n(\theta)\, d\theta = A_\nu x. \tag{1.69}$$

Since

$$X((\zeta_o + re^{i\eta_\mu^n})I - X)^{-1} =$$
$$-I + (\zeta_o + re^{i\eta_\mu^n})((\zeta_o + re^{i\eta_\mu^n})I - X)^{-1},$$

then

$$X\left(\frac{r^{-\nu}}{2\pi} \int_0^{2\pi} \varphi_n(\theta)\, d\theta\right)(x) =$$
$$-\frac{r^{-\nu}}{2\pi} \sum_\mu (\theta_{\mu+1}^n - \theta_\mu^n) e^{-i\nu\eta_\mu^n} x$$
$$+\frac{r^{-\nu}}{2\pi} \sum_\mu (\theta_{\mu+1}^n - \theta_\mu^n) e^{-i\nu\eta_\mu^n} \times$$
$$(\zeta_o + re^{i\eta_\mu^n})((\zeta_o + re^{i\eta_\mu^n})I - X)^{-1}x.$$

Since moreover the limits when $n \to +\infty$ of the two summands in the latter equality both exist and are equal, respectively, to

$$\frac{1}{2\pi i} \int_\gamma (\zeta - \zeta_0)^{-(\nu+1)}\, d\zeta = \delta_{\nu,0}$$

and to

$$A_{\nu-1} + \zeta_o A_\nu,$$

then

$$\lim_{n \to +\infty} X\left(\frac{r^{-\nu}}{2\pi} \int_0^{2\pi} \varphi_n(\theta)\, d\theta\right) = \tag{1.70}$$
$$= -\delta_{\nu,0} x + A_{\nu-1} x + \zeta_o A_\nu x.$$

Since X is closed, the existence of this limit and (1.70) imply that $A_\nu x \in \mathcal{D}(X)$ for all $x \in \mathcal{E}$ (which completes the proof of (1.68)), and that, moreover,

$$XA_\nu x = -\delta_{\nu,0} x + A_{\nu-1} x + \zeta_o A_\nu x,$$

for all $x \in \mathcal{E}$, i.e.

$$(\zeta_o I - X)A_\nu = \delta_{\nu,0} I - A_{\nu-1} \tag{1.71}$$

on $\mathcal{E}$ for all $\nu \in \mathbf{Z}$.

We consider now the operators $A_{-\nu}$ with $\nu > 0$.

Let γ_1 and γ_2 be two circles with center ζ_o and radii r_1 and r_2, with $0 < r_1 < r_2 < r$.

If $\zeta_1, \zeta_2 \in r(X)$, then

$$(\zeta_2 I - X)^{-1} - (\zeta_1 I - X)^{-1} = (\zeta_1 - \zeta_2)(\zeta_1 I - X)^{-1}(\zeta_2 I - X)^{-1}. \tag{1.72}$$

Hence, if $\zeta_1 \in \gamma_1$ and $\zeta_2 \in \gamma_2$, for $\nu > 0$, $\mu > 0$, Fubini's theorem and (1.66) yield

$$A_{-\nu}A_{-\mu} =$$
$$\frac{1}{(2\pi i)^2} \int_{\gamma_1} \int_{\gamma_2} (\zeta_1 - \zeta_o)^{\nu-1}(\zeta_2 - \zeta_o)^{\mu-1}(\zeta_1 I - X)^{-1}(\zeta_2 I - X)^{-1} d\zeta_1 d\zeta_2$$
$$= \frac{1}{(2\pi i)} \int_{\gamma_2} (\zeta_2 - \zeta_o)^{\mu-1} \left(\frac{1}{2\pi i} \int_{\gamma_1} \frac{(\zeta_1 - \zeta_o)^{\nu-1}}{\zeta_1 - \zeta_2} d\zeta_1 \right) (\zeta_2 I - X)^{-1} d\zeta_2 +$$
$$\frac{1}{2\pi i} \int_{\gamma_1} (\zeta_1 - \zeta_o)^{\nu-1} \left(\frac{1}{2\pi i} \int_{\gamma_2} \frac{(\zeta_2 - \zeta_o)^{\mu-1}}{\zeta_2 - \zeta_1} d\zeta_2 \right) (\zeta_1 I - X)^{-1} d\zeta_1$$
$$= \frac{1}{2\pi i} \int_{\gamma_1} (\zeta_1 - \zeta_o)^{\mu+\nu-2}(\zeta_1 I - X)^{-1} d\zeta_1,$$

i. e.,

$$A_{-\nu}A_{-\mu} = A_{-(\nu+\mu-1)}, \tag{1.73}$$

because

$$\int_{\gamma_1} \frac{(\zeta_1 - \zeta_o)^{\nu-1}}{\zeta_1 - \zeta_2} d\zeta_1 = 0$$

by the Cauchy integral theorem, and

$$\int_{\gamma_2} \frac{(\zeta_2 - \zeta_o)^{\mu-1}}{\zeta_2 - \zeta_1} d\zeta_2 = (\zeta_1 - \zeta_o)^{\mu-1}$$

by the Cauchy integral formula.

By (1.73) the operator $P := A_{-1}$ is a projector ($P^2 = P$) for which

$$PA_{-\nu} = A_{-\nu}P = A_{-\nu} \text{ for } \nu = 1, 2, \dots .$$

For $x \in \mathcal{R}(P) \subset \mathcal{D}(X)$, (1.71) yields

$$Xx = \zeta_o x + A_{-2}x$$

for all $x \in \mathcal{D}(X)$. This proves the following lemma (which can be established also as a consequence of the closed graph theorem).

Lemma 1.14.1 *The restriction of X to $\mathcal{R}(P)$ is continuous.*

It follows from (1.73) that

$$A_{-\nu} = (X - \zeta_o I)^{\nu-1} P \tag{1.74}$$

for $\nu = 1, 2, \dots$, and, from (1.67), that, whenever $\nu > 0$,

$$A_{-\nu}x = P(X - \zeta_o I)^{\nu-1}x \tag{1.75}$$

for all $x \in \mathcal{D}((X - \zeta_o I)^{\nu-1})$, that is, for all $x \in \mathcal{D}(X)$ such that $(X - \zeta_o I)^\mu \in \mathcal{D}(X)$ for $\mu = 1, 2, \dots, \nu - 2$.

In view of (1.71)

$$P = I + (X - \zeta_o I)A_0$$

and

$$A_\nu = (X - \zeta_o I)A_{\nu+1} \tag{1.76}$$

for $\nu \geq 0$.

Lemma 1.14.2 *If $\nu \geq 0$,*

$$A_\nu = (-1)^\nu A_0{}^{\nu+1}.$$

Proof It suffices to show that

$$A_{\nu+1} = -A_0 A_\nu$$

for $\nu \geq 0$.

Choosing γ, γ_1, γ_2 as before, (1.66), (1.72) and Fubini's theorem yield

$$A_0 A_\nu =$$

$$= \frac{1}{(2\pi i)^2} \int_{\gamma_1} \int_{\gamma_2} \frac{1}{(\zeta_1 - \zeta_o)(\zeta_2 - \zeta_o)^{\nu+1}}(\zeta_1 I - X)^{-1}(\zeta_2 I - X)^{-1}d\zeta_1\,d\zeta_2$$

$$= \frac{1}{(2\pi i)^2}\left(\int_{\gamma_1} \frac{1}{\zeta_1 - \zeta_o}\left[\int_{\gamma_2} \frac{1}{(\zeta_2 - \zeta_o)^{\nu+1}(\zeta_2 - \zeta_1)}d\zeta_2\right](\zeta_1 I - X)^{-1}d\zeta_1 - \right.$$

$$\left. \int_{\gamma_2} \frac{1}{(\zeta_2 - \zeta_o)^{\nu+1}}\left[\int_{\gamma_1} \frac{1}{(\zeta_1 - \zeta_o)(\zeta_2 - \zeta_1)}d\zeta_1\right](\zeta_2 I - X)^{-1}d\zeta_2\right).$$

Since $|\zeta_1 - \zeta_o| < |\zeta_2 - \zeta_o|$, then

$$\frac{1}{\zeta_2 - \zeta_1} = \frac{1}{\zeta_2 - \zeta_o - (\zeta_1 - \zeta_o)} = \sum_{\mu=0}^{+\infty} \frac{(\zeta_1 - \zeta_o)^\mu}{(\zeta_2 - \zeta_o)^{\mu+1}}, \qquad (1.77)$$

the convergence being absolute and uniform when ζ_2 varies in γ_2. Hence

$$\int_{\gamma_2} \frac{1}{(\zeta_2 - \zeta_o)^{\nu+1}(\zeta_2 - \zeta_1)}d\zeta_2 = \sum_{\mu=0}^{+\infty}(\zeta_1 - \zeta_o)^\mu \int_{\gamma_2} \frac{1}{(\zeta_2 - \zeta_o)^{\mu+\nu+2}}d\zeta_2 = 0.$$

Since the function $\zeta \mapsto \frac{1}{\zeta_2 - \zeta}$ is holomorphic in a neighbourhood of the closed disc bounded by γ_1, the Cauchy integral formula yields

$$\frac{1}{2\pi i} \int_{\gamma_1} \frac{1}{(\zeta_1 - \zeta_o)(\zeta_2 - \zeta_1)}d\zeta_1 = \frac{1}{\zeta_2 - \zeta_o}. \qquad (1.78)$$

Thus,

$$A_0 A_\nu = -\frac{1}{2\pi i} \int_{\gamma_2} \frac{1}{(\zeta_2 - \zeta_o)^{\nu+2}}(\zeta_2 I - X)^{-1}d\zeta_2 = -A_{\nu+1}.$$

∎

By a similar computation, (1.72), (1.77) and (1.78) yield

$$A_o P = P A_o =$$

$$\frac{1}{(2\pi i)^2} \int_{\gamma_2} \int_{\gamma_1} \frac{1}{\zeta_2 - \zeta_o}(\zeta_2 I - X)^{-1}(\zeta_1 I - X)^{-1}d\zeta_1 d\zeta_2$$

$$= \frac{1}{(2\pi i)^2} \int_{\gamma_2} \int_{\gamma_1} \sum_{\mu=0}^{+\infty} \frac{(\zeta_1 - \zeta_o)^\mu}{(\zeta_2 - \zeta_o)^{\mu+2}}[(\zeta_1 I - X)^{-1} - (\zeta_2 I - X)^{-1}]d\zeta_1 d\zeta_2 = 0.$$

Hence, by (1.67) and Lemma 1.14.2,

$$A_\mu A_{-\nu} = A_{-\nu} A_\mu = 0$$

whenever $\mu \geq 0$ and $\nu \geq 1$.

In view of (1.74) and of Lemma 1.14.2, (1.65) can now be written, for $\zeta \in \Delta(\zeta_o, r_o)$,

$$(\zeta I - X)^{-1} = \sum_{\nu=1}^{+\infty} (\zeta - \zeta_o)^{-\nu}(X - \zeta_o I)^{\nu-1} P$$

$$+ A_0 \sum_{\nu=0}^{+\infty} ((\zeta_o - \zeta)A_o)^\nu,$$

and therefore

$$(\zeta I - X)^{-1} = \sum_{\nu=1}^{+\infty} (\zeta - \zeta_o)^{-\nu}(X - \zeta_o I)^{\nu-1} P + A_0(I + (\zeta - \zeta_o)A_0)^{-1}$$

$$(1.79)$$

when ζ varies in a small neighbourhood of ζ_o.

By Lemma 1.14.2 and by (1.68), (1.76), if $\nu \geq 0$, then

$$A_0^{\nu+1} = (\zeta_o I - X) A_0^{\nu+2}.$$

Thus, if A_0 were invertible, with $A_0^{-1} \in \mathcal{L}(\mathcal{E})$, then also $\zeta_o I - X$ should be invertible, with $(\zeta_o I - X)^{-1} \in \mathcal{L}(\mathcal{E})$ contradicting the hypothesis whereby $\zeta_o \in \sigma(X)$. Thus $0 \in \sigma(A_0)$, and therefore also

$$0 \in \sigma(A_\nu) \quad \text{for all} \ \nu \geq 0.$$

We consider now the part of (1.79) corresponding to the negative powers of $(\zeta - \zeta_0)$, and we restrict it to the space $\mathcal{R}(P)$, on which $X - \zeta_o I$ is continuous, by Lemma 1.14.1. If $0 < r < \min(r_o, 1)$, letting $\tau = \frac{1}{\zeta - \zeta_o}$, then

$$\lim_{n \to +\infty} \tau^n (\zeta_o I - X)^n x = 0$$

for all $x \in \mathcal{R}(P)$, whenever $|\tau| > \frac{1}{r}$. Thus $\{\|\tau^n(\zeta_o I - X)_{|\mathcal{R}(P)}{}^n\| : n = 1, 2, \dots\}$ is bounded whenever $|\tau| > \frac{1}{r}$. By Lemma 1.6.6, $(\zeta_o I - X)_{\mathcal{R}(P)}$ is topologically nilpotent, i.e.

Lemma 1.14.3 *The spectrum of the restriction of $\zeta_o I - X$ to $\mathcal{R}(P)$ is* $\{0\}$.

Let ζ_o be a pole of $(\bullet I - X)^{-1}$, i.e. let $A_{-\nu} = 0$ when $\nu \gg 0$. The *order* of the pole is the positive integer ν_o such that $A_{-\nu_o} \neq 0$ while $A_{-\nu} = 0$ for all $\nu > \nu_o$. Then ν_o is uniquely defined by the condition

$$(\zeta_o I - X)^{\nu_o - 1} P \neq 0, \ (\zeta_o I - X)^{\nu_o} P = 0, \tag{1.80}$$

and $A_{-\nu_o}$ is given by

$$A_{-\nu_o} = \lim_{\zeta \to \zeta_o} (\zeta - \zeta_o)^{\nu_o} (\zeta I - X)^{-1}.$$

Hence

Lemma 1.14.4 *If ζ_o is a pole of $(\zeta I - X)^{-1}$, then ζ_o is an eigenvalue of X.*

Examples The following example, taken from [40] and whose details may be worked out as an exercise, shows that the converse is false. Let $\mathcal{E} = l^2(\mathbf{N}^*)$ and let $X \in \mathcal{L}(\mathcal{E})$ be defined on $(x_1, x_2, \dots) \in \mathcal{E}$ by

$$X(x_1, x_2, x_3, \dots) = \left(x_2, \frac{x_3}{2}, \frac{x_4}{3}, \dots \right).$$

The operator X is topologically nilpotent; 0 is an eigenvalue of X and an essential singularity of $(\bullet I - X)^{-1}$.

Consider now the following examples, found again in [40].

Let $\mathcal{F} = C[0,1]$, $\mathcal{G} = \{x \in \mathcal{F} : x(0) = 0\}$ and let X be the continuous linear operator defined on $\mathcal{F}$ and on $\mathcal{G}$ by the integral

$$Xx(t) = \int_0^t x(s)\,ds,$$

where $0 \leq t \leq 1$.

In both cases $\sigma(X) = \{0\}$ and 0 is an essential singularity of $(\bullet I - X)^{-1}$. When $X \in \mathcal{L}(\mathcal{F})$, then $0 \in r\sigma(X)$; if $X \in \mathcal{L}(\mathcal{G})$, then $0 \in c\sigma(X)$.

Since P is a projector then

$$x \in \ker P \Leftrightarrow x = x - Px,$$

i.e.
$$\ker P = \mathcal{R}(I - P).$$

Since, for $\nu > 1$,

$$(X - \zeta_o I)^\nu A_{\nu-1} = (X - \zeta_o I)^{\nu-1} A_{\nu-2},$$

then

$$(X - \zeta_o I)^\nu A_{\nu-1} = P - I \qquad (1.81)$$

whenever $\nu > 0$.

For $\nu > 0$, let

$$\mathcal{K}_\nu = \ker((\zeta_o I - X)^\nu), \ \mathcal{R}_\nu = \mathcal{R}((\zeta_o I - X)^\nu).$$

If $x \in \mathcal{K}_\nu$ for some $\nu \geq 1$, then

$$0 = (X - \zeta_o I)^\nu A_{\nu-1} x = Px - x,$$

showing that

$$\mathcal{K}_\nu \subset \mathcal{R}(P) \ \text{ for all } \ \nu \geq 1. \qquad (1.82)$$

If $x \in \mathcal{R}(P)$, then

$$(X - \zeta_o I)^\nu x = (X - \zeta_o I)^\nu Px = A_{-(\nu+1)} x$$

for all $\nu \geq 1$. Hence, if ζ_o is a pole of order $\nu_0 > 0$, then $(X - \zeta_o I)^\nu x = 0$, i.e., $x \in \mathcal{K}_\nu$, for all $\nu \geq \nu_o$.

That proves the first part of the following lemma.

Lemma 1.14.5 *If ζ_o is a pole of order $\nu_0 > 0$, then*

$$\ker(\zeta_o I - X)^\nu = \mathcal{R}(P)$$

and

$$\mathcal{R}(\zeta_o I - X)^\nu = \ker P$$

for all $\nu \geq \nu_o$.

Proof In view of (1.81),

$$\ker P = \mathcal{R}(P - I) = \mathcal{R}(X - \zeta_o I)^\nu A_{\nu-1}$$
$$\subset \mathcal{R}(X - \zeta_o I)^\nu = \mathcal{R}_\nu.$$

For $\nu \geq \nu_0$ let $x \in \mathcal{R}_\nu \cap \mathcal{K}_\nu$, i.e.

$$x = (\zeta_o I - X)^\nu y \text{ for some } y \in \mathcal{E},$$

and

$$(\zeta_o I - X)^\nu x = 0.$$

Thus $y \in \mathcal{K}_{2\nu}$, and therefore, by the first part of this lemma, $y \in \mathcal{K}_\nu$, whence $x = 0$, showing that

$$\mathcal{K}_\nu \cap \mathcal{R}_\nu = \{0\} \text{ for all } \nu \geq \nu_0. \tag{1.83}$$

To complete the proof we show now that $\mathcal{R}_\nu \subset \ker P$.

For $x \in \mathcal{R}_\nu$, let $x_1 = (I - P)x$ and $x_2 = Px$. Thus $x_1 \in \mathcal{R}(I - P) = \ker P \subset \mathcal{R}_\nu$, and therefore $x_2 = x - x_1 \in \mathcal{R}_\nu$. Since, on the other hand, $x_2 \in \mathcal{R}(P) = \mathcal{K}_\nu$, then, by (1.83), $x_2 = 0$, and therefore $x = x_1 \in \ker P$. ∎

Being

$$\mathcal{E} = \mathcal{R}(P) \oplus \mathcal{R}(I - P),$$

the facts established above are summarized by the following theorem.

Theorem 1.14.6 *If ζ_o is a pole of order $\nu_0 > 0$, then*

$$\mathcal{R}(P) = \ker(\zeta_o I - X)^\nu,$$
$$\ker P = \mathcal{R}(\zeta_o I - X)^\nu,$$
$$\mathcal{E} = \ker(\zeta_o I - X)^\nu \oplus \mathcal{R}(\zeta_o I - X)^\nu$$

for all $\nu \geq \nu_o$.

Since

$$\ker(\zeta_o I - X) \subset \ker(\zeta_o I - X)^\nu$$

for all $\nu > 0$, then, by Lemma 1.14.5,

$$\dim_{\mathbf{C}} \mathcal{R}(P) \geq \dim_{\mathbf{C}} \ker(\zeta_o I - X).$$

Furthermore the sequence of strict inclusions

$$\{0\} = (\zeta_o I - X)^{\nu_o}\mathcal{R}(P) \subset (\zeta_o I - X)^{\nu_o-1}\mathcal{R}(P) \subset \cdots \subset$$
$$\subset (\zeta_o I - X)\mathcal{R}(P) \subset \mathcal{R}(P)$$

implies that

$$0 < \dim_{\mathbf{C}}(\zeta_o I - X)^{\nu_o-1}\mathcal{R}(P) \leq \dim_{\mathbf{C}}\mathcal{R}(P) - (\nu_o - 1),$$

whence

$$\max\{\nu_0, \dim_{\mathbf{C}}\ker(\zeta_o I - X)\} \leq \dim_{\mathbf{C}}\mathcal{R}(P). \tag{1.84}$$

As a consequence, if ζ_o is a pole and $\mathcal{R}(P)$ has finite dimension, the eigenspace $\ker(\zeta_o I - X)$ has finite dimension. This conclusion is improved by the following theorem.

Theorem 1.14.7 If $\dim_{\mathbf{C}}\mathcal{R}(P) < \infty$, ζ_o is a pole.

Proof In view of (1.74), the theorem is equivalent to the existence of some positive integer α such that

$$(X - \zeta_o I)^{\alpha} P = 0 \quad \text{on} \quad \mathcal{R}(P).$$

Let $\{x_1, \dots, x_N\}$ be a basis of $\mathcal{R}(P)$. Since $\mathcal{R}(P)$ has finite dimension and is invariant under the action of X, for every x_j ($j = 1, 2, \dots, N$), the vectors $x_j, Xx_j, \dots, X^N x_j$ are linearly dependent. Thus, there exists a monic polynomial p_j with complex coefficients, such that

$$p_j(X)(x_j) = 0.$$

Let p be the monic polynomial $p = p_1 \cdots p_N$. Since $p(X)(x_j) = 0$ for $j = 1, 2, \dots, N$, then
$$p(X)(x) = 0$$
for all $x \in \mathcal{R}(P)$.

Let $\zeta_1, \dots, \zeta_m$ be the roots of the polynomial p.

To prove the theorem, assume that for all $\alpha \in \mathbf{N}^*$ there is some $y \in \mathcal{R}(P)$ such that
$$(X - \zeta_o I)^{\alpha} \neq 0.$$

We shall show that this assumption leads to a contradiction.

There exists a polynomial q with complex coefficients and a root ζ_β of p, for which

$$q(X)(X - \zeta_o I)^\alpha y \neq 0,$$
$$(X - \zeta_\beta I)q(X)(X - \zeta_o I)^\alpha y = 0.$$

In view of our assumption, $(X - \zeta_o)_{|\mathcal{R}(P)}$ is injective, and therefore (being $\dim_{\mathbf{C}} P < \infty$) bijective. Thus, $\zeta_\beta \neq \zeta_o$. The vector $z = q(X)(X - \zeta_o I)^\alpha y \neq 0$ is an eigenvector of X with eigenvalue ζ_β, and, for every $\zeta \in \mathbf{C}$,

$$(\zeta I - X)z = \zeta z - Xz = (\zeta - \zeta_\beta)z.$$

Hence, if $\zeta \notin \sigma(X)$, then

$$z = (\zeta - \zeta_\beta)(\zeta I - X)^{-1}z.$$

Since $z \in \mathcal{R}(P)$, then

$$z = Pz = \frac{1}{2\pi i} \int_\gamma (\zeta I - X)^{-1} z \, d\zeta$$
$$= \frac{1}{2\pi i} \left(\int_\gamma \frac{1}{\zeta - \zeta_\beta} \, d\zeta \right) z.$$

Choosing the radius $r > 0$ of γ so small that ζ_β lies outside the closed disc bounded by γ, shows that

$$\int_\gamma \frac{1}{\zeta - \zeta_\beta} d\zeta = 0,$$

and therefore $z = 0$. Contradiction. ■

We prove now that, if ζ_o is a pole and $\dim_{\mathbf{C}} \ker(\zeta_o I - X) < \infty$, then $\dim_{\mathbf{C}} \mathcal{R}(P) < \infty$.

Proposition 1.14.8 Let $A \in \mathcal{L}(\mathcal{E})\backslash\{0\}$ be such that $A^n = 0$ for some integer $n > 1$. If $\dim_{\mathbf{C}} \mathcal{E} = \infty$, then $\dim_{\mathbf{C}} \ker A = \infty$.

Proof Let $\mathcal{F} = \ker A$. If $\dim_{\mathbf{C}} \mathcal{F} < \infty$, there exists a closed linear subspace $\mathcal{G}$ of $\mathcal{E}$ such that

$$\mathcal{E} = \mathcal{F} \oplus \mathcal{G}. \tag{1.85}$$

The operator A is represented by the matrix

$$\begin{pmatrix} A_{11} & A_{12} \\ A_{21} & A_{22} \end{pmatrix}$$

where $A_{11} \in \mathcal{L}(\mathcal{F})$, $A_{12} \in \mathcal{L}(\mathcal{G}, \mathcal{F})$, $A_{21} \in \mathcal{L}(\mathcal{F}, \mathcal{G})$, $A_{22} \in \mathcal{L}(\mathcal{G})$.

a) Let $n = 2$. For $x = (u, v) \in \mathcal{E}$, with $u \in \mathcal{F}$ and $v \in \mathcal{G}$,

$$Ax = \begin{pmatrix} A_{11}u + A_{12}v \\ A_{21}u + A_{22}v \end{pmatrix}.$$

Being,

$$\ker A = \{x \in \mathcal{E} : v = 0\},$$

then $A_{11}u = 0$, $A_{21}u = 0$ $\forall u \in \mathcal{F}$, i.e.,

$$A_{11} = 0, \quad A_{21} = 0. \tag{1.86}$$

Since

$$A^2 = 0 \Leftrightarrow \mathcal{R}(A) \subset \ker A$$
$$\Leftrightarrow A_{22}v = 0 \quad \forall v \in \mathcal{G}$$
$$\Leftrightarrow A_{22} = 0,$$

whence

$$A = \begin{pmatrix} 0 & A_{12} \\ 0 & 0 \end{pmatrix}. \tag{1.87}$$

Since $\dim_{\mathbf{C}} \mathcal{R}(A_{12}) < \infty$ and $\dim_{\mathbf{C}} \mathcal{E} = \infty$, then $\ker A_{12} \neq \{0\}$, i.e., there is $v \in \mathcal{G} \backslash \{0\}$ for which $A_{12}v = 0$. Thus, $x := (0, v) \in \ker A = \mathcal{F}$, contradicting (1.85).

b) Proceeding by induction on n, let $n > 2$, and suppose the proposition to have been established for $n - 1$, so that

$$A^{n-1} = \begin{pmatrix} 0 & A_{12}A_{22}{}^{n-2} \\ 0 & A_{22}{}^{n-1} \end{pmatrix}$$

Thus,

$$A^n = 0 \Leftrightarrow \mathcal{R}(A^{n-1}) \subset \ker A$$
$$A_{22}{}^{n-1}v = 0 \ \forall v \in \mathcal{G}$$
$$\Leftrightarrow A_{22}{}^{n-1} - 0.$$

By the inductive hypothesis, that implies that $\dim_{\mathbf{C}} \ker A_{22} = \infty$. Since

$$A_{|\{0\} \oplus \ker A_{22}} = \begin{pmatrix} 0 & A_{12| \ker A_{22}} \\ 0 & 0 \end{pmatrix},$$

from here the proof proceeds as at the end of a). ∎

Theorem 1.14.9 *If ζ_o is a pole and if $\dim_{\mathbf{C}} \ker(\zeta_o - X) < \infty$, then $\dim_{\mathbf{C}} \mathcal{R}(P) < \infty$.*

Proof Since $\mathcal{R}(P)$ is invariant under the action of $\zeta_o I - X$, if $\nu_0 > 0$ is the order of the pole ζ_o then $((\zeta_o I - X)_{|\mathcal{R}(P)})^{\nu_0} = \{0\}$.

Since $(\zeta_o I - X)_{|\mathcal{R}(P)} \in \mathcal{L}(\mathcal{R}(P))$, (1.82) and Proposition 1.14.8 yield the conclusion. ∎

Thus, (1.75) implies

Corollary 1.14.10 *If ζ_o is a pole, then $\dim_{\mathbf{C}} \mathcal{R}(P) < \infty$ if, and only if, $\dim_{\mathbf{C}} \ker(\zeta_o I - X) < \infty$.*

Let ζ_o be a pole of order ν_o and let $\dim_{\mathbf{C}} \mathcal{R}(P) < \infty$.

The operator $(X - \zeta_o I)_{|\mathcal{R}(P)}$ is represented - with respect to a suitable basis for $\mathcal{R}(P)$ - by a finite direct sum

$$(X - \zeta_o I)_{|\mathcal{R}(P)} = J_1 \oplus O_1 \oplus J_2 \oplus O_2 \oplus \cdots \tag{1.88}$$

of zero matrices O_m of order p_m and of h Jordan matrices $J_r = (J_{r;\alpha,\beta})$, $(r = 1, \dots, h)$ of order $q(r) \geq 2$, whose entries are all zero except $J_{r;\alpha,\alpha+1} = 1$ for $\alpha = 2, \dots, q(r)$.

Since $\ker J_r$ is one-dimensional, (1.82) and (1.88) yield

$$\dim_{\mathbf{C}} \ker(\zeta_o I - X) = h + \sum p_m. \tag{1.89}$$

Let N be the order of the matrix (1.88). Being $J_r^{q(r)-1} \neq 0$, $J_r^{q(r)} = 0$ and $(\zeta_o I - X)^{\nu_0-1} \neq 0$, $(\zeta_o I - X)^{\nu_0} = 0$, then $q(r) \leq \nu_0$ for all r, and there exists some $r = 1, \ldots, h$ for which $q(r) = \nu_0$. Thus, in view of (1.89),

$$\dim_{\mathbf{C}} \mathcal{R}(P) = N \leq h\nu_0 + \sum p_m$$
$$\leq \nu_0 \dim_{\mathbf{C}} \ker(\zeta_o I - X).$$

Setting $0 \cdot \infty = \infty$, and assuming $\nu_0 = \infty$ when ζ_o is an essential singularity of $(\bullet I - X)^{-1}$, the following theorem has been established.

Theorem 1.14.11 *If ζ_o is an isolated point of $\sigma(X)$, then*

$$\max\{\nu_0, \dim_{\mathbf{C}} \ker(\zeta_o I - X)\} \leq \dim_{\mathbf{C}} \mathcal{R}(P) \leq \nu_0 \dim_{\mathbf{C}} \ker(\zeta_o I - X).$$

The complex dimension of $\mathcal{R}(P)$ is called the *algebraic multiplicity* of ζ_o. The complex dimension of the eigenspace $\ker(\zeta_o I - X)$ is called the *geometric multiplicity* of ζ_o.

1.15 Commutative Banach algebras

A unital algebra $\mathcal{A}$ is called a *division algebra* if $\mathcal{A}\backslash\{0\}$ is a group for the product in $\mathcal{A}$.

Theorem 1.15.1 *(Theorem of Gelfand-Mazur) A complex normed division algebra is canonically isomorphic to the complex field.*

Proof Let $\mathcal{A}$ be a complex normed division algebra. If $x \in \mathcal{A}\backslash\{0\}$, then $0 \notin \sigma(x)$. Hence, if $\zeta \in \sigma(x)$, then $\zeta 1 - x = 0$, since, otherwise, $\zeta 1 - x$ would be invertible in $\mathcal{A}$. Thus $x = \zeta 1$, and therefore $\sigma(x) = \{\zeta\}$. The map $x \mapsto \zeta$ defines an isomorphism of $\mathcal{A}$ onto $\mathbf{C}$ because, if $x_1 = \zeta_1 1$, $x_2 = \zeta_2 I$, then $x_1 + x_2 = (\zeta_1 + \zeta_2) 1$, $x_1 x_2 = \zeta_1\zeta_2 1$. ∎

Remark The homomorphism $x \mapsto \zeta$ is unique. In fact, if $\phi : \mathcal{A} \to \mathbf{C}$ is an isomorphism of $\mathcal{A}$ onto $\mathbf{C}$, then the unit 1 of $\mathcal{A}$ is $\phi^{-1}(1)$, and therefore, for $x \in \mathcal{A}$,

$$\phi(x - \phi(x)1) = \phi(x) - \phi(x) = 0,$$

whence $x - \phi(x)1 = 0$. Thus, $\sigma(x) = \phi(x)$.

Let $\mathcal{E}$ be a normed space, let $\mathcal{F}$ be a linear closed subspace of $\mathcal{E}$ and let $\pi : \mathcal{E} \to \mathcal{E}/\mathcal{F}$ be the canonical projection of $\mathcal{E}$ onto the quotient vector space $\mathcal{E}/\mathcal{F}$. For $z \in \mathcal{E}/\mathcal{F}$ let

$$\|z\| := \inf\{\|x\| : x \in \pi^{-1}(z)\}.$$

Lemma 1.15.2 *The function $z \mapsto \|z\|$ is a norm in $\mathcal{E}/\mathcal{F}$, for which π is continuous. If $\mathcal{E}$ is complete, $\mathcal{E}/\mathcal{F}$ is complete. If $\mathcal{E}$ is a commutative normed algebra and $\mathcal{F}$ is a closed ideal, then $\mathcal{E}/\mathcal{F}$ is a normed algebra.*

Proof I. For $z, z_1, z_2 \in \mathcal{E}$, $a \in \mathbf{C}$,

$$\begin{aligned}
\|z_1 + z_2\| &= \inf\{\|x\| : x \in \pi^{-1}(z_1 + x_2)\} \\
&= \inf\{\|x\| : x \in \pi^{-1}(z_1) + \pi^{-1}(z_2)\} \\
&\leq \inf\{\|x\| : x \in \pi^{-1}(z_1)\} + \inf\{\|x\| : x \in \pi^{-1}(z_2)\} \\
&= \|z_1\| + \|z_2\|,
\end{aligned}$$

$$\begin{aligned}
\|az\| &= \inf\{\|x\| : x \in \pi^{-1}(az)\} \\
&= \inf\{\|ax\| : x \in \pi^{-1}(z)\} \\
&= |a|\,\|z\|,
\end{aligned}$$

and that proves that $z \mapsto \|z\|$ is a seminorm. If $x \in \mathcal{F}$, then $(\pi(x) = 0$ and therefore) $x = 0$. Viceversa, if $\|z\| = 0$ for some $z \in \mathcal{E}/\mathcal{F}$, for any fixed $x \in \pi^{-1}(z)$, there is a sequence $\{y_\nu\}$ of points $y_\nu \in \mathcal{F}$ such that $\lim_{\nu \to +\infty} \|x + y_\nu\| = 0$, and therefore $\lim_{\nu \to +\infty} y_\nu = -x$, whence $x \in \mathcal{F}$. Thus $\|z\| = 0 \Rightarrow \pi^{-1}(z) \subset \mathcal{F}$, and therefore $z = 0$. That proves that $\mathcal{E}/\mathcal{F}$ is a normed space.

Since $\|\pi(x)\| \leq \|x\|$ for all $x \in \mathcal{E}$, then π is continuous, and $\|\pi\| \leq 1$.

II. If $\{z_\nu\}$ is a Cauchy sequence in $\mathcal{E}/\mathcal{F}$, there is a subsequence $\{z'_\nu\}$ such that $\|z'_{\nu+1} - z'_\nu\| < \frac{1}{2^{\nu+1}}$ for all indices $\nu \geq 1$. For every $\nu \geq 1$ there is some $x'_\nu \in \pi^{-1}(z_\nu)$ such that $\|x'_{\nu+1} - x'_\nu\| \leq \frac{1}{2^\nu}$. Thus $\{x'_\nu\}$ is a Cauchy sequence. If $\mathcal{E}$ is complete, $\{x'_\nu\}$ converges to some $x \in \mathcal{E}$. The continuity of π implies that

$$\pi(x) = \lim_{\nu \to +\infty} x'_\nu = \lim_{\nu \to +\infty} z'_\nu.$$

Since $\{z'_\nu\}$ converges and $\{z_\nu\}$ is a Cauchy sequence, then $\{z_\nu\}$ converges.

III. Let $\mathcal{E}$ be a commutative, normed algebra and let $\mathcal{F}$ be a closed ideal. Since $\pi(x_1\, x_2) = \pi(x_1)\, \pi(x_2)$ for all $x_1, x_2 \in \mathcal{E}$, $\mathcal{E}/\mathcal{F}$ inherits from $\mathcal{E}$ the structure of a commutative algebra, for which π is an algebra homomorphism. If $z_1, z_2 \in \mathcal{E}/\mathcal{F}$, then

$$\begin{aligned}
\|z_1\, z_2\| &= \inf\{\|x_1\, x_2\| : x_1 \in \pi^{-1}(z_1), x_2 \in \pi^{-1}(z_2)\} \\
&\leq \inf\{\|x_1\|\, \|x_2\| : x_1 \in \pi^{-1}(z_1), x_2 \in \pi^{-1}(z_2)\} \\
&\leq \inf\{\|x_1\| : x_1 \in \pi^{-1}(z_1)\}\, \inf\{\|x_2\| : x_2 \in \pi^{-1}(z_2)\} \\
&= \|z_1\|\, \|z_2\|.
\end{aligned}$$

■

Lemma 1.15.3 *If $\mathcal{A}$ is a unital commutative algebra, $x \in \mathcal{A}$ is invertible if, and only if, x is contained in no proper ideal of $\mathcal{A}$.*

Proof If $x \in \mathcal{A}$ is invertible and if an ideal $\mathcal{I}$ of $\mathcal{A}$ contains x, then $1 = x^{-1}\, x \in \mathcal{I}$, and therefore $\mathcal{I} = \mathcal{A}$.

If $x \in \mathcal{A}$ is not invertible, the ideal generated by x does not coincide with $\mathcal{A}$, since, otherwise, there would exist some $x \in \mathcal{A}$ such that $1 = yx$, and therefore x would be invertible. ■

Henceforth, $\mathcal{A}$ will be a unital, commutative, complex Banach algebra. The identity element 1 is contained in the open set $\mathcal{A}^{-1}$ of all invertible elements of $\mathcal{A}$. Hence, if $\mathcal{I}$ is a proper ideal of $\mathcal{A}$, $1 \notin \overline{\mathcal{I}}$, the closure of $\mathcal{I}$. That implies

Lemma 1.15.4 *The closure of any proper ideal of $\mathcal{A}$ is a proper ideal.*

We order all ideals of $\mathcal{A}$ by inclusion.

Corollary 1.15.5 *Every maximal proper ideal of $\mathcal{A}$ is closed.*

The union of all proper ideals containing the proper ideal $\mathcal{I}$ is an ideal which does not contain 1, and therefore is a proper ideal of $\mathcal{A}$. The Zorn lemma yelds then

Proposition 1.15.6 *Every proper ideal of $\mathcal{A}$ is contained in a proper maximal ideal.*

If $\mathcal{A}$, is any algebra, a non-identically vanishing linear form $\chi : \mathcal{A} \to \mathbf{C}$ such that $\chi(xy) = \chi(x) \chi(y)$ for all $x, y \in \mathcal{A}$, is called a *character* of $\mathcal{A}$. Since $\chi(1) = \chi(1^2) = \chi(1)^2$, then, either $\chi(1) = 1$ or $\chi(1) = 0$, in which latter case $\chi(x) = \chi(1x) = 0$ for all $x \in \mathcal{A}$, contradicting the assumption $\chi \neq 0$.

Example Let K be a compact Hausdorff space, and let $C(K)$ be the vector space of all continuous, complex valued functions on K. This space is a Banach algebra for the operations of point-wise sums and products and for the uniform norm:

$$\|x\| = \max\{|x(t)| : t \in K\}, \quad x \in C(K). \tag{1.90}$$

For any $t \in K$, the map $x \mapsto x(t)$ is a character of $C(K)$.

If χ is a character of $\mathcal{A}$, then $\mathcal{I} := \ker \chi$ is a proper ideal of $\mathcal{A}$ for which

$$\mathcal{A}/\mathcal{I} \cong \mathcal{R}(\chi) = \mathbf{C},$$

and therefore is a proper maximal ideal of $\mathcal{A}$.

Viceversa, if $\mathcal{I}$ is a proper maximal ideal of a commutative Banach algebra $\mathcal{A}$ (hence closed, by Corollary 1.15.5), the Banach algebra $\mathcal{A}/\mathcal{I}$ does not contain any proper ideal different from zero, and therefore is a division algebra. By Theorem 1.15.1 $\mathcal{A}/\mathcal{I} \simeq \mathbf{C}$, and the projection

$$\chi : \mathcal{A} \to \mathcal{A}/\mathcal{I} \cong \mathbf{C}$$

Thus, the following theorem holds.

Theorem 1.15.7 *Every proper maximal ideal of $\mathcal{A}$ is the kernel of a character, and viceversa.*

Since a linear form on a normed space is continuous if, and only if, its kernel is closed, all characters of $\mathcal{A}$ are continuous. It will be useful in the following obtaining this result by a different argument.

If χ is a character of $\mathcal{A}$, for all $x \in \mathcal{A}$

$$\chi(x - \chi(x)\,1) = \chi(x) - \chi(x) = 0,$$

i.e.,

$$x - \chi(x)\,1 \in \ker\chi.$$

By Lemma 1.15.3, $x - \chi(x)\,1$ is not invertible, and therefore $\chi(x) \in \sigma(x)$. Thus,

$$|\chi(x)| \leq \rho(x) \leq \|x\| \quad \forall x \in \mathcal{A}. \tag{1.91}$$

Hence χ is continuous; since, moreover, $\chi(1) = 1$, then

$$\|\chi\| = 1. \tag{1.92}$$

If $\zeta \in \sigma(x)$, $x - \zeta 1$ is not invertible, that is, by Lemma 1.15.3, is contained in a proper ideal of $\mathcal{A}$. Thus, by Proposition 1.15.6 and Theorem 1.15.7, there is a character χ of $\mathcal{A}$ such that $\chi(x - \zeta 1) = 0$, i.e., $\zeta = \chi(x)$. That proves

Theorem 1.15.8 *For every $x \in \mathcal{A}$, the spectrum $\sigma(x)$ is the set of the values taken on x by all characters of $\mathcal{A}$.*

Corollary 1.15.9 *An element $x \in \mathcal{A}$ is invertible if, and only if, there is no character of $\mathcal{A}$ vanishing on x.*

By (1.92), the set $M(\mathcal{A})$ of all characters of $\mathcal{A}$ is contained in the closure $\overline{B'}$ of the open unit disc B' of the topological dual $\mathcal{A}'$ of $\mathcal{A}$. Since

$$\begin{aligned} M(\mathcal{A}) = \{\chi \in \mathcal{A}' : \chi(1) = 1, \\ \chi(xy) = \chi(x)\,\chi(y) \ \ \forall x, y \in \mathcal{A}\}, \end{aligned}$$

and since, for every $x \in \mathcal{A}$, the map $\mathcal{A}' \to \mathbf{C}$ defined by $\chi \mapsto \chi(x)$, is continuous for the weak-star topology, $M(\mathcal{A})$ - as intersection of zero-sets of continuous functions - is closed in $\overline{B'}$ for the weak-star topology. The Banach-Alaoglu theorem yields then

Theorem 1.15.10 *The set $M(\mathcal{A})$ of all characters of a complex, commutative, unital Banach algebra $\mathcal{A}$ is compact for the weak-star topology.*

For $x \in \mathcal{A}$, let $\hat{x} : M(\mathcal{A}) \to \mathbf{C}$ be the function defined by $\hat{x}(\chi) = \chi(x)$.

Proposition 1.15.11 *The topology defined on $M(\mathcal{A})$ by the weak-star topology in $\mathcal{A}'$, is the weakest topology for which all functions $\hat{x}$ are continuous.*

Proof Exercise.

The set $M(\mathcal{A})$, endowed with the topology considered above is called the *space of maximal ideals*, or *space of characters*, or *Gelfand space*, or *Gelfand spectrum* of $\mathcal{A}$, and its topology is called the *Gelfand topology*.

Example Consider again the unital commutative Banach algebra $C(K)$ of all continuous, complex valued functions on the compact Hausdorff space K, endowed with the uniform norm (1.90).

Lemma 1.15.12 *The set of all functions in $C(K)$ vanishing at a point of K is a maximal ideal of $C(K)$, and viceversa.*

Proof For any $t_o \in K$, the set $\mathcal{I}$ of all functions in $C(K)$ vanishing at t_o is a proper ideal of $C(K)$. To show that it is a maximal ideal, choose any $y \in C(K)$ such that $y(t_o) \neq 0$. For all $t \in K$ and all $z \in C(K)$,

$$z(t) = \frac{z(t_o)}{y(t_o)}\, y(t) + \left(z(t) - \frac{z(t_o)}{y(t_o)}\, y(t) \right).$$

Since

$$\left(z - \frac{z(t_o)}{y(t_o)}\, y \right)(t_o) = 0,$$

then

$$z - \frac{z(t_o)}{y(t_o)}\, y \in \mathcal{I}.$$

As a consequence, 1 and y generate the whole algebra $C(K)$, showing that $\mathcal{I}$ (has codimension one, and therefore) is a maximal ideal of $C(K)$.

Now, let $\mathcal{I}$ be an ideal in $C(K)$ such that, for every $t \in K$, there is some $x \in K$ for which $x(t) \neq 0$. Since K is compact and all elements of $C(K)$ are continuous, there exist $x_1, \ldots, x_n$ in $\mathcal{I}$ such that the function $z \in C(K)$ defined by

$$z(t) = x_1(t)\overline{x_1(t)} + \cdots + x_n(t)\overline{x_n(t)} \quad (t \in K),$$

is strictly positive at each point $t \in K$. Since z is continuous and K is compact, then the function $t \mapsto \frac{1}{z(t)}$ belongs to $C(K)$. Being $z \in \mathcal{I}$ and $z(t) \frac{1}{z(t)} = 1$ for all $t \in K$, then $1 \in \mathcal{I}$, and in conclusion $\mathcal{I} = C(K)$. ∎

By this lemma, the set $M(C(K))$ of all maximal ideals of $C(K)$ is the set of all *Dirac measures*

$$\delta_t : x \mapsto x(t),$$

when t varies in K. The Gelfand topology on $M(C(K))$ is the weakest topology for which all functions $x \in C(K)$ are continuous. Since $C(K)$ separates the points of K, the following theorem has been proved.

Theorem 1.15.13 *Every compact Hausdorff space K is (canonically homeomorphic to) the space $M(C(K))$ of all maximal ideals of $C(K)$, endowed with the Gelfand topology.*

Going back to the general case of the unital, commutative, Banach algebra $\mathcal{A}$, the map

$$x \mapsto \hat{x} \tag{1.93}$$

is called *Gelfand transform*.

For $a, b \in \mathbf{C}$ and $x, y \in \mathcal{A}$,

$$\widehat{ax + by} = a\hat{x} + b\hat{y},$$
$$\widehat{xy} = \hat{x}\,\hat{y},$$
$$\hat{1} = 1.$$

Thus, the map (1.93) is a homomorphism. Since furthermore, by Theorem 1.15.8,

$$\|\hat{x}\|_{C(K)} = \rho(x) \leq \|x\|_{\mathcal{A}} \tag{1.94}$$

for all $x \in \mathcal{A}$, the following theorem holds

Theorem 1.15.14 *The Gelfand transform is a continuous homomorphism of the Banach algebra $\mathcal{A}$ into the Banach algebra $C(M(\mathcal{A}))$.*

If $\chi_1, \chi_2 \in M(\mathcal{A})$ and $\chi_1 \neq \chi_2$, there is some $x \in \mathcal{A}$ such that $\chi_1(x) \neq \chi_2(x)$. That proves

Lemma 1.15.15 *The image of the Gelfand transform separates the points of $M(\mathcal{A})$.*

Lemma 1.6.2 and (1.94) yield

Lemma 1.15.16 *The Gelfand transform is an isometry if, and only if, $\|x^2\| = \|x\|^2$ for all $x \in \mathcal{A}$.*

The kernel of the Gelfand transform - which is called the *radical* of $\mathcal{A}$ - is the set of all $x \in \mathcal{A}$ for which $\chi(x) = 0$ for every choice of the character χ of $\mathcal{A}$. Hence

Lemma 1.15.17 *The radical is the intersection of all proper maximal ideals of $\mathcal{A}$.*

In view of (1.94), the radical is the set of all quasi-nilpotent elements of $\mathcal{A}$.

Theorem 1.15.18 *The radical of $\mathcal{A}$ is zero and the image of $\mathcal{A}$ by the Gelfand transform is closed in $C(M(\mathcal{A}))$ if, and only if, there exists a finite constant $c > 0$ such that*

$$\|x\|^2 \leq c \, \|x^2\| \tag{1.95}$$

for all $x \in \mathcal{A}$.

Proof If the Gelfand transform has kernel zero and a closed image, then the transform is a continuous isomorphism of $\mathcal{A}$ onto its image, which - as a closed subspace of $C(M(\mathcal{A}))$ - is a Banach space. Thus, the Gelfand transform is invertible, and, by the Banach open mapping theorem, its inverse is continuous. In other words, there is a finite constant $c > 0$ such that

$$\|x\| \leq c \, \|\hat{x}\|_{C(M(\mathcal{A}))}, \tag{1.96}$$

i.e., by (1.94),

$$\|x\| \leq c \, \rho(x) \tag{1.97}$$

for all $x \in \mathcal{A}$. As a consequence,

$$\|x\|^2 \leq c^2 \rho(x)^2 = c^2 \rho(x^2) \leq \|x^2\|.$$

To establish the converse, note first that, if (1.95) holds, for all $x \in \mathcal{A}$ and for every positive integer n,

$$\|x\| \leq c^{\frac{1}{2}} \|x^2\|^{\frac{1}{2}} \leq c^{\frac{1}{2}} \left(c^{\frac{1}{2}} \|x^4\|^{\frac{1}{2}} \right)^{\frac{1}{2}}$$

$$= c^{\frac{1}{2}+\frac{1}{2^2}} \|x^{2^2}\|^{\frac{1}{2^2}}$$

$$\leq c^{\frac{1}{2}+\frac{1}{2^2}} \left(c^{\frac{1}{2}} \|(x^{2^2})^2\|^{\frac{1}{2}} \right)^{\frac{1}{2^2}}$$

$$= c^{\frac{1}{2}+\frac{1}{2^2}+\frac{1}{2^3}} \|x^{2^3}\|^{\frac{1}{2^3}} \leq \cdots$$

$$\leq c^{\frac{1}{2}+\frac{1}{2^2}+\cdots+\frac{1}{2^n}} \|x^{2^n}\|^{\frac{1}{2^n}}.$$

Since

$$\sum_{n=1}^{+\infty} \frac{1}{2^n} = 1$$

and

$$\lim_{n \to +\infty} \|x^{2^n}\|^{\frac{1}{2^n}} = \rho(x),$$

then (1.97) holds, which, by (1.94), yields (1.96) and completes the proof of the theorem. ∎

Note that, under the hypotheses of Theorem 1.15.18, there exists a finite constant $k > 0$ such that

$$k\|x\| \leq \|\hat{x}\|_{C(M(\mathcal{A}))} \leq \|x\|$$

for all $x \in \mathcal{A}$.

Commutative Banach algebras whose radical is reduced to zero are called *semisimple*.

1.16 Continuous linear forms and characters

We have seen that every character of a commutative Banach algebra is a continuous linear form. Under which conditions a continuous linear form is a character? This section is devoted to answering this question, following [32], [43], [84] and [70].

We begin with some preliminaries of linear algebra.

Let $\mathcal{A}$ be a unital algebra and let λ be a homomorphism of $\mathcal{A}$ into $\mathbf{C}$ such that $\lambda(1) = 1$. Let $\mathcal{N} = \ker \lambda$.

Lemma 1.16.1 *If*

$$\xi, \eta \in \mathcal{N} \implies \xi\eta \in \mathcal{N}, \tag{1.98}$$

then

$$\lambda(uv) = \lambda(u)\lambda(v) \ \ \forall\, u, v \in \mathcal{A}. \tag{1.99}$$

Proof If $u \in \mathcal{A}$, then $u - \lambda(u)1 \in \mathcal{N}$, that is to say,

$$u = \xi + \lambda(u)1 \ \text{ for some } \xi \in \mathcal{N}. \tag{1.100}$$

Similarly, for $v \in \mathcal{A}$,

$$v = \eta + \lambda(v)1 \ \text{ for some } \eta \in \mathcal{N},$$

whence

$$uv = \xi\eta + \lambda(u)\eta + \lambda(v)\xi + \lambda(u)\lambda(v)1. \tag{1.101}$$

Applying λ to both sides and using (1.98) we obtain (1.99). ∎

Corollary 1.16.2 *If*

$$\xi \in \mathcal{N} \implies \xi^2 \in \mathcal{N}, \tag{1.102}$$

then

$$\lambda(u^2) = \lambda(u)^2 \ \ \forall\, u \in \mathcal{A}.$$

Lemma 1.16.3 *If (1.102) holds, then $\mathcal{N}\mathcal{A} \subset \mathcal{A}$, i.e., $\mathcal{N}$ is a right ideal.*

Proof In view of Corollary 1.16.2, for all $u, v \in \mathcal{A}$,

$$\lambda((u + v)^2) = \lambda(u + v)^2$$
$$= \lambda(u)^2 + 2\lambda(u)\lambda(v) + \lambda(v)^2,$$

i.e.,

$$\lambda(u^2) + \lambda(uv + vu) + \lambda(v^2) = \lambda(u)^2 + 2\lambda(u)\lambda(v) + \lambda(v)^2,$$

that is to say,

$$\lambda(uv + vu) = 2\lambda(u)\lambda(v).$$

Thus,

$$u \in \mathcal{N}, v \in \mathcal{A} \Longrightarrow uv + vu \in \mathcal{N}, \tag{1.103}$$

and therefore also

$$u \in \mathcal{N}, v \in \mathcal{A} \Longrightarrow (uv + vu)^2 \in \mathcal{N}.$$

The identity

$$(uv + vu)^2 + (uv - vu)^2 = 2[u(vuv) + (vuv)u]$$

yields then

$$u \in \mathcal{N} \Longrightarrow (uv - vu)^2 \in \mathcal{N} \ \forall \, v \in \mathcal{A}.$$

Since, by Corollary 1.16.2,

$$\lambda((uv - vu)^2) = \lambda(uv - vu)^2,$$

then

$$u \in \mathcal{N} \Longrightarrow (uv - vu) \in \mathcal{N} \ \forall \, v \in \mathcal{A}.$$

This inclusion, together with (1.103), shows that

$$u \in \mathcal{N} \Longrightarrow uv \in \mathcal{N} \ \forall \, v \in \mathcal{A},$$

and proves the lemma. ∎

Assume now that $\mathcal{A}$ is a unital Banach algebra, and let λ be a (not necessarily continuous) linear form on $\mathcal{A}$ such that $\lambda(1) = 1$ and that $\mathcal{N} = \ker \lambda$ contains no invertible elements of $\mathcal{A}$.

Then, for all $\xi \in \mathcal{N}$,

$$\|1 - \xi\| \geq 1,$$

and therefore

$$\|\zeta - \xi\| \geq |\zeta| \ \forall \zeta \in \mathbf{C}, \ \forall \xi \in \mathcal{N}.$$

Thus, writing u as in (1.100),

$$|\lambda(u)| \leq \|\xi + \lambda(u)1\| = \|u\| \ \forall u \in \mathcal{A},$$

proving that λ is continuous, and $\|\lambda\| \leq 1$. Since $\lambda(1) = 1$, then

$$\|\lambda\| = 1$$

We need a lemma from the theory of entire functions.

Lemma 1.16.4 *Let $f : \mathbf{C} \to \mathbf{C}$ be an entire function such that $f(0) = 1$, $f'(0) = 0$ and $0 < |f(\zeta)| \leq e^{|\zeta|}$. Then $f \equiv 1$.*

Proof Since $f(\zeta) \neq 0$ for all ζ, there is an entire function g such that

$$f(\zeta) = e^{g(\zeta)}.$$

Hence $g(0) = g'(0) = 0$ and the inequality

$$\Re g(\zeta) \leq |\zeta| \tag{1.104}$$

holds for all ζ.

Thus, if $r > 0$ and $|\zeta| \leq r$, then

$$|2r - g(\zeta)|^2 = 4r^2 - 4r\Re g(\zeta) + |g(\zeta)|^2$$
$$\geq 4r(r - |\zeta|) + |g(\zeta)|^2,$$

and therefore, again by (1.104),

$$|g(\zeta)| \leq |2r - g(\zeta)| \quad \text{whenever } |\zeta| \leq r. \tag{1.105}$$

The function h_r defined for $|\zeta| \leq 2r$ by

$$h_r(\zeta) = \frac{r^2 g(\zeta)}{\zeta^2 (2r - g(\zeta))} \tag{1.106}$$

is holomorphic on $\Delta(0, 2r) = \{\zeta \in \mathbf{C} : |\zeta| < 2r\}$ and, by (1.105), is such that $|h_r(\zeta)| \leq 1$ if $|\zeta| = r$. By the maximum principle,

$$|h_r(\zeta)| \leq 1 \quad \text{whenever } |\zeta| \leq r. \tag{1.107}$$

Keep ζ fixed and let $r \to +\infty$. If $g \not\equiv 0$, (1.106) implies that $|h_r(\zeta)| \to +\infty$, contradicting (1.107) ∎

We prove now that (1.102) holds.

For any $\xi \in \mathcal{N}\backslash\{0\}$ consider the holomorphic map $\mathbf{C} \to \mathcal{A}$ defined by

$$\zeta \mapsto \exp(\zeta\xi) = \sum_0^{+\infty} \frac{\zeta^n}{n!}\xi^n.$$

Since λ is continuous, the complex-valued function

$$f : \zeta \mapsto \lambda(\exp(\zeta\xi)) = \sum_0^{+\infty} \frac{\zeta^n}{n!}\lambda(\xi^n)$$

is an entire function. Since $\exp(\zeta\xi)$ is invertible, then

$$f(\zeta) = \lambda(\exp(\zeta\xi)) \neq 0 \ \forall \, \zeta \in \mathbf{C}.$$

Since, moreover, $f(0) = 1$ and $f'(0) = \lambda(\xi) = 0$, by Lemma 1.16.4 $f \equiv 1$, and therefore $\lambda(\xi^2) = f''(0) = 0$, *i.e.* $\xi^2 \in \mathcal{N}$.

In conclusion, the following theorem has been established.

Theorem 1.16.5 *[84],[70]* [7] *If $\mathcal{A}$ is a unital Banach algebra, and if λ is a linear form on $\mathcal{A}$ such that $\lambda(1) = 1$ and $\lambda(x) \neq 0$ for every invertible element $x \in \mathcal{A}$, then λ is a continuous character of $\mathcal{A}$.*

As a consequence, $\ker \lambda$ is a closed bilateral ideal of $\mathcal{A}$.

Theorem 1.16.6 *[32] Let $\mathcal{B}$ be a linear subspace of codimension one in the unital Banach algebra $\mathcal{A}$. Then, $\mathcal{B}$ is a closed bilateral ideal of $\mathcal{A}$ if, and only if, all its elements are not invertible in $\mathcal{A}$.*

---

[7]In the commutative case, this theorem was first established by A. Gleason in [32]. See also, Theorem 1.4.4, pp. 52-53, of [9], and also [71].

Proof Suppose that all elements of $\mathcal{B}$ are not invertible in $\mathcal{A}$. Since the identity element 1 is not in the closure $\overline{\mathcal{B}}$ of $\mathcal{B}$ and since codim $\mathcal{B} = 1$, then $\mathcal{B}$ is closed and, by the Hahn-Banach theorem, there is a continuous linear form λ on $\mathcal{A}$ such that $\mathcal{B} = \ker\lambda$ and $\lambda(1) = 1$. Thus, $\lambda(x) \neq 0$ for every invertible element of $\mathcal{A}$, and, as a consequence of Theorem 1.16.5, λ is a continuous character of $\mathcal{A}$.

Viceversa, if $\mathcal{B}$ is a closed bilateral ideal, then $1 \notin \mathcal{B}$, since otherwise $\mathcal{B} = \mathcal{A}$. But, if $y \in \mathcal{B}$ were invertible, then $1 \in \mathcal{B}$. ∎

Theorem 1.16.7 *Any linear form χ on the unital Banach algebra $\mathcal{A}$ such that $\chi(x) \in \sigma(x)$ for all $x \in \mathcal{A}$ is a continuous character of $\mathcal{A}$.*

Proof Since $\chi(1) \in \sigma(1) = \{1\}$ and since $0 \notin \sigma(x)$ whenever x is invertible in $\mathcal{A}$ the conclusion folllows from Theorem 1.16.5. ∎

Remark If $\mathcal{A}$ is commutative, this theorem (which in the commutative case is due to J.P.Kahane and W.Zelazko, [43]), together with Theorem 1.15.8 characterizes the characters among all linear forms on $\mathcal{A}$. In [43], J.P.Kahane and W.Zelazko have extended Theorem 1.16.7 to linear maps, in the following form.

Theorem 1.16.8 *Let $\mathcal{A}$ and $\mathcal{C}$ be unital, commutative Banach algebras, and let $A : \mathcal{A} \to \mathcal{C}$ be a linear map.*
a) If $\mathcal{C}$ is semisimple, and if

$$\sigma(Ax) \subset \sigma(x) \ \forall\, x \in \mathcal{A}, \tag{1.108}$$

then A is multiplicative, i.e.,

$$A(uv) = Au\, Av \ \forall\, u, v \in \mathcal{A}. \tag{1.109}$$

b) If A is multiplicative, and if

$$A\, 1_{\mathcal{A}} = 1_{\mathcal{C}}, \tag{1.110}$$

then (1.108) holds.

Proof a) If χ is a character of $\mathcal{C}$, then $\lambda := \chi \circ A$ is a continuous linear form on $\mathcal{A}$ such that

$$\lambda(x) = \chi(Ax) \in \sigma(Ax) \subset \sigma(x)$$

for all $x \in \mathcal{A}$. By Theorem 1.16.7, λ is a character of $\mathcal{A}$. Thus, for any character χ of $\mathcal{C}$,

$$\chi(A(uv)) = \lambda(uv) = \lambda(u)\,\lambda(v)$$
$$= \chi(Au)\,\chi(Av) \ \ \forall\, u, v \in \mathcal{A}.$$

Since $\mathcal{A}$ is semisimple, that implies (1.109).

b) In view of (1.110), if $x \in \mathcal{A}$ is invertible, then

$$1_\mathcal{C} = A(x\,x^{-1}) = A(x)\,A(x^{-1})$$
$$= A(x^{-1}\,x) = A(x^{-1})\,A(x).$$

Hence $A(x)$ is invertible. Thus, if $y \in \mathcal{A}$ and if $\zeta \notin \sigma(y)$, *i.e.*, if $y - \zeta 1$ is invertible in $\mathcal{A}$, then

$$A(y - \zeta 1) = Ay - \zeta A 1_\mathcal{A} = Ay - \zeta 1_\mathcal{C}$$

is invertible in $\mathcal{C}$. Therefore $\zeta \notin \sigma(Ay)$. ∎

1.17 Dual operators

Let $X : \mathcal{D}(X) \subset \mathcal{E} \to \mathcal{E}$ be a linear operator with dense domain, and let $\mathcal{D}(X')$ be the linear subspace of $\mathcal{E}'$ consisting of all continuous linear forms φ on $\mathcal{E}$ such that the linear form $x \mapsto\, < Xx, \varphi >$ is continuous on $\mathcal{D}(X)$. Since $\overline{\mathcal{D}(X)} = \mathcal{E}$, the linear form $< X\bullet, \varphi >$ has a unique extension to a continuous linear form on $\mathcal{E}$. Denoting this latter form by $X'\varphi$, the map $\varphi \mapsto X'\varphi$ defines a linear operator $X' : \mathcal{D}(X') \subset \mathcal{E}' \to \mathcal{E}'$, that is called the *dual* or the *adjoint* and for which

$$< Xx, \varphi > = < x, X'\varphi > \quad \forall x \in \mathcal{D}(X) \ \forall \varphi \in \mathcal{D}(\mathcal{E}'). \tag{1.111}$$

Lemma 1.17.1 *The operator X' is closed.*

Proof If $\Gamma(X')$ is the graph of X' [8], then

$$(\varphi, \psi) \in \Gamma(X') \Leftrightarrow \varphi \in \mathcal{D}(X'), \text{ and } \psi = X'\varphi$$
$$\Leftrightarrow \varphi \in \mathcal{D}(X') \text{ and}$$
$$< Xx, \varphi >=< x, \psi > \forall x \in \mathcal{D}(X).$$

If $\varphi, \psi \in \mathcal{E}'$ are such that $< Xx, \varphi >=< x, \psi >$ for all $x \in \mathcal{D}(X)$, the linear form $x \mapsto < Xx, \varphi >$ is continuous on $\mathcal{D}(X)$, and therefore $\varphi \in \mathcal{D}(X')$. Thus,

$$(\varphi, \psi) \in \Gamma(X') \Leftrightarrow < Xx, \varphi >=< x, \psi > \quad \forall x \in \mathcal{D}(X).$$

If $(\varphi, \psi) \in \overline{\Gamma(X')}$ there is a sequence $\{(\varphi_\nu, \psi_\nu)\}$ in $\Gamma(X')$ converging to (φ, ψ). Since

$$| < Xx, \varphi > - < x, \psi > | \leq$$
$$| < Xx, \varphi - \varphi_\nu > | + | < x, \psi - \psi_\nu > |$$
$$\leq \|Xx\| \, \|\varphi - \varphi_\nu\| + \|x\| \, \|\psi - \psi_\nu\| \to 0$$

as $\nu \to +\infty$, then $< Xx, \varphi > - < x, \psi >= 0$. ■

If $X \in \mathcal{L}(\mathcal{E})$, then $\mathcal{D}(X') = \mathcal{E}'$ and X' is closed. By the closed graph theorem, $X' \in \mathcal{L}(\mathcal{E}')$. That proves the first part of the following

Lemma 1.17.2 *If $X \in \mathcal{L}(\mathcal{E})$, then $X' \in \mathcal{L}(\mathcal{E}')$ and $\|X\| = \|X'\|$.*

Proof

$$\|X'\| = \sup\{\|X'\varphi\| : \varphi \in \mathcal{E}', \|\varphi\| \leq 1\}$$
$$= \sup\{\sup\{| < x, X'\varphi > | : x \in \mathcal{E}, \|x\| \leq 1\} : \varphi \in \mathcal{E}', \|\varphi\| \leq 1\}$$
$$= \sup\{\sup\{| < x, X'\varphi > | : \varphi \in \mathcal{E}', \|\varphi\| \leq 1\} : x \in \mathcal{E}, \|x\| \leq 1\}$$
$$= \sup\{\sup\{| < Xx, \varphi > | : \varphi \in \mathcal{E}', \|\varphi\| \leq 1\} : x \in \mathcal{E}, \|x\| \leq 1\}$$
$$= \sup\{\|Xx\| : x \in \mathcal{E}, \|x\| \leq 1\} = \|X\|.$$

■

---

[8]with the norm $\|(\varphi, \psi)\| = \|\varphi\| + \|\psi\|$.

Lemma 1.17.3 *If X is closed and densely defined, then $r(X) \subset r(X')$ and*

$$(\zeta I - X')^{-1} = ((\zeta I - X)^{-1})' \quad \forall \zeta \in r(X). \qquad (1.112)$$

Proof We begin by showing that, if $\zeta \in r(X)$, $\zeta I - X'$ is injective. Let $\varphi \in \mathcal{D}(X')$ be such that $(\zeta I - X')\varphi = 0$, *i.e.*,

$$
\begin{aligned}
0 &= <x, (\zeta I - X')\varphi> = <x, (\zeta I - X)'\varphi> \\
&= <(\zeta I - X)x, \varphi> \quad \forall x \in \mathcal{D}(X).
\end{aligned}
$$

Thus, $\varphi = 0$ because $\mathcal{R}(\zeta I - X) = \mathcal{E}$.

By Lemma 1.17.2, $((\zeta I - X)^{-1})' \in \mathcal{L}(\mathcal{E}')$.

If $x \in \mathcal{E}$ and $\varphi \in \mathcal{D}(X')$,

$$
\begin{aligned}
<x, \varphi> &= <(\zeta I - X)(\zeta I - X)^{-1}x, \varphi> \\
&= <(\zeta I - X)^{-1}x, (\zeta I - X')\varphi> \\
&= <x, ((\zeta I - X)^{-1})'(\zeta I - X')\varphi>,
\end{aligned}
$$

whence

$$((\zeta I - X)^{-1})'(\zeta I - X') = I \quad \text{on } \mathcal{D}(X').$$

If $x \in \mathcal{D}(X)$ and $\varphi \in \mathcal{E}'$,

$$
\begin{aligned}
<x, \varphi> &= <(\zeta I - X)^{-1}(\zeta I - X)x, \varphi> \\
&= <(\zeta I - X)x, ((\zeta I - X)^{-1})'\varphi>. \qquad (1.113)
\end{aligned}
$$

The map

$$
\begin{aligned}
x \mapsto <Xx, ((\zeta I - X)^{-1})'\varphi> &= \\
-<(\zeta I - X)x, ((\zeta I - X)^{-1})'\varphi> &+ \zeta <x, ((\zeta I - X)^{-1})'\varphi> \\
= -<(\zeta I - X)^{-1}(\zeta I - X)x, \varphi> &+ \zeta <x, ((\zeta I - X)^{-1})'\varphi> \\
= -<x, \varphi> &+ \zeta <x, ((\zeta I - X)^{-1})'\varphi>
\end{aligned}
$$

is continuous on $\mathcal{D}(X)$. As a consequence, $((\zeta I - X)^{-1})'\varphi \in \mathcal{D}(X')$ and, by (1.113),

$$<x, \varphi> = <x, (\zeta I - X')((\zeta I - X)^{-1})'\varphi>$$

$$(\zeta I - X')((\zeta I - X)^{-1})' = I \quad \text{on } \mathcal{E}'.$$

That proves (1.112). ∎

Lemma 1.17.4 *If X is densely defined and $\zeta \in r\sigma(X)$, then $\zeta \in p\sigma(X')$.*

Proof By our hypothesis, there exists $\varphi \in \mathcal{E}'\backslash\{0\}$ such that

$$< (\zeta I - X)x, \varphi >= 0 \ \forall \ x \in \mathcal{D}(X). \tag{1.114}$$

The linear form $x \mapsto < Xx, \varphi >= \zeta < x, \varphi >$ is continuous on $\mathcal{D}(X)$. Hence, $\varphi \in \mathcal{D}(X)$ and

$$< x, (\zeta I - X')\varphi >= 0 \ \forall \ x \in \mathcal{D}(X).$$

Since $\overline{\mathcal{D}(X)} = \mathcal{E}$, then

$$X'\varphi = \zeta \varphi.$$

∎

Lemma 1.17.5 *If X is densely defined and $\zeta \in p\sigma(X')$, then $\zeta \in p\sigma(X) \cup r\sigma(X)$.*

Indeed, if $X'\varphi = \zeta\varphi$ for some $\varphi \in \mathcal{D}(X')\backslash\{0\}$, then (1.114) implies that

$$\overline{\mathcal{R}(X - \zeta I)} \neq \mathcal{E}. \tag{1.115}$$

Hence, either $X - \zeta I$ is injective - in which case $\zeta \in r\sigma(X)$ - or $\ker(X - \zeta I) \neq \{0\}$, i. e. $\zeta \in p\sigma(X)$.

The set of points $\zeta \in \mathbf{C}$ for which (1.115) holds is called the *compression spectrum* of X, [38], and is denoted by $k\sigma(X)$[9]. Note that $r\sigma(X) \subset k\sigma(X)$, but that $p\sigma(X)$ is not always disjoint from $k\sigma(X)$.

The above argument proves

Lemma 1.17.6 *If X is densely defined, the compression spectrum of X is $p\sigma(X')$.*

Proposition 1.17.7 *If $\mathcal{E}$ is reflexive and X is closed and densely defined, X' is densely defined.*

---

[9]According to some authors, *e. g.* [12], [58], $k\sigma(X)$ is called residual spectrum.

Proof Since $\mathcal{E}$ is reflexive, if $\mathcal{D}(X')$ were not dense in $\mathcal{E}'$, there would exist some $x_o \in \mathcal{E}\backslash\{0\}$ such that

$$< x_o, \varphi >= 0 \ \forall \ \varphi \in \mathcal{D}(X'). \tag{1.116}$$

Since $(0, x_o) \notin \Gamma(X) = \overline{\Gamma(X)}$, by the Hahn-Banach theorem there exist $\varphi_1, \varphi_2 \in \mathcal{E}$ such that

$$< Xx, \varphi_1 >=< x, \varphi_2 > \ \forall x \in \mathcal{D}(X) \tag{1.117}$$

and

$$< x_o, \varphi_1 >\neq< 0, \varphi_2 >= 0. \tag{1.118}$$

On the other hand, (1.117) implies that $\varphi_1 \in \mathcal{D}(X')$, and (1.118) contradicts (1.116). ∎

Exercise Prove that, under the hypotheses of Proposition 1.17.7, $X" = (X')' = X$.

If $\mathcal{H}$ is a complex Hilbert space with inner product $(\ |\ )$, the map $x \mapsto (\bullet|x)$ defines a $\overline{\mathbf{C}}$-isomorphism of $\mathcal{H}$ onto $\mathcal{H}'$. If X is a densely defined linear operator on $\mathcal{H}$ and X^* is its Hilbert space adjoint, (1.111) reads now

$$(Xx|y) = (x|X^*y) \ \forall x \in \mathcal{D}(X), \ \ \forall y \in \mathcal{D}(X^*),$$

where $\mathcal{D}(X^*)$ consists of all $y \in \mathcal{H}$ such that the linear form $x \mapsto (Xx|y)$ is continuous on $\mathcal{D}(X)$. Standard references concerning the geometry of Hilbert spaces are, *e.g.*, [3], [72].

1.18 Introduction to the spectral theory for non linear operators

As before, $\mathcal{E}$ will indicate a complex Banach space; $2^{\mathcal{E}}$ will stand for the set of all subsets of $\mathcal{E}$.

Any subset Γ of $\mathcal{E} \times \mathcal{E}$ defines two subsets

$$\mathcal{D}(\hat{\Gamma}) \subset \mathcal{E} : \{x \in \mathcal{E} : \exists y \in \mathcal{E} \text{ such that } (x, y) \in \Gamma\},$$

$$\mathcal{R}(\hat{\Gamma}) \subset \mathcal{E} : \{y \in \mathcal{E} : \exists x \in \mathcal{E} \text{ such that } (x,y) \in \Gamma\},$$

and a set-valued map

$$\hat{\Gamma} : \mathcal{D}(\hat{\Gamma}) \to 2^{\mathcal{E}} \backslash \{\emptyset\}$$

defined on $x \in \mathcal{D}(\hat{\Gamma})$ by

$$x \mapsto \hat{\Gamma}(x) = \{y \in \mathcal{E} : (x,y) \in \Gamma\}.$$

Viceversa, given a subset E of $\mathcal{E}$ and a set-valued map

$$A : E \to 2^{\mathcal{E}} \backslash \{\emptyset\},$$

the subset of $\mathcal{E} \times \mathcal{E}$ defined by

$$G_A : \{(x,y) : x \in E, y \in A(x)\}$$

is is the graph of A, *i.e.*,

$$A = \widehat{G_A}.$$

Note In a similar way, given a complex Banach space $\mathcal{F}$, one defines set-valued maps $E \to 2^{\mathcal{F}} \backslash \{\emptyset\}$ and their graphs.

Let $\Upsilon(\mathcal{E})$ be the set of all set-valued maps

$$\mathcal{E} \supset E \to 2^{\mathcal{E}} \backslash \{\emptyset\}.$$

For any $A \in \Upsilon(\mathcal{E})$, the *resolvent set* $r(A)$ of A is, by definition, the set of all $\zeta \in \mathbf{C}$ such that:

$\zeta I - A$ has a dense image in $\mathcal{E}$. That is to say, for any $y \in \mathcal{E}$ and any $\epsilon > 0$ there are $x \in \mathcal{D}(A)$ and $z \in A(x)$ such that

$$\|(\zeta x - z) - y\| < \epsilon;$$

$\zeta I - A$ is injective;

$$(\zeta I - A)^{-1} \in \text{Lip}(\mathcal{M}_\zeta), \tag{1.119}$$

where

$$\mathcal{M}_\zeta = \cup\{\zeta I - A(x) : x \in \mathcal{D}(A)\}.$$

The set $\sigma(A) := \mathbf{C} \backslash r(A)$ is called the *spectrum of A*.[10]

Let $\zeta \in r(A)$. In view of (1.119), there exists $M > 0$ such that

$$\left\| (\zeta I - A)^{-1}(y_1) - (\zeta I - A)^{-1}(y_2) \right\| \leq M \left\| y_1 - y_2 \right\| \ \ \forall \, y_1, y_2 \in \mathcal{M}_\zeta.$$

Letting

$$x_j = (\zeta I - A)^{-1}(y_j) \in \mathcal{D}(A) \ \ (j = 1, 2),$$

there exist $z_j \in A(x_j)$ for which

$$y_j = \zeta x_j - z_j.$$

For every $y \in \mathcal{E}$ there are sequences $\{x_\nu\}$ and $\{z_\nu\}$, with $x_\nu \in \mathcal{D}(A)$ and $z_\nu \in A(x_\nu)$, such that

$$\lim_{\nu \to \infty} (\zeta x_\nu - z_\nu) = y. \tag{1.120}$$

Since

$$\|x_\nu - x_\mu\| \leq M \left\| (\zeta x_\nu - z_\nu) - (\zeta x_\mu - z_\mu) \right\|,$$

$\{x_\nu\}$ is a Cauchy sequence. By (1.120), also $\{z_\nu\}$ is a Cauchy sequence. If

$$x = \lim_{\nu \to \infty} x_\nu, \quad z = \lim_{\nu \to \infty} z_\nu,$$

then

$$y = \zeta x - z.$$

If the graph $G_{\zeta I - A}$ of $\zeta I - A$ is closed, then $(x, z) \in G_{\zeta I - A}$, that is, $x \in \mathcal{D}(A)$ and $z \in \xi x - A(x)$.

Since, if G_A is closed, $G_{\zeta I - A}$ is closed, that proves

Lemma 1.18.1 *If A is closed, then $r(A)$ consists of all $\zeta \in \mathbf{C}$ for which: $\mathcal{M}_\zeta = \mathcal{E}$, $\zeta I - A$ is injective, and*

$$(\zeta I - A)^{-1} \in \mathrm{Lip}(\mathcal{E}).$$

The proof of the lemma yields the following

---

[10]These definitions are taylored on the ones given in n. 1.8 in the case of a linear operator A, and differ from those given, *e.g.*, in [14].

Lemma 1.18.2 *If* $\overline{\mathcal{M}_\zeta} = \mathcal{E}$ *for some* $\zeta \in r(A)$, *then* $\zeta I - \overline{A}$ *is surjective.*

Proof Exercise.

In order to show that $\zeta I - \overline{A}$ is injective, let $x_1, x_2 \in \mathcal{D}(\overline{A})$ and $z_1 \in \overline{A}(x_1)$, $z_2 \in \overline{A}(x_2)$ be such that

$$\zeta x_1 - z_1 = \zeta x_2 - z_2. \tag{1.121}$$

Since

$$G_{\zeta I - \overline{A}} = \overline{G_{\zeta I - A}},$$

there exist sequences $\{x_1^\nu\}$, $\{x_2^\nu\}$ in $\mathcal{D}(A)$ and sequences $\{z_1^\nu\}$, $\{z_2^\nu\}$, with $z_1^\nu \in A(x_1^\nu)$, $z_2^\nu \in A(x_2^\nu)$, converging respectively to x_1, x_2 and z_1, z_2.

The inequalities

$$\|x_1^\nu - x_2^\nu\| \le M\|(\zeta x_1^\nu - z_1^\nu) - (\zeta x_2^\nu - z_2^\nu)\| \tag{1.122}$$

and (1.121) imply that $x_1 = x_2$ and $z_1 = z_2$.

Thus, for any $\zeta \in r(A)$, $\zeta I - \overline{A}$ is bijective.

We show now that the map $(\zeta I - \overline{A})^{-1}$ is lipschitz.

For any choice of $x_1, x_2 \in \mathcal{D}(\overline{A})$ and $z_1 \in \overline{A}(x_1)$, $z_2 \in \overline{A}(x_2)$, let $\{x_1^\nu\}$, $\{x_2^\nu\}$ in $\mathcal{D}(A)$ and $\{z_1^\nu\}$, $\{z_2^\nu\}$, with $z_1^\nu \in A(x_1^\nu)$, $z_2^\nu \in A(x_2^\nu)$, be sequences converging respectively to x_1, x_2 and z_1, z_2. The equation (1.122) implies that

$$\|x_1 - x_2\| \le M\|(\zeta x_1 - z_1) - (\zeta x_2 - z_2)\|,$$

proving thereby that

$$(\zeta I - \overline{A})^{-1} \in \mathrm{Lip}(\mathcal{E}).$$

Summing up, $\zeta \in r(\overline{A})$.

On the other hand, if $\zeta \in r(\overline{A})$, for every $y \in \mathcal{E}$ there exist a unique $x \in \mathcal{D}(\overline{A})$ and a unique $z \in \overline{A}(x)$ such that $y = \zeta x - z$. Furthermore, the map $y \mapsto x$ is lipschitz. Hence, $\zeta \in r(A)$, and, in conclusion, the following theorem has been established.

Theorem 1.18.3 *For any $A \in \Upsilon(\mathcal{E})$,*

$$r(A) = r(\overline{A}),$$

or, equivalently,

$$\sigma(A) = \sigma(\overline{A}).$$

Lemma 1.18.4 *If $\mathcal{R}(\zeta I - A) = \mathcal{E}$ for some $\zeta \in r(A)$, A is closed.*

Proof By Theorem 1.18.3, $\zeta \in r(\overline{A})$. If $\mathcal{D}(A) \overset{\subseteq}{\neq} \mathcal{D}(\overline{A})$, $\overline{A}$ is not injective. ∎

Let $A \in \Upsilon(\mathcal{E})$, $\zeta_o \in r(A)$ and

$$s_o = \frac{1}{p_L((\zeta_o I - A)^{-1})}.$$

Proposition 1.18.5 *One has*

$$\Delta(\zeta_o, s_o) \subset r(A),$$

and, for any $\zeta \in \Delta(\zeta_o, s_o)$,

$$(\zeta I - \overline{A})^{-1} = (\zeta_o I - \overline{A})^{-1} \circ \left(I + (\zeta - \zeta_o)(\zeta_o I - \overline{A})^{-1}\right)^{-1}.$$

Proof Being

$$|\zeta - \zeta_o| \, p_L((\zeta_o I - \overline{A})^{-1}) < 1, \tag{1.123}$$

then, by Proposition 1.2.6,

$$\left(I + (\zeta - \zeta_o)(\zeta_o I - \overline{A})^{-1}\right)^{-1} \in \mathrm{Lip}(\mathcal{E}).$$

Letting

$$F = (\zeta_o I - \overline{A})^{-1} \circ \left(I + (\zeta - \zeta_o)(\zeta_o I - \overline{A})^{-1}\right)^{-1} \in \mathrm{Lip}(\mathcal{E}),$$

we have to show that $F = (\zeta I - \overline{A})^{-1}$; that is to say, for all $x \in \mathcal{E}$ the equation

$$\zeta u - x \in \overline{A}(u) \tag{1.124}$$

has the unique solution $u = F(x)$.

To establish uniqueness, assume that $\zeta u - x \in \overline{A}(u)$, and let $v = \zeta u - x$. Because $v \in \overline{A}(u)$, then

$$(\zeta_o I - \overline{A})^{-1}(\zeta_o u - v) = u,$$

i.e.,

$$(\zeta_o I - \overline{A})^{-1}(x + (\zeta_o - \zeta)u) = u.$$

Thus u is fixed by the map $C : \mathcal{E} \to \mathcal{E}$ defined by

$$y \mapsto C(y) = (\zeta_o I - \overline{A})^{-1}(x + (\zeta_o - \zeta)y).$$

Since, by (1.123), $p_L(C) < 1$, Theorem 1.2.2 implies that u is the unique fixed point of C.

We are left to prove that $u = F(x)$. Letting

$$z = \left(I + (\zeta - \zeta_o)(\zeta_o I - \overline{A})^{-1}\right)^{-1}(x),$$

so that

$$(\zeta_o I - \overline{A})^{-1}(z) = F(x),$$

then

$$
\begin{aligned}
\zeta F(x) - x &= \zeta(\zeta_o I - \overline{A})^{-1}(z) - \left(I + (\zeta - \zeta_o)(\zeta_o I - \overline{A})^{-1}\right)(z) \\
&= \zeta_o(\zeta_o I - \overline{A})^{-1}(z) - z \in \overline{A}\left((\zeta_o I - \overline{A})^{-1}(z)\right) \\
&= \overline{A}(F(x)).
\end{aligned}
$$

Thus, $u = F(x)$ solves the equation (1.124). $\blacksquare$

Corollary 1.18.6 *For any $A \in \Upsilon(\mathcal{E})$, $r(A)$ is open.*

Equivalently, $\sigma(A)$ is closed.

Theorem 1.18.7 *For any $x \in \mathcal{E}$ the function*

$$r(A) \ni \zeta \mapsto (\zeta I - \overline{A})^{-1}(x)$$

is locally lipschitz.

Proof For $\zeta, \zeta_o \in r(A)$, $\zeta \neq \zeta_o$,

$$(\zeta I - \overline{A})^{-1}(x) - (\zeta_o I - \overline{A})^{-1}(x) = (\zeta I - \overline{A})^{-1}(x) - \\ -(\zeta I - \overline{A})^{-1}(x + (\zeta - \zeta_o)(\zeta_o I - \overline{A})^{-1}(x)).$$

Hence

$$\|(\zeta I - \overline{A})^{-1}(x) - (\zeta_o I - \overline{A})^{-1}(x)\| \leq \\ \leq |\zeta - \zeta_o| \, p_L((\zeta I - \overline{A})^{-1}) \, \|(\zeta_o I - \overline{A})^{-1}(x)\| \,.$$

On the other hand,

$$p_L((\zeta I - \overline{A})^{-1}) = \sup \left\{ \frac{1}{\|x - y\|} \|(\zeta I - \overline{A})^{-1}(x) - (\zeta I - \overline{A})^{-1}(y)\| : \right.$$

$$\left. x, y \in \mathcal{E}, x \neq y \right\} = \sup \left\{ \frac{1}{\|x - y\|} \|(\zeta_o I - \overline{A})^{-1} \times \right.$$

$$\times (I + (\zeta - \zeta_o)(\zeta_o I - \overline{A})^{-1})^{-1}(x) - \\ (\zeta_o I - \overline{A})^{-1}(I + (\zeta - \zeta_o)(\zeta_o I - \overline{A})^{-1})^{-1}(y)\| :$$

$$\left. x, y \in \mathcal{E}, x \neq y \right\}$$

$$\leq \|(\zeta_o I - \overline{A})^{-1}\| \, p_L \left(\left(I + (\zeta - \zeta_o)(\zeta_o I - \overline{A})^{-1} \right)^{-1} \right)$$

$$\leq \frac{\|(\zeta_o I - \overline{A})^{-1}\|}{1 - |\zeta - \zeta_o| \, p_L(\zeta_o I - \overline{A})^{-1})},$$

because of Proposition 1.2.6, of the fact that $(\zeta_o I - \overline{A})^{-1}$ is injective, and because

$$|\zeta - \zeta_o| \, p_L(\zeta_o I - \overline{A})^{-1}) < 1.$$

Hence,

$$\|(\zeta I - \overline{A})^{-1}(x) - (\zeta_o I - \overline{A})^{-1}(x)\| \leq \\ \leq \frac{|\zeta - \zeta_o| \, \|(\zeta_o I - \overline{A})^{-1}\|}{1 - |\zeta - \zeta_o| \, p_L(\zeta_o I - \overline{A})^{-1})} \, \|(\zeta_o I - \overline{A})^{-1}(x)\|.$$

That proves the theorem. ∎

Let $f : \mathcal{E} \to \mathcal{E}$ and $g : \mathcal{E} \to \mathcal{E}$ be such that $r(f) \cap r(g) \neq \emptyset$, and let $\zeta \in r(f) \cap r(g)$.

For all $x \in \mathcal{E}$

$$(\zeta I - g)(x) = (\zeta I - f)(x) + (f - g)(x).$$

Hence, if $x = (\zeta I - f)^{-1}(y)$,

$$(\zeta I - g)((\zeta I - f)^{-1}(y)) = y + (f - g)((\zeta I - f)^{-1}(y)),$$

i.e.,

$$(\zeta I - f)^{-1}(y) = (\zeta I - g)^{-1}\left(y + (f - g)\left((\zeta I - f)^{-1}(y)\right)\right) \qquad (1.125)$$

for all $y \in \mathcal{E}$.

Let $A \in \Upsilon(\mathcal{E})$ with $r(A) \neq \emptyset$. For $\zeta \in r(A)$ and $x \in \mathcal{E}$, let

$$H(\zeta, x) = (\zeta I - \overline{A})^{-1}(x).$$

Proposition 1.18.8 *If H is Fréchet differentiable at all points $x \in \mathcal{E}$ and for all $\zeta \in r(A)$, $H(\bullet, x)$ is holomorphic in $r(A)$. Furthermore,*

$$\frac{d}{d\zeta}H(\zeta, x) = -d_x H(\zeta, x)(H(\zeta, x))$$

for any $x \in \mathcal{E}$ and any $\zeta \in r(A)$.

Proof For $x, y \in \mathcal{E}$, with $y \neq 0$, and $\zeta \in r(A)$, let

$$h(y) := H(\zeta, x + y) - H(\zeta, x) - d_x H(\zeta, x)(y).$$

Then
$$\lim_{y \to 0} \frac{\|h(y)\|}{\|y\|} = 0.$$

If $\kappa \in \mathbf{C}$ is such that $\zeta + \kappa \in r(A)$, then, choosing $g = \overline{A}$ and $f = -\kappa I + \overline{A}$, (1.125) yields

$$
\begin{aligned}
H(\zeta + \kappa, x) &= ((\zeta + \kappa)I - \overline{A})^{-1}(x) \\
&= (\zeta I - \overline{A})^{-1}\left(x + ((-\kappa I + \overline{A} - \overline{A})\left((\zeta + \kappa)I - \overline{A})^{-1}(x)\right)\right) \\
&= (\zeta I - \overline{A})^{-1}\left(x - \kappa\left((\zeta + \kappa)I - \overline{A})^{-1}(x)\right) \\
&= H(\zeta, x - \kappa H(\zeta + \kappa, x)),
\end{aligned}
$$

and therefore

$$H(\zeta + \kappa, x) - H(\zeta, x) = H(\zeta, x - \kappa H(\zeta + \kappa, x)) - H(\zeta, x)$$
$$= -\kappa d_x H(\zeta, x)(H(\zeta + \kappa, x)) + h(-\kappa H(\zeta + \kappa, x)).$$

Since $H(\bullet, x)$ is locally lipshitz, the conclusion follows. ∎

Theorem 1.18.9 *Let K be a non-empty, closed and connected subset of* **C**. *Let $A \in \Upsilon(\mathcal{E})$, let $\zeta_o \in K$ and let $\phi : K \to \mathbf{R}_+$ be continuous and such that:*

$$\|x_1 - x_2\| \leq \phi(\zeta)\|\zeta(x_1 - x_2) - (z_1 - z_2)\| \qquad (1.126)$$

for all $x_1, x_2 \in \mathcal{D}(A)$, all $z_1 \in A(x_1)$, $z_2 \in A(x_2)$ and all $\zeta \in K$, and that

$$\overline{(\zeta_o I - A)(\mathcal{D}(A))} = \mathcal{E}. \qquad (1.127)$$

Then:

$$K \subset r(A); \qquad (1.128)$$

$$p_L((\zeta I - A)^{-1}) \leq \phi(\zeta) \ \forall \zeta \in K, \qquad (1.129)$$

and

$$\overline{(\zeta I - A)(\mathcal{D}(A))} = \mathcal{E} \ \forall \zeta \in K. \qquad (1.130)$$

Proof Note first of all that (1.126) holds, by continuity, when A is replaced by $\overline{A}$, i.e.,

$$\|x_1 - x_2\| \leq \phi(\zeta)\|\zeta(x_1 - x_2) - (z_1 - z_2)\| \qquad (1.131)$$

for all $x_1, x_2 \in \mathcal{D}(\overline{A})$, all $z_1 \in \overline{A}(x_1)$, $z_2 \in \overline{A}(x_2)$ and all $\zeta \in K$.
    We begin by showing that, for all $y \in \mathcal{E}$, there exists a unique $x \in \mathcal{D}(\overline{A})$ solving the equation

$$\zeta_o x - y \in \overline{A}(x). \qquad (1.132)$$

By (1.131), there exist sequences $\{x_\nu\}$, with $x_\nu \in \mathcal{D}(A)$, and $\{z_\nu\}$, with $z_\nu \in A(x_\nu)$, such that, letting

$$\zeta_o x_\nu - z_\nu = y_\nu,$$

then

$$\lim_{\nu \to \infty} y_\nu = y.$$

It follows from (1.126) that

$$\|x_\nu - x_\mu\| \leq \phi(\zeta_o)\|y_\nu - y_\mu\|$$

for all indices ν and μ.

Since $\mathcal{E}$ is complete, $\{x_\nu\}$ converges to some $x \in \mathcal{D}(\overline{A})$. Hence, $\{z_\nu\}$ converges to some $z \in \overline{A}(x)$ for which

$$\zeta_o x - z = y.$$

Thus, (1.132) holds.

Let $x' \in \mathcal{D}(\overline{A})$ and $z' \in \overline{A}(x')$ be such that

$$\zeta_o x' - z' = y.$$

Since

$$\|x - x'\| \leq \phi(\zeta_o)\|(\zeta_o x - z) - (\zeta_o x' - z')\|$$
$$= \|y - y\| = 0,$$

then $x = x'$, *i.e.*, the solution of the equation (1.132) is unique.

Since, furthermore,

$$p_L((\zeta_o I - \overline{A})^{-1}) \leq \phi(\zeta_o), \tag{1.133}$$

then

$$\zeta_o \in r(A) \cap K.$$

The intersection $K_o = K \cap r(A)$ is non-empty and open in K.

To establish (1.128), we prove now that K_o is also closed.

Let $\zeta \in \overline{K_o}$, and let $\{\zeta_\nu\}$, with $\zeta_\nu \in K_o$, be a sequence converging to ζ. By Proposition 1.18.5 and by (1.126),

$$\Delta\left(\zeta_\nu, \frac{1}{\phi(\zeta_\nu)}\right) \cap K \subset K_o.$$

If $\epsilon > 0$ is such that $\phi(\zeta_\nu) < \frac{1}{\epsilon}$, then

$$\Delta(\zeta_\nu, \epsilon) \cap K \subset K_o,$$

and therefore $\zeta \in K_o$. Since K_o is open in K, closed and non-empty, and since K is connected, then $K = K_o \subset r(A)$.

Because

$$p_L((\zeta I - A)^{-1}) \leq p_L((\zeta I - \overline{A})^{-1}),$$

(1.131), which now reads

$$p_L((\zeta I - \overline{A})^{-1}) \leq \phi(\zeta) \ \forall \zeta \in K,$$

yields (1.129).

We complete the proof of the theorem showing that (1.130) holds. Indeed, if $\lambda \in \mathcal{E}'$ is such that

$$\langle \zeta x - y, \lambda \rangle = 0 \ \forall x \in \mathcal{D}(A); \forall y \in A(x),$$

the same condition is satisfied whenever $x \in \mathcal{D}(\overline{A})$ and $y \in \overline{A}(x)$. But, since $\zeta \in r(\overline{A})$, then $\lambda = 0$. ∎

The following theorems will concern lipschitz functions on $\mathcal{E}$.

Theorem 1.18.10 *If $f \in \mathrm{Lip}(\mathcal{E})$, then*

$$\dot{\sigma}(f) \subset \overline{\Delta(0, p_L(f))}.$$

If $|\zeta| > p_L(f)$,

$$(\zeta I - f)^{-1} = \left(I - \frac{1}{\zeta}f\right)^{-1} \circ \frac{1}{\zeta},$$

and

$$p_L((\zeta I - f)^{-1}) \leq \frac{1}{|\zeta| - p_L(f)}. \tag{1.134}$$

Proof If $\zeta \neq 0$ and $y \in \mathcal{M}_\zeta$, the equation

$$\zeta x - f(x) = y$$

is equivalent to

$$x - \frac{1}{\zeta} f(x) = \frac{1}{\zeta} y.$$

If $|\zeta| > p_L(f)$,

$$p_L \left(\frac{1}{\zeta} f \right) < 1.$$

By Proposition 1.2.6, $I - \frac{1}{\zeta} f$ is injective, and

$$p_L \left(\left(I - \frac{1}{\zeta} f \right)^{-1} \right) \leq \frac{1}{1 - p_L \left(\frac{1}{\zeta} f \right)},$$

which implies (1.134).

The rest of the proof is trivial. ∎

Theorem 1.18.11 *Let* $f \in \mathrm{Lip}(\mathcal{E})$. *If* $(\zeta I - \overline{f})^{-1}(x)$ *is Fréchet differentiable with respect to x for all $x \in \mathcal{E}$ and all $\zeta \in r(f)$, the function*

$$\zeta \mapsto (\zeta I - \overline{f})^{-1}(x) - (\zeta I - \overline{f})^{-1}(y) \tag{1.135}$$

is holomorphic in $r(f)$ and bounded at infinity for all $x, y \in \mathcal{E}$. Furthermore, $\sigma(f) \neq \emptyset$.

Proof The holomorphy of the function (1.135) is stated in Proposition 1.18.8. Boundedness at infinity follows from the inequalities

$$\|(\zeta I - \overline{f})^{-1}(x) - (\zeta I - \overline{f})^{-1}(y)\| \leq p_L((\zeta I - \overline{f})^{-1}) \|x - y\|$$
$$\leq \frac{\|x - y\|}{|\zeta| - p_L(\overline{f})}$$

holding whenever $|\zeta| > p_L(\overline{f})$.

Let $x \neq y$. If $\sigma(f) = \emptyset$, the holomorphic function (1.135) on $\mathbf{C}$ is equal to a constant $k \in \mathcal{E}$, by the Liouville theorem. As a consequence of the latter inequality,

$$\|k\| \leq \frac{\|x - y\|}{|\zeta| - p_L(\overline{f})} \to 0$$

as $|\zeta| \to \infty$. Hence $k = 0$, and therefore $x = y$. ∎

By (1.7), Theorems 1.18.10 and 1.18.11 imply Theorem 1.10.1, (1.38) and Lemma 1.10.3.

Theorem 1.18.12 *Let $f : \mathcal{D}(f) \subset \mathcal{E} \to \mathcal{E}$ and let $f_\nu : \mathcal{D}(f_\nu) \subset \mathcal{E} \to \mathcal{E}$ ($\nu = 1, 2, \dots$) satisfy the following conditions:*

$$\mathcal{D}(f) \subset \mathcal{D}(f_\nu) \ \forall \nu;$$

$$r(f) \cap \{r(f_\nu) : \nu = 1, 2, \dots\} \neq \emptyset;$$

there is a set $\mathcal{D} \subset \mathcal{D}(f)$ which is dense in $\mathcal{E}$ and such that

$$\lim_{\nu \to \infty} f_\nu(x) = f(x) \ \forall x \in \mathcal{D};$$

for every $\zeta \in r(f) \cap \{r(f_\nu) : \nu = 1, 2, \dots\}$, $(\zeta I - f)^{-1} \in \mathrm{Lip}(\mathcal{E})$.

Under these hypotheses,

$$\lim_{\nu \to \infty} (\zeta I - f_\nu)^{-1}(x) = (\zeta I - f)^{-1}(x)$$

for all $x \in \mathcal{E}$ and all $\zeta \in r(f) \cap \{r(f_\nu) : \nu = 1, 2, \dots\}$.

Proof For $\zeta \in r(f) \cap \{r(f_\nu) : \nu = 1, 2, \dots\}$ let $k > 0$ be such that

$$p_L((\zeta I - f)^{-1}) \leq k.$$

By (1.125),

$$
\begin{aligned}
(\zeta I - f_\nu)^{-1}(x) - (\zeta I - f)^{-1}(x) = (\zeta I - f_\nu)^{-1}(x) - \\
(\zeta I - f_\nu)^{-1}\left(x - (f - f_\nu)((\zeta I - f)^{-1}(x))\right)
\end{aligned}
$$

for all $x \in \mathcal{E}$.

Therefore

$$\left\|(\zeta I - f_\nu)^{-1}(x) - (\zeta I - f)^{-1}(x)\right\| \leq k \left\|(f - f_\nu)((\zeta I - f)^{-1}(x))\right\| \to 0$$

as $\nu \to \infty$. ∎

Note A large part of the results in this section can be found in [14].

Chapter 2

Strongly continuous semigroups

This chapter will be devoted to the general theory of strongly continuous semigroups of continuous linear operators acting on Banach spaces. Standard references will be [12], [15], [35], [40], [63], [81].

2.1   Semigroups

As before, $\mathcal{E}$ will be a complex Banach space (which - to avoid trivialities - will always be assumed to have positive dimension); $\mathcal{E}'$ will denote the strong dual of $\mathcal{E}$, that is, the Banach space of all linear continuous forms $\lambda : \mathcal{E} \to \mathbf{C}$ with the norm $\|\lambda\| = \sup\{| < x, \lambda > | : x \in \mathcal{E}, \|x\| \leq 1\}$, where $< x, \lambda >$ is the value of λ at the point $x \in \mathcal{E}$.

Let $T : \mathbf{R}_+ \to \mathcal{L}(\mathcal{E})$ be a *semigroup*, that is, a map of $\mathbf{R}_+$ into $\mathcal{L}(\mathcal{E})$ such that

$$T(0) = I \quad T(t_1 + t_2) = T(t_1)T(t_2) \tag{2.1}$$

for all t_1, t_2 in $\mathbf{R}_+ = \{t \in \mathbf{R} : t \geq 0\}$, where I is the identity operator in $\mathcal{E}$.

Different topologies in $\mathcal{L}(\mathcal{E})$ generate different continuity conditions for T.

The strong operator topology in $\mathcal{L}(\mathcal{E})$ defines the *strong continuity*

118

for T, that is, the continuity of the function

$$t \mapsto T(t)x \tag{2.2}$$

from $\mathbf{R}_+$ to $\mathcal{E}$, for all $x \in \mathcal{E}$. If T is strongly continuous, then

$$\lim_{t \downarrow 0} T(t)x = x \quad \text{for all } x \in \mathcal{E}. \tag{2.3}$$

This condition will be shown to be also sufficient for the function $\mathbf{R}_+ \times \mathcal{E} \to \mathcal{E}$, defined by

$$(t, x) \mapsto T(t)x, \tag{2.4}$$

to be continuous, that is, for the semigroup T to be strongly continuous.

Strongly continuous semigroups will be also called C_o semigroups.

The semigroup T will be said to be *weakly continuous* if, for every $x \in \mathcal{E}$, the function (2.2) is continuous for the weak topology, i.e., if the function $t \mapsto < T(t)x, \lambda >$ from $\mathbf{R}_+$ to $\mathbf{C}$ is continuous for all $x \in \mathcal{E}$ and for every continuous linear form λ on $\mathcal{E}$. That implies, in particular, that

$$\text{w} - \lim_{t \downarrow 0} T(t)x = x \quad \text{for all } x \in \mathcal{E}, \tag{2.5}$$

i.e.

$$\lim_{t \downarrow 0} < T(t)x, \lambda > = < x, \lambda >$$

for all $x \in \mathcal{E}$ and all $\lambda \in \mathcal{E}'$.

This condition will be shown to imply the strong continuity of T.

The semigroup T will be said to be *uniformly continuous* or also *norm-continuous* if the function $t \mapsto \|T(t)\|$, from $\mathbf{R}_+$ to $\mathbf{R}_+$, is continuous.

Suppose now that $T : \mathbf{R} \to \mathcal{L}(\mathcal{E})$ is group, that is to say, the second condition in (2.1) holds for all t_1, $t_2 \in \mathbf{R}$ (in which case $T(t)$ is invertible, with $T(t)^{-1} = T(-t) \in \mathcal{L}(\mathcal{E})$ for all $t \in \mathbf{R}$). In a similar way to the one described above, one defines *strongly continuous, weakly continuous, uniformly continuous* or *norm-continuous groups*. We shall come back on groups in the next section.

Theorem 2.1.1 *If $T \to \mathcal{L}(\mathcal{E})$ is a semigroup, (2.3) holds if, and only if, the map (2.4) is continuous. If T is strongly continuous, there exist two finite real constants a and $M \geq 1$ such that*

$$\|T(t)\| \leq Me^{at} \ \forall t \in \mathbf{R}_+. \tag{2.6}$$

Proof For $n \in \mathbf{N}^*$, let

$$c_n = \sup\{\|T(t)\| : 0 \leq t \leq \frac{1}{n}\}.$$

We begin by showing that, if (2.3) holds, c_n is finite for some $n \geq 1$. Indeed, if that were not the case, for every integer $n \geq 1$ there would exist some $t_n \in (0, \frac{1}{n}]$ for which $\|T(t_n)\| \geq n$. But then, by the theorem of Banach-Steinhaus, there would exist some $x \in \mathcal{E}$ for which

$$\limsup_{n \to +\infty} \|T(t_n)x\| = +\infty,$$

contradicting (2.3).

Let $n \in \mathbf{N}^*$ be such that c_n is finite. Setting $c = c_n{}^n$, then $\|T(t)\| \leq c$ for all $t \in [0,1]$, and $c \geq \|T(0)\| = \|I\| = 1$. For $t > 0$ let $[t]$ be the integral part of t. Then

$$\|T(t)\| \leq c^{[t]+1} \leq c^{t+1} = ce^{t\log c}.$$

We complete the proof by showing that, if (2.3) holds, the map (2.4) is continuous.

If $\{t_\nu\}$ and $\{x_\nu\}$ are sequences in $\mathbf{R}_+$ and in $\mathcal{E}$ converging to t and to x, then

$$
\begin{aligned}
\|T(t_\nu)x_\nu - T(t)x\| &= \|T(t_\nu)x_\nu - T(t_\nu)x + T(t_\nu)x - T(t)x\| \\
&\leq \|T(t_\nu)(x - x_\nu)\| + \|(T(t_\nu) - T(t))x\| \\
&\leq \|T(t_\nu)\| \, \|x - x_\nu\| + \|(T(t_\nu) - T(t))x\| \\
&\leq Me^{at_\nu}\|x - x_\nu\| + \|(T(t_\nu) - T(t))x\|.
\end{aligned}
$$

According as $t_\nu \leq t$ or $t_\nu \geq t$ we can write, respectively,

$$
\begin{aligned}
\|(T(t_\nu) - T(t))x\| &= \|(T(t_\nu) - T(t - t_\nu + t_\nu))x\| \\
&= \|T(t_\nu)(x - T(t - t_\nu)x)\| \\
&\leq \|T(t_\nu)\| \, \|T(t - t_\nu)x - x\| \\
&\leq Me^{at_\nu}\|T(t - t_\nu)x - x\|,
\end{aligned}
$$

or

$$\begin{aligned}
\|(T(t_\nu) - T(t))x\| &= \|(T(t_\nu - t + t) - T(t))x\| \\
&= \|T(t)(T(t_\nu - t)x - x)\| \\
&\leq \|T(t)\| \, \|T(t_\nu - t)x - x\| \\
&\leq Me^{at}\|T(t_\nu - t)x - x\|.
\end{aligned}$$

Thus, in both cases,

$$\|(T(t_\nu) - T(t))x\| \leq Me^{\min\{at, at_\nu\}}\|T(|t - t_\nu|)x - x\|,$$

and therefore

$$\begin{aligned}
\lim_{\nu \to +\infty} \|T(t_\nu)x_\nu - T(t)x\| &\leq \lim_{\nu \to +\infty} \left(Me^{at_\nu}\|x - x_\nu\| \right. \\
&\quad \left. + Me^{\min\{at, at_\nu\}}\|T(|t - t_\nu|)x - x\| \right) \\
&= 0.
\end{aligned}$$

∎

Corollary 2.1.2 *The semigroup* $T : \mathbf{R}_+ \to \mathcal{L}(\mathcal{E})$ *is strongly continuous if, and only if, the map (2.4) is continuous.*

Theorem 2.1.3 *If* $T : \mathbf{R}_+ \to \mathcal{L}(\mathcal{E})$ *is such that*

$$T(0) = I \text{ and } T(t_1 + t_2) = T(t_1)T(t_2)$$

for all $t_1 > 0$, $t_2 > 0$, *and that (2.5) holds, then* T *is a strongly continuous semigroup.*

Proof The proof will consist in showing that

$$\lim_{t \to t_o} T(t)x_o = T(t_o)x_o$$

for all $x_o \in \mathcal{E}$ and for every choice of t_o in $\mathbf{R}_+$ (with the *proviso* that, if $t_o = 0$, the limit should be computed from the right).

For $x_o \in \mathcal{E}$, let $x : \mathbf{R}_+ \to \mathcal{E}$ be defined by $x(t) = T(t)x_o$.

1. For every $t_o \geq 0$, x is weakly continuous from the right at t_o. because

$$\begin{aligned} \text{w}-\lim_{t\downarrow t_o} x(t) &= \text{w}-\lim_{t\downarrow t_o} T(t)x_o = \text{w}-\lim_{h\downarrow 0} T(h+t_o)x_o = \\ &= \text{w}-\lim_{h\downarrow 0} T(h)T(t_o)x_o = T(t_o)x_o = x(t_o). \end{aligned}$$

Hence, for every $\lambda \in \mathcal{E}'$, the scalar-valued function

$$t \mapsto\, < T(t)x_o, \lambda > \tag{2.7}$$

is continuous from the right at t_o.

We show now that the function $t \mapsto \|T(t)\|$ is bounded on a right neighbourhood of 0.

Indeed, if that were not the case, there would exist a sequence $\{t_n\}$ in $\mathbf{R}_+$, with $t_n \downarrow 0$, such that $\limsup_{n\to+\infty} \|T(t_n)\| = \infty$. By the theorem of Banach-Steinhaus, $\limsup_{n\to+\infty} \|T(t_n)x_o\| = \infty$ for some $x_o \in \mathcal{E}$. Again by the same theorem, there would exist some $\lambda \in \mathcal{E}'$, with $\|\lambda\| = 1$, for which $\limsup_{n\to+\infty} < T(t_n)x_o, \lambda >= \infty$, contradicting the continuity from the right of the function (2.7).

Thus, let $\delta > 0$ and $k > 0$ be such that $\|T(s)\| \leq k$ for all $s \in [0, \delta]$. If $a \geq 0$, every $t \in [a, a + \delta]$ can be written as $t = a + s$ with $s \in [0, \delta]$. Thus

$$\begin{aligned} \|T(t)\| = \|T(a + s)\| &= \|T(a)T(s)\| \\ &\leq \|T(a)\| \, |\, T(s)\| \leq k\|T(a)\| \end{aligned}$$

for every $t \in [a, a + \delta]$. Hence the function $t \mapsto \|T(t)\|$ is bounded on all compact subsets of $\mathbf{R}_+$. As a consequence, for every $x_o \in \mathcal{E}$, the function x is bounded on all compact subsets of $\mathbf{R}_+$.

2. We order now in a sequence $\{t_n\}$ all rational numbers in $\mathbf{R}_+$ and we order in a sequence $\{\alpha_n\}$ all complex numbers whose real parts and imaginary coefficients are rational.

Denoting by $\mathcal{C}$ the closure in $\mathcal{E}$ of the set of all finite linear combinations $\sum \alpha_j x(t_j)$, we show now that

$$\{x(t) : t \geq 0\} \subset \mathcal{C}. \tag{2.8}$$

If that were not the case, there would exist some $r_o \in \mathbf{R}_+$ for which $x(r_o) \notin \mathcal{C}$. On the other hand, the linear subspace $\mathcal{C}$, closed in $\mathcal{E}$ for the strong topology, is also weakly closed. Hence, if $\{s_n\}$ is a subsequence of $\{t_n\}$ such that $s_n \downarrow r_o$, then

$$x(r_o) = \mathrm{w} - \lim_{n \to +\infty} x(s_n) \in \mathcal{C}.$$

As a consequence, the range, $x(\mathbf{R}_+)$, of x is separable.

3. The proof of the following lemma of function theory will be left as an exercise:

Any function $\mathbf{R}_+ \to \mathbf{R}$ which is continuous from the right is Lebesgue measurable.

As a consequence, $t \mapsto x(t)$ is weakly (Lebesgue) measurable, that is, $t \mapsto < x(t), \lambda >$ is Lebesgue measurable for every $\lambda \in \mathcal{E}'$. Thus, by Pettis' theorem [1], x is strongly Lebesgue measurable.

We will show now that x is continuous. We begin by proving the continuity on $\mathbf{R}_+^*$. Let $t_o > 0$, $\epsilon \in (0, t_o)$, and let a, $b \in \mathbf{R}$ be such that $0 \le a < b < t_o - \epsilon$. Since for every $t \in (0, t_o)$,

$$x(t_o) = T(t_o)x = T(t)T(t_o - t)x_o = T(t)x(t_o - t),$$

then

$$(b - a)x(t_o) = \int_a^b x(t_o)dt = \int_a^b T(t)x(t_o - t)dt,$$

$$(b - a)(x(t_o \pm \epsilon) - x(t_o)) = \int_a^b T(t)(x(t_o \pm \epsilon - t) - x(t_o - t))dt,$$

whence

$$(b - a)\|x(t_o \pm \epsilon) - x(t_o)\|$$

$$\le \sup\{\|T(t)\| : a \le t \le b\} \int_a^b \|x(t_o \pm \epsilon - t) - x(t_o - t)\|dt$$

$$= \sup\{\|T(t)\| : a \le t \le b\} \int_{t_o - b}^{t_o - a} \|x(t \pm \epsilon) - x(t)\|dt.$$

---

[1] *A function x from an interval E in $\mathbf{R}$ to $\mathcal{E}$ is strongly measurable (i.e., there exists a sequence of finitely-valued functions strongly convergent almost everywhere to x) if, and only if, the function is weakly measurable and its range is almost everywhere separable (i.e., there exists a measurable set K in E, of measure zero, such that $\{x(t) : t \in E \backslash K\}$ is separable). Cf. e.g. [81], pp. 130-132.*

If f is an elementary function with a finite number of values, then

$$\lim_{\epsilon \downarrow 0} \int_{t_o-b}^{t_o-a} \|f(t \pm \epsilon) - f(t)\|dt = 0.$$

The dominated convergence theorem implies then that

$$\lim_{\epsilon \downarrow 0} \int_{t_o-b}^{t_o-a} \|x(t \pm \epsilon) - x(t)\|dt = 0,$$

whence

$$\lim_{\epsilon \downarrow 0} x(t_o \pm \epsilon) = x(t_o)$$

whenever $t_o > 0$.

4. To complete the proof of the theorem we are left to show that x is continuous at $0 \in \mathbf{R}_+$.

For every $t \geq 0$ and for any rational number $t_n > 0$,

$$T(t)x(t_n) = T(t)T(t_n)x_o = T(t+t_n)x_o = x(t+t_n).$$

In view of what has been seen in 3.,

$$\lim_{\delta \downarrow 0} T(t)x(t_n) = \lim_{t \downarrow 0} x(t+t_n) = x(t_n).$$

With the same notations as in 2., if x_ν is a finite linear combination of the $x(t_n)$,

$$\lim_{t \downarrow 0} T(t)x_\nu = x_\nu. \tag{2.9}$$

Since for any $t \in [0,1]$,

$$\|x(t) - x_o\| \leq \|T(t)x_o - T(t)x_\nu\| + \|T(t)x_\nu - x_\nu\| + \|x_\nu - x_o\|$$
$$\leq (1 + \sup\{\|T(t)\| : 0 \leq t \leq 1\})\|x_o - x_\nu\| + \|T(t)x_\nu - x_\nu\|.$$

For any $\epsilon > 0$, let x_ν be such that $\|x_o - x_\nu\| < \epsilon$. By (2.9) there exists $\delta > 0$ such that, if $0 < t < \delta$, then $\|T(t)x_\nu - x_\nu\| < \epsilon$, and therefore

$$\|x(t) - x_o\| < (2 + \sup\{\|T(t)\| : 0 \leq t \leq 1\})\epsilon.$$

That establishes the continuity of x at 0, and completes the proof of the theorem. ∎

124

Note The proof follows the argument esposed in [81], pp.233-234.

Let $T : \mathbf{R}_+ \to \mathcal{E}$ be a strongly continuous semigroup. The function $\phi : \mathbf{R}_+ \to [-\infty, +\infty)$ defined by

$$\phi(t) = \log \|T(t)\|$$

satisfies all the hypotheses of Theorem 1.5.2 of Chapter I. Hence

$$\lim_{t \to +\infty} \frac{1}{t} \log \|T(t)\|$$

exists, and

$$\omega := \lim_{t \to +\infty} \frac{1}{t} \log \|T(t)\| = \inf\{\frac{\log \|T(t)\|}{t} : t > 0\} \in [-\infty, +\infty).$$

$$(2.10)$$

Since, moreover, for $t \geq 0$ and $n \in \mathbf{N}^*$,

$$\|T(t)^n\|^{\frac{1}{n}} = \|T(nt)\|^{\frac{1}{n}} = e^{\frac{\phi(nt)}{n}} = e^{t\frac{\phi(nt)}{nt}},$$

the following theorem holds

Theorem 2.1.4 *For any $t \in \mathbf{R}_+$ the spectral radius $\rho(T(t))$ of $T(t)$ is*

$$\rho(T(t)) = e^{\omega t},$$

where ω is given by (2.10).

Let $a \in \mathbf{R}$ and $M \geq 1$ be chosen in such a way that (2.6) holds, and therefore

$$\frac{\log \|T(t)\|}{t} \leq \frac{M}{t} + a$$

for all $t > 0$. Then (2.10) implies that

$$\omega = \inf\{a \in \mathbf{R} : \text{ there exists } M \in \mathbf{R}_+ \text{ such that (2.6) holds}\}. \quad (2.11)$$

The real number ω, which will be also denoted by $\omega(T)$, is called the *growth bound* or *type* of the strongly continuous semigroup T.

Example Let Υ be a locally compact, Hausdorff space, and let $C_o(\Upsilon)$ be the vector space of all continuous functions $\Upsilon \to \mathbf{C}$ which vanish at infinity, endowed with the uniform norm $\|x\| = \sup\{|x(v)| : v \in \Upsilon\}$. Let $\varphi : \Upsilon \to \mathbf{C}$ be a continuous function, and let M_φ be the *multiplication operator* acting on $C_o(\Upsilon)$, defined by

$$(M_\varphi x)(v) = \varphi(v)\, x(v). \qquad (2.12)$$

The maximal domain of M_φ is the linear subspace of $C_o(\Upsilon)$

$$\mathcal{D}(M_\varphi) = \{x \in C_o(\Upsilon) : \varphi x \in C_o(\Upsilon)\}.$$

As an exercise, show that $\mathcal{D}(\Upsilon)$ is dense in $C_o(\Upsilon)$.

Let now $\{x_\nu\}$ be a sequence in $\mathcal{D}(M_\varphi)$ converging to some $x \in C_o(\Upsilon)$ and such that $\{\varphi x_\nu\}$ converges to some $y \in C_o(\Upsilon)$. The sequence $\{x_\nu\}$ converges to x uniformly on all compact subsets of Υ, and therefore also the sequence $\{\varphi x_\nu\}$ converges to φx on all compact subsets of Υ. Thus, $\varphi x \in C_o(\Upsilon)$, and therefore $x \in \mathcal{D}(M_\varphi)$ and $y = \varphi x$, showing thereby that the operator M_φ is closed.

For $x \in C_o(\Upsilon)$ and $t \geq 0$, let $T(t)x$ be the continuous function $\Upsilon \to \mathbf{C}$ defined by

$$(T(t)x)(v) = e^{t\,\varphi(v)}x(v). \qquad (2.13)$$

As an exercise, show that, if $t > 0$, then $T(t)x \in C_o(\Upsilon)$ for all $x \in C_o(\Upsilon)$ if, and only if,

$$\sup\{\Re\varphi(v) : v \in \Upsilon\} < \infty. \qquad (2.14)$$

That proves

Proposition 2.1.5 *The semigroup $T : \mathbf{R}_+ \to \mathcal{L}(C_o(\Upsilon))$ defined by (2.13) is strongly continuous if, and only if, (2.14) holds. In which case, the left-hand side of (2.14) is the growth bound of T.*

Exercise Replace the topological space Υ by a σ − finite measure space $\{\Upsilon, \mu\}$ and $C_o(\Upsilon)$ by the Banach space $\mathcal{E} := L^p(\Upsilon, \mu)$ of all (complex valued) L^p − functions on Υ, $1 \leq p < \infty$. Let $\varphi : \Upsilon \to \mathbf{C}$ be a measurable function, and, for $x \in \mathcal{E}$, let $M_\varphi x$ be the measurable function defined by (2.12), the (maximal) domain of the linear operator M_φ being $\mathcal{D}(M_\varphi) = \{x \in \mathcal{E} : \varphi x \in \mathcal{E}\}$.

Show that $\mathcal{D}(M_\varphi)$ is dense in $\mathcal{E}$, that M_φ is closed, and prove

Proposition 2.1.6 *The semigroup $T : \mathbf{R}_+ \to \mathcal{E}$ defined by (2.13) is strongly continuous if, and only if,*

$$\operatorname{ess\,sup}\{\Re\varphi(v) : v \in \Upsilon\} < \infty.$$

If this is the case, the left-hand side is the growth bound of T.

2.2 The infinitesimal generator

Let $T : \mathbf{R}_+ \to \mathcal{L}(\mathcal{E})$ be a strongly continuous semigroup, and let X be the linear operator defined on the linear submanifold of $\mathcal{E}$

$$\mathcal{D}(X) = \{x \in \mathcal{E} : \lim_{t \downarrow 0} \frac{1}{t}(T(t) - I)x \text{ exists in } \mathcal{E}\}$$

by

$$Xx = \lim_{t \downarrow 0} \frac{1}{t}(T(t) - I)x.$$

The linear operator X is called the *infinitesimal generator* of the semigroup T.

Lemma 2.2.1 *The linear manifold $\mathcal{D}(X)$ is dense in $\mathcal{E}$, invariant under the action of $T(t)$ and such that*

$$T(t)Xx = XT(t)x \tag{2.15}$$

for all $t \geq 0$ and all $x \in \mathcal{D}(X)$.

Proof For $t > 0$ and $x \in \mathcal{E}$, let

$$x(t) = \int_0^t T(s)x\,ds.$$

If $h > 0$,

$$\frac{1}{h}(T(h)x(t) - x(t)) = \frac{1}{h}[T(h)\int_0^t T(s)x\,ds - \int_0^t T(s)x\,ds]$$

$$= \frac{1}{h}[\int_0^t T(s+h)x\,ds - \int_0^t T(s)x\,ds]$$

$$= \frac{1}{h}[\int_h^{t+h} T(s)x\,ds - \int_0^t T(s)x\,ds]$$

$$= \frac{1}{h}[(\int_h^0 + \int_0^t + \int_t^{t+h} - \int_0^t)T(s)x\,ds]$$

$$= \frac{1}{h}[\int_t^{t+h} T(s)x\,ds - \int_0^h T(s)x\,ds],$$

and therefore $x(t) \in \mathcal{D}(X)$ and $Xx(t) = T(t)x - x$, i.e., $\int_0^t T(s)x\,ds \in \mathcal{D}(X)$ and

$$X\int_0^t T(s)x\,ds = T(t)x - x \qquad (2.16)$$

for all $t \geq 0$ and all $x \in \mathcal{E}$.

Since

$$\lim_{t \downarrow 0} \frac{1}{t}x(t) = \lim_{t \downarrow 0} \frac{1}{t}\int_0^t T(s)x\,ds = x,$$

then $\overline{\mathcal{D}(X)} = \mathcal{E}$.

If $t \geq 0$, $h > 0$ and $x \in \mathcal{D}(X)$, then

$$\frac{1}{h}(T(h) - I)T(t)x = T(t)\frac{1}{h}(T(h) - I)x. \qquad (2.17)$$

Thus $T(t)x \in \mathcal{D}(X)$, and (2.15) holds. ■

It follows from (2.17) that the function $t \mapsto T(t)x$ is differentiable from right for all $x \in \mathcal{D}(X)$, and the right derivative, $D^+T(t)x$ is given by

$$D^+T(t)x = XT(t)x = T(t)Xx. \qquad (2.18)$$

Before discussing the differentiability *tout court* of the function $T(\bullet)x$ we review a few facts of elementary calculus. Let $\varphi : [a,b] \to \mathbf{R}$ be a continuous function with a right derivative at each point of the open interval (a,b).

I. *If $D^+\varphi(t) > 0$ for all $t \in (a,b)$, then φ is increasing on $[a,b]$.*

Proof. Let $a < t_1 < t_2 < b$ and let $t_o \in [t_1, t_2]$ be such that $\varphi(t_o) \geq \varphi(t)$ for all $t \in [t_1, t_2]$. If $t_o < t_2$, then $D^+\varphi(t_o) \leq 0$. Thus $t_o = t_2$, and therefore $\varphi(t_1) \leq \varphi(t_2)$. Since $D^+\varphi(t_1) > 0$, there is some $t \in (t_1, t_2)$ such that $\varphi(t) > \varphi(t_1)$. Hence, if $\varphi(t_1) = \varphi(t_2)$, $\varphi(t_o) > \varphi(t_1) = \varphi(t_2)$, contradicting the fact that $t_o = t_2$. Hence $\varphi(t_1) < \varphi(t_2)$. ∎

II. *If $D^+\varphi(t) = 0$ for all $t \in (a,b)$, then φ is constant on $[a,b]$.*

Proof Let $\epsilon > 0$, and let $\varphi_\epsilon(t) = \varphi(t) + \epsilon(t-a)$. Since $D^+\varphi_\epsilon(t) = \epsilon$ at all $t \in (a,b)$, in view of I. φ_ϵ is increasing on $[a,b]$ for all $\epsilon > 0$. As a consequence, φ, which is the lower envelope of the functions φ_ϵ, is non-decreasing on $[a,b]$. Similarly, since the function ψ defined by $\psi(t) = -\varphi(t) + \epsilon(t-a)$ is increasing on $[a,b]$, the function $-\varphi$ is non-decreasing on $[a,b]$. In conclusion, φ is constant. ∎

Applying II. to the real part and to the imaginary coefficient, one proves that

III. *If $\varphi : [a,b] \to \mathbf{C}$ is continuous, with right-derivative at each point of (a,b), and if $D^+\varphi(t) = 0$ for all $t \in (a,b)$, then φ is constant on $[a,b]$.*

Lemma 2.2.2 *If $x \in \mathcal{D}(X)$, then*

$$T(t)x - x = \int_0^t T(x)Xx\,ds$$

for all $t \geq 0$.

Proof Let λ be a continuous linear form on $\mathcal{E}$, and let $\varphi : \mathbf{R}_+ \to \mathbf{C}$ be the continuous function defined by

$$\varphi(t) = <T(t)x - x - \int_0^t T(s)Xx\,ds, \lambda>.$$

By (2.18),

$$D^+\varphi(t) = \lim_{h\downarrow 0} < \frac{1}{h}(T(t+h) - T(t))x - \frac{1}{h}\int_t^{t+h} T(s)Xx\,ds, \lambda >$$
$$= < D^+T(t)x - T(t)Xx, \lambda > = < T(t)Xx - T(t)Xx, \lambda > = 0,$$

and, by III., $\varphi(t) = 0$ at every $t \in \mathbf{R}_+$. The Hahn-Banach theorem yields the conclusion. ∎

If $t > 0$, the left derivative $D^-T(t)x$, for $x \in \mathcal{D}(X)$, is given by

$$D^-T(t)x = \lim_{h\downarrow 0} \frac{1}{t}(T(t)x - T(t-h)x),$$

provided that this limit exists. If $0 < h < t$, Lemma 2.2.2 implies:

$$\frac{1}{h}(T(t) - T(t-h))x = \frac{1}{h}(T(h)T(t-h) - T(t-h))x$$
$$= \frac{1}{h}\int_0^h T(s)XT(t-h)x\,ds$$
$$= \frac{1}{h}\int_0^h T(s+t\quad h)Xx\,ds$$
$$= \frac{1}{h}\int_{t-h}^t T(s)Xx\,ds,$$

showing that the limit exists, and that

$$D^-T(t)x = T(t)Xx \qquad (2.19)$$

for all $x \in \mathcal{D}(X)$ and all $t > 0$.

Comparison of (2.18) and (2.19) yields

Theorem 2.2.3 *For all $x \in \mathcal{D}(X)$, the function $t \mapsto T(t)x$ from $\mathbf{R}_+$ to $\mathcal{E}$ is differentiable of class C^1, and*

$$\frac{d}{dt}T(t)x = T(t)Xx = XT(t)x.$$

Theorem 2.2.4 *Let X be the infinitesimal generator of the strongly continuous semigroup T, and, for some $a > 0$, let $x : [0, a] \to \mathcal{D}(X)$ be a function of class C^1 such that*

$$\dot{x}(t) = Xx(t)$$

for all $t \in [0, a]$. Then $x(a) = T(a)x(0)$.

Proof Let $\lambda \in \mathcal{E}'$ and let $\psi : [0, a] \to \mathbf{C}$ be the continuous function defined by

$$\psi(t) = < T(t)x(a - t), \lambda > .$$

If $0 \le t < a$, $D^+\psi(t)$ is given by

$$D^+\psi(t) = \lim_{h \downarrow 0} \frac{1}{h} < T(t + h)x(a - t - h) - T(t)x(a - t), \lambda >$$

$$= \lim_{h \downarrow 0} [< T(t + h)\frac{1}{h}(x(a - t - h) - x(a - t)), \lambda >$$

$$+ < \frac{1}{h}(T(t + h) - T(t))x(a - t), \lambda >]$$

$$= < T(t)\frac{dx}{dt}(a - t), \lambda > + < XT(t)x(a - t), \lambda >$$

$$= < -T(t)Xx(a - t), \lambda > + < XT(t)x(a - t), \lambda > = 0.$$

If $0 < t \le a$, $D^-\psi(t)$ is given by

$$D^-\psi(t) = \lim_{h \downarrow 0} \frac{1}{h} < T(t)x(a - t) - T(t - h)x(a - t + h), \lambda >$$

$$= \lim_{h \downarrow 0} [< \frac{1}{h}(T(t) - T(t - h))x(a - t), \lambda >$$

$$+ < T(t - h)\frac{1}{h}(x(a - t) - x(a - t + h)), \lambda >]$$

$$= < T(t)Xx(a - t), \lambda > + < T(t)\frac{d}{dt}x(a - t), \lambda >$$

$$= < XT(t)x(a - t), \lambda > + < -T(t)Xx(a - t), \lambda > = 0.$$

Hence, $\psi(t) = \psi(0)$, i.e. $< T(t)x(a - t), \lambda > = < x(a), \lambda >$, for all $t \in [0, a]$. In particular, $< T(a)x(0), \lambda > = < x(a), \lambda >$ for all $\lambda \in \mathcal{E}'$. The Hahn-Banach theorem yields the conclusion. ∎

Corollary 2.2.5 *Strongly continuous semigroups having the same infinitesimal generator coincide.*

If $X \in \mathcal{L}(\mathcal{E})$, setting, for $\zeta \in \mathbf{C}$,

$$\exp\zeta X = I + \zeta X + \frac{\zeta^2}{2}X^2 + \frac{\zeta^3}{3!}X^3 + \cdots,$$

defines a function $\exp \bullet X \in \mathrm{Hol}(\mathbf{C}, \mathcal{L}(\mathcal{E}))$ such that

$$\frac{d}{d\zeta}\exp\zeta X = X \exp\zeta X. \tag{2.20}$$

Since

$$\exp(\zeta_1 + \zeta_2)X = \exp\zeta_1 X \cdot \exp\zeta_2 X$$

for all $\zeta_1, \zeta_2 \in \mathbf{C}$, the map $\zeta \mapsto \exp\zeta X$ is an example of a *holomorphic semigroup* (on which we shall come back in Chapter 4). In particular, setting, for $t \in \mathbf{R}$,

$$G(t) = \exp tX, \tag{2.21}$$

defines a norm-continuous group $G : \mathbf{R} \to \mathcal{L}(\mathcal{E})$. In view of (2.20), the infinitesimal generator of the semigroup $T = G_{|\mathbf{R}_+}$ is X. The fact that the linear operator X is bounded characterizes the uniformly continuous semigroups, as the following theorem shows.

Theorem 2.2.6 *The strongly continuous semigroup $T : \mathbf{R}_+ \to \mathcal{L}(\mathcal{E})$ is uniformly continuous if, and only if, its infinitesimal generator X is a continuous linear operator on $\mathcal{E}$. If that is the case, T is the restriction to $\mathcal{R}_+$ of the group $G : \mathbf{R} \to \mathcal{L}(\mathcal{E})$ defined by (2.21).*

Proof. If T is uniformly continuous, then

$$\lim_{t\downarrow 0}\left\|\frac{1}{t}\int_0^t T(s)ds - I\right\| = \lim_{t\downarrow 0}\left\|\frac{1}{t}\int_0^t (T(s) - I)ds\right\|$$

$$\leq \lim_{t\downarrow 0}\frac{1}{t}\int_0^t \|T(s) - I\|ds = 0$$

because $\lim_{s\downarrow 0}\|T(s) - I\| = 0$. Hence there is some $k > 0$ such that the operator $\int_0^t T(s)ds \in \mathcal{L}(\mathcal{E})$ is invertible in $\mathcal{L}(\mathcal{E})$ for every $t \in (0, k]$.

For $0 < t \leq k$ and $0 < h \leq k$,

$$\frac{1}{h}(T(h) - I) \int_0^t T(s)ds = \frac{1}{h}\left(\int_h^{t+h} T(s)ds - \int_0^t T(s)ds\right)$$

$$= \frac{1}{h}\left(\int_h^0 + \int_0^t + \int_t^{t+h} - \int_0^t\right)T(s)ds$$

$$= \frac{1}{h}\left(\int_t^{t+h} T(s)ds - \int_0^h T(s)ds\right),$$

and therefore

$$\lim_{h \to 0} \left\|\frac{1}{h}(T(h) - I) \int_0^t T(s)ds - (T(t) - I)\right\| = \qquad (2.22)$$

$$= \lim_{h \downarrow 0} \left\|\frac{1}{h}\left(\int_t^{t+h} - \int_0^h\right)T(s)ds - (T(t) - I)\right\|$$

$$\leq \lim_{h \downarrow 0} \frac{1}{h}\left(\left\|\int_t^{t+h} (T(s) - T(t))ds\right\| + \left\|\int_0^h (T(s) - I)ds\right\|\right)$$

$$\leq \lim_{h \downarrow 0} \frac{1}{h}\left(\int_t^{t+h} \|T(s) - T(t)\|ds + \int_0^h \|T(s) - I\|ds\right)$$

$$= 0$$

because

$$\lim_{s \downarrow t} \|T(s) - T(t)\| = \lim_{s \downarrow t} \|T(t + s - t) - T(t)\|$$

$$= \lim_{h \downarrow 0} \|T(t)(T(h) - I)\|$$

$$\leq \|T(t)\| \lim_{h \downarrow 0} \|T(h) - I\| = 0,$$

where $h = s - t$. Let $Q = \int_0^t T(s)ds$ and $h > 0$. Since Q is invertible, with $Q^{-1} \in \mathcal{L}(\mathcal{E})$, then

$$\left\|\frac{1}{h}(T(h) - I) - (T(t) - I)Q^{-1}\right\| =$$

$$\left\|\frac{1}{h}(T(h) - I)Q\,Q^{-1} - (T(t) - I)Q^{-1}\right\|$$

$$\leq \left\|\frac{1}{h}(T(h) - I)Q - (T(t) - I)\right\| \|Q^{-1}\|.$$

Thus, by (2.22), the infinitesimal generator of T is

$$X := \lim_{h \downarrow 0} \frac{1}{h}(T(h) - I) = (T(t) - I)Q^{-1} \in \mathcal{L}(\mathcal{E}).$$

Corollary 2.2.5 implies then that T is the restriction to $\mathbf{R}_+$ of the group G defined by (2.21). ∎

The differentiablity properties established so far will now be used to characterize strongly continuous groups.

Let $G : \mathbf{R} \to \mathcal{L}(\mathcal{E})$ be a strongly continuous group. The restriction $G_{|\mathbf{R}_+}$ is a strongly continuous semigroup. Similarly, the semigroup $\mathbf{R}_+ \ni t \mapsto G(-t)$ is strongly continuous. If X is the infinitesimal generator of the first, $-X$ generates the second. This fact characterizes strongly continuous groups, as the following theorem shows.

Theorem 2.2.7 *If X and $-X$ are the infinitesimal generators of two strongly continuous semigroups T and S, the map $G : \mathbf{R} \to \mathcal{L}(\mathcal{E})$ defined by $G(t) = T(t)$ if $t \geq 0$, $G(t) = S(-t)$ if $t \leq 0$ is a strongly continuous group.*

Proof The function G is strongly continuous, and moreover

$$\lim_{t \to 0} \frac{1}{t}(G(t) - I)x = Xx$$

for all $x \in \mathcal{D}(X)$. Furthermore, $G(0) = I$ and

$$G(t_1 + t_2) = G(t_1)\,G(t_2) \tag{2.23}$$

whenever t_1 and t_2 are both non-negative or both non-positive. What is left to prove is that (2.23) holds also when t_1 and t_2 have opposite signs. One checks easily that it suffices to prove that

$$S(t)\,T(t) = T(t)\,S(t) = I \tag{2.24}$$

for all $t \geq 0$.

For $x \in \mathcal{D}(X)$ and $t \geq 0$, let $x(t) = S(t)\,T(t)x$. Then $x(t) \in \mathcal{D}(X)$ for all $t \in \mathbf{R}_+$, and, if $t > 0$,

$$
\begin{aligned}
\frac{d}{dt}x(t) &= \lim_{h \to 0} \frac{1}{h}(S(t+h)\,T(t+h) - S(t)\,T(t))x \\
&= \lim_{h \to 0} S(t+h)\frac{1}{h}(T(t+h) - T(t))x \\
&\quad + \lim_{h \to 0}\frac{1}{h}(S(t+h) - S(t))T(t)x \\
&= S(t)\,T(t)\,Xx - S(t)\,XT(t)x = 0.
\end{aligned}
$$

Thus,

$$
S(t)\,T(t)x = x \tag{2.25}
$$

for all $t \geq 0$ and all $x \in \mathcal{D}(X)$. Since $\mathcal{D}(X)$ is dense in $\mathcal{E}$, and the linear operator $S(t)\,T(t)$ is continuous on $\mathcal{E}$, (2.25) holds for all $x \in \mathcal{E}$. In a similar way, one stablishes the second part of (2.24). ∎

The operator X will be called the *infinitesimal generator* of the group G.

Let X be the infinitesimal generator of the strongly continuous semi-group T.

Theorem 2.2.8 *The operator X is closed.*

Proof Let $\{x_\nu\}$ be a sequence of points $x_\nu \in \mathcal{D}(X)$ converging to $x \in \mathcal{E}$ and such that $\{Xx_\nu\}$ converges to $y \in \mathcal{E}$. By Lemma 2.2.2, if $t > 0$,

$$
\begin{aligned}
T(t)x - x &= \lim_{\nu \to +\infty}(T(t)x_\nu - x_\nu) \\
&= \lim_{\nu \to +\infty}\int_0^t T(s)Xx_\nu ds = \int_0^t T(s)y ds,
\end{aligned}
$$

whence $x \in \mathcal{D}(X)$ and $Xx = y$. ∎

We introduce now on $\mathcal{D}(X)$ the graph norm $\|\|\ \|\|$:

$$
\|\|x\|\| = \|x\| + \|Xx\|,
$$

and we denote this normed space by $\mathcal{E}_1$. Since X is closed, $\mathcal{E}_1$ is complete, and the map $x \mapsto j(x) := (x, Xx)$ defines a surjective isometry of $\mathcal{E}_1$ onto the graph $\Gamma(X)$ of X. The restriction $T_1(t) := T(t)_{|\mathcal{E}_1}$ is bounded for all $t \geq 0$. Furthermore, by Lemma 2.2.1, $T_1 : t \mapsto T_1(t)$ is a semigroup in $\mathcal{E}_1$. If $\hat{T} : \mathbf{R}_+ \to \mathcal{L}(\mathcal{E} \oplus \mathcal{E})$ is defined by

$$\hat{T}(t)(x, y) = (T(t)x, T(t)y),$$

$\hat{T}$ is a strongly continuous semigroup. Since

$$j \circ T_1(t) = \hat{T}(t) \circ j$$

on $\mathcal{D}(X)$, the following lemma holds.

Lemma 2.2.9 *The semigroup* $T_1 : \mathbf{R}_+ \to \mathcal{E}_1$ *is strongly continuous for the norm* $\| \ \|$.

Here is a direct proof of this lemma.

For $x \in \mathcal{D}(X)$ and $t \geq 0$ we have, by Lemma 2.2.1,

$$\|T(t)x - x\| = \|T(t)x - x\| + \|X(T(t)x - x)\|$$
$$= \|T(t)x - x\| + \|T(t)Xx - Xx\|.$$

As a consequence

$$\lim_{t \downarrow 0} \|T(t)x - x\| = 0$$

if, and only if,

$$\lim_{t \downarrow 0} \|T(t)x - x\| = \lim_{t \downarrow 0} \|T(t)Xx - Xx\| = 0.$$

That proves Lemma 2.2.9.

We will find now the domain $\mathcal{D}(X_1)$ of the infinitesimal generator X_1 of the semigroup T_1. For $t > 0$ and $x \in \mathcal{D}(X)$,

$$\frac{1}{t}\|T(t)x - x\| = \frac{1}{t}(\|T(t)x - x\| + \|X(T(t)x - x)\|)$$
$$= \frac{1}{t}(\|T(t)x - x\| + \|T(t)Xx - Xx\|).$$

Hence $x \in \mathcal{D}(X_1)$ if, and only if, $x \in \mathcal{D}(X)$ and $Xx \in \mathcal{D}(X)$. Thus,

$$\mathcal{D}(X_1) = \{x \in \mathcal{D}(X) : Xx \in \mathcal{D}(X)\} = \mathcal{D}(X^2),$$

and, if $x \in \mathcal{D}(X_1)$, then $X_1 x = X x$.

Remark Since $\mathcal{D}(X_1)$ is dense in $\mathcal{D}(X)$ for the norm $\| \ \|$, and $\mathcal{D}(X)$ is dense in $\mathcal{E}$ for the norm $\| \ \|$, then $\mathcal{D}(X_1)$ is dense in $\mathcal{E}$ (for the norm $\| \ \|$).

Theorem 2.2.10 *If the linear space $\mathcal{D}$ is dense in $\mathcal{D}(X)$ and such that $T(t)\mathcal{D} \subset \mathcal{D}$ for all $t \geq 0$, then $\mathcal{D}$ is a core of X.*

Proof Denoting by $\overline{\mathcal{D}}^{\| \ \|}$ the closure of $\mathcal{D}$ for the norm $\| \ \|$, we have to prove that $\overline{\mathcal{D}}^{\| \ \|} = \mathcal{D}(X)$.

For any $x \in \mathcal{D}(X)$ there is a sequence $\{x_\nu\}$ in $\mathcal{D}$ for which

$$\lim_{\nu \to +\infty} \|x - x_\nu\| = 0$$

Since the function $t \mapsto T_1(t)x_\nu$ is continuous for the norm $\| \ \|$, then

$$\int_0^t T(s)x_\nu ds \in \overline{\mathcal{D}}^{\| \ \|}$$

for all $t \geq 0$. Furthermore, by (2.16),

$$\left\| \int_0^t T(s)x_\nu ds - \int_0^t T(s)x ds \right\| =$$

$$= \left\| \int_0^t T(s)(x_\nu - x)ds \right\| + \left\| X \int_0^t T(s)(x_\nu - x)ds \right\|$$

$$= \left\| \int_0^t T(s)(x_\nu - x)ds \right\| + \|T(t)(x_\nu - x) - (x_\nu - x)\|$$

$$\leq \left\| \int_0^t T(s)(x_\nu - x)ds \right\| + \|T(t)(x_\nu - x)\| + \|x_\nu - x\|.$$

Hence

$$\lim_{\nu \to +\infty} \left\| \int_0^t T(s)x_\nu ds - \int_0^t T(s)x ds \right\| \leq$$

$$\leq \lim_{\nu \to +\infty} \left\{ \left\| \int_0^t T(s)(x_\nu - x)ds \right\| + \|T(t)(x_\nu - x)\| + \|x_\nu - x\| \right\},$$

and therefore

$$\int_0^t T(s)x\,ds \in \overline{\mathcal{D}}^{\|\cdot\|}. \tag{2.26}$$

Again by (2.16), for $t > 0$,

$$\left\| \frac{1}{t} \int_0^t T(s)x\,ds - x \right\| = \left\| \frac{1}{t} \int_0^t T(s)x\,ds - x \right\|$$

$$+ \left\| \frac{1}{t} X \int_0^t T(s)x\,ds - Xx \right\|$$

$$= \left\| \frac{1}{t} \int_0^t T(s)x\,ds - x \right\|$$

$$+ \left\| \frac{1}{t}(T(t)x - x) - Xx \right\|.$$

Since $x \in \mathcal{D}(X)$,

$$\lim_{t \downarrow 0} \left\| \frac{1}{t} \int_0^t T(s)x\,ds - x \right\| = 0,$$

and therefore $x \in \overline{\mathcal{D}}^{\|\cdot\|}$, by (2.26). ∎

In line with Theorem 2.1.3, the following theorem clarifies the re-lationship between the infinitesimal generator and the weak topology. As before, the complex Banach space $\mathcal{E}'$ is the topological dual of $\mathcal{E}$.

Theorem 2.2.11 *Let $\mathcal{F}'$ be a dense linear subspace of $\mathcal{E}'$. If, for $x, y \in \mathcal{E}$ there exists in $\mathbf{R}_+^*$ a sequence $\{t_\nu\}$ converging to zero, such that*

$$\lim_{\nu \to +\infty} \frac{< T(t_\nu)x - x, \lambda >}{t_\nu} = < y, \lambda >$$

for every $\lambda \in \mathcal{F}'$, then $x \in \mathcal{D}(X)$ and $y = Xx$.

Proof For $\lambda \in \mathcal{F}'$, the function $f : \mathbf{R}_+ \to \mathbf{C}$ defined by

$$f(t) = < T(t)x - x - \int_0^t T(s)y\,ds, \lambda >$$

is continuous, and $f(0) = 0$. Furthermore,

$$\frac{f(t+t_\nu) - f(t)}{t_\nu} =$$

$$\frac{1}{t_\nu} < (T(t+t_\nu) - T(t))x - \int_t^{t+t_\nu} T(s)y\,ds, \lambda >$$

$$= <T(t)\left(\frac{1}{t_\nu}(T(t_\nu) - I)x\right), \lambda > -\frac{1}{t_\nu} < \int_t^{t+t_\nu} T(s)y\,ds, \lambda >$$

$$= <\left(\frac{1}{t_\nu}(T(t_\nu) - I)x\right), T(t)'\lambda > -\frac{1}{t_\nu} < \int_t^{t+t_\nu} T(s)y\,ds, \lambda >,$$

where $T(t)'\lambda$ is the image of λ by the dual $T(t)'$ of $T(t)$. Hence

$$\lim_{\nu \to +\infty} \frac{f(t+t_\nu) - f(t)}{t_\nu} = <y, T(t)'\lambda > - <T(t)y, \lambda >$$

$$= <T(t)y, \lambda > - <T(t)y, \lambda >$$

$$= 0. \tag{2.27}$$

For $k > 0$, the set

$$L_k := \{t \in \mathbf{R}_+ : |f(t)| \le kt\}$$

is closed and contains 0. If $s \in L_k$, for all $\epsilon > 0$ there exist, by (2.27), some $t \in L_k$ such that $s < t < s + \epsilon$. Thus, L_k is open, and therefore $L_k = \mathbf{R}_+$. Hence $f(t) = 0$ for all $t \ge 0$, i.e.,

$$<T(t)x - x, \lambda > = <\int_0^t T(s)y\,ds, \lambda >$$

for all $t \in \mathbf{R}_+$ and all $\lambda \in \mathcal{F}'$, and therefore, being $\overline{\mathcal{F}'} = \mathcal{E}'$, for all $\lambda \in \mathcal{E}'$. By the Hahn-Banach theorem,

$$T(t)x - x = \int_0^t T(s)y\,ds \quad \forall t \in \mathbf{R}_+.$$

Consequently,

$$\lim_{t \downarrow 0} \frac{1}{t}(T(t)x - x) = \lim_{t \downarrow 0} \frac{1}{t}\int_0^t T(s)y\,ds = y,$$

that is to say, $x \in \mathcal{D}(X)$ and $Xx = y$. ■

2.3 Examples

Let $C_o(\mathbf{R})$ be the Banach space of all complex-valued continuous functions on $\mathbf{R}$ which vanish at infinity, endowed with the sup-norm. Setting, for $t \in \mathbf{R}$ and $x \in C_o(\mathbf{R})$,

$$(G(t)x)(r) = x(r + t) \quad (r \in \mathbf{R}), \tag{2.28}$$

defines a strongly continuous group $G : \mathbf{R} \to \mathcal{L}(C_o(\mathbf{R}))$. Let X be its infinitesimal generator.

Lemma 2.3.1 *The domain $\mathcal{D}(X)$ of X is given by*

$$\mathcal{D}(X) = \{x \in C_o(\mathbf{R}) : \; x \in C^1(\mathbf{R}) \text{ and } \dot{x} \in C_o(\mathbf{R})\},$$

and, for all $x \in \mathcal{D}(X)$,

$$Xx(r) = \dot{x}(r) \quad (r \in \mathbf{R}). \tag{2.29}$$

Proof If $x \in C_o(\mathbf{R})$ is contained in the domain $\mathcal{D}(X)$ of the infinitesimal generator X of the group G, then the limit $\lim_{h \to 0} \frac{1}{h}(T(h) - I)x$ exists, and

$$\lim_{h \to 0} \frac{1}{h}(T(h)x - x)(r) = \lim_{h \to 0} \frac{x(r + h) - x(r)}{h} = \dot{x}(r).$$

Hence $x \in C^1(\mathbf{R})$, and $\dot{x} \in C_o(\mathbf{R})$.

Viceversa, let $x \in C_o(\mathbf{R}) \cap C^1(\mathbf{R})$, with $\dot{x} \in C_o(\mathbf{R})$. For all $r \in \mathbf{R}$ and $h \neq 0$,

$$\left| \frac{x(r + h) - x(r)}{h} - \dot{x}(r) \right| = \left| \frac{1}{h} \int_r^{r+h} (\dot{x}(s) - \dot{x}(r))ds \right|$$

$$\leq \frac{1}{|h|} \int_r^{r+h} |\dot{x}(s) - \dot{x}(r)|ds,$$

and

$$\lim_{h \to 0} \frac{1}{|h|} \int_r^{r+h} |\dot{x}(s) - \dot{x}(r)|ds = 0$$

uniformly with respect to r. Thus, $x \in \mathcal{D}(X)$ and $Xx = \dot{x}$. ∎

Let now $T : \mathbf{R} \to \mathcal{L}(L^p(\mathbf{R}))$ be the strongly continuous group defined by (2.28) in the Banach space $L^p(\mathbf{R})$, with $1 \leq p < \infty$, for the Lebesgue measure.

Lemma 2.3.2 *The infinitesimal generator X is defined by (2.29) on the domain*

$$\mathcal{D} = \{x \in L^p(\mathbf{R}) : x \text{ absolutely continuous, and } \dot{x} \in L^p(\mathbf{R})\}. \quad (2.30)$$

Proof Let x be an element of the domain of the operator X defined by (2.29), and let

$$y = \lim_{h \downarrow 0} \frac{1}{h}(T(h) - I)x \in L^p(\mathbf{R}).$$

For $a < b$ and $h \neq 0$, the interval $(a + h, b + h)$ is the image of (a, b) by the traslation defined by h. Hence,

$$\frac{1}{h}\left(\int_b^{b+h} x(r)dr - \int_a^{a+h} x(r)dr\right) = \int_a^b \frac{x(r + h) - x(r)}{h}dr.$$

By the dominated convergence theorem,

$$\lim_{h \to 0} \int_a^b \frac{x(r + h) - x(r)}{h}dr = \int_a^b y(r)dr;$$

On the other hand, for almost all a and b in $\mathbf{R}$,

$$\lim_{h \to 0} \frac{1}{h}\int_a^{a+h} x(r)dr = x(a), \quad \lim_{h \to 0} \frac{1}{h}\int_b^{b+h} x(r)dr = x(b),$$

whence, for almost all a and b in $\mathbf{R}$,

$$x(b) - x(a) = \int_a^b y(r)dr.$$

In other words, modifying x on a set of measure zero, we have

$$x(s) = \int_a^s y(r)dr + x(a)$$

for all $s \in \mathcal{R}$. That shows that x is absolutely continuous, and $\dot{x} = y \in L^p(\mathbf{R})$ almost everywhere.

Viceversa, let $x \in L^p(\mathbf{R})$ be absolutely continuous, with $\dot{x} \in L^p(\mathbf{R})$. For $r \in \mathbf{R}$,

$$\lim_{h \to 0} \int_{\mathbf{R}} \left| \frac{x(r+h) - x(r)}{h} - \dot{x}(r) \right|^p dr =$$

$$\lim_{h \to 0} \int_{\mathbf{R}} \left| \frac{1}{h} \int_0^h (\dot{x}(r+h) - \dot{x}(r)) ds \right|^p dr$$

$$= \lim_{h \to 0} \int_{\mathbf{R}} \left| \int_0^1 (\dot{x}(r+th) - \dot{x}(r)) dt \right|^p dr$$

$$\leq \lim_{h \to 0} \int_{\mathbf{R}} \int_0^1 |\dot{x}(r+th) - \dot{x}(r)|^p dt\, dr$$

$$= \lim_{h \to 0} \int_0^1 \int_{\mathbf{R}} |\dot{x}(r+th) - \dot{x}(r)|^p dr\, dt$$

$$= \int_0^1 (\lim_{h \to 0} \int_{\mathbf{R}} |\dot{x}(r+th) - \dot{x}(r)|^p dr) dt$$

$$= 0,$$

and therefore x is contained in the domain of X and $Xx = \dot{x}$. ∎

Exercise Going back to the Propositions 2.1.5 and 2.1.6, show that, if the semigroup defined by (2.6) is strongly continuous, then its infinitesimal generator is M_φ with its maximal domain.

Strongly continuous semigroups whose values are linear operators of particular type are necessarily uniformly continuous. Here is an example. Let $\mathcal{A}$ be a complex unital Banach algebra. Every $a \in \mathcal{A}$ defines a linear operator $L_a \in \mathcal{L}(\mathcal{A})$ acting on $\mathcal{A}$ by $L_a : x \mapsto ax$. As was noticed before, the map $a \mapsto L_a$ is a linear isometric homomorphism of $\mathcal{A}$ into $\mathcal{L}(\mathcal{A})$, or - in other words - a faithful linear representation of the Banach algebra $\mathcal{A}$ in the Banach space $\mathcal{A}$. Let $\mathcal{M}(\mathcal{A})$ be the image of $\mathcal{A}$ in $\mathcal{L}(\mathcal{A})$. The following theorem holds, ([75], [77])

Theorem 2.3.3 *Let* $T : \mathbf{R}_+ \to \mathcal{L}(\mathcal{A})$ *be a strongly continuous semigroup. If there is* $t_o > 0$ *such that,* $T(t) \in \mathcal{M}(\mathcal{A})$ *for any* $t \in (0, t_o)$, *then* T *is uniformly continuous, and there exists* $a \in \mathcal{A}$ *such that*

$$T(t)x = (\exp ta)x \tag{2.31}$$

for all $t \geq 0$ and all $x \in \mathcal{A}$.

Proof The set $\mathcal{A}^{-1}$ of all invertible elements of $\mathcal{A}$ is open and non-empty. Hence, if $\mathcal{D}(X)$ is the domain of the infinitesimal generator X of T, then $\mathcal{D}(X) \cap \mathcal{A}^{-1} \neq \emptyset$. For any $y \in \mathcal{D}(X) \cap \mathcal{A}^{-1}$ the limit $\lim_{t \downarrow 0} \frac{1}{t}(T(t) - I)y$ exists. Since, for $t \in (0, t_o)$,

$$\frac{1}{t}(T(t) - I)1 = \frac{1}{t}(T(t) - I)(y\,y^{-1})$$
$$= \left(\frac{1}{t}(T(t) - I)y \right) y^{-1},$$

and since the product in $\mathcal{A}$ is continuous, the limit

$$a := \lim_{t \downarrow 0} \frac{1}{t}(T(t) - I)1 \tag{2.32}$$

exists in $\mathcal{A}$. Hence, for all $x \in \mathcal{D}(X)$, the limit

$$\lim_{t \downarrow 0} \frac{1}{t}(T(t) - I)x = \lim_{t \downarrow 0} \frac{1}{t}(T(t) - I)(1\,x)$$
$$= \left(\lim_{t \downarrow 0} \frac{1}{t}(T(t) - I)1 \right) x$$

exists in $\mathcal{A}$. Thus, $x \in \mathcal{D}(X)$, and therefore $\mathcal{A} = \mathcal{D}(X)$. Since X is closed, the closed graph theorem implies that $X \in \mathcal{L}(\mathcal{A})$. Hence T is uniformly continuous, and (2.32) yields (2.31). ∎

Similar considerations, [75], lead to the following slightly more general result.

Lemma 2.3.4 *Assume that there are $t_0 < t_1$ in $\mathbf{R}_+$ such that*

$$T((t_0, t_1)) \subset \mathcal{M}(\mathcal{A}).$$

If $T(t_0)\mathcal{A}$ is closed in $\mathcal{A}$ ($T(t_0)$ is invariant with respect to $T(t)$ for all $t \geq 0$ and) the map $t \mapsto T(t)_{|T(t_0)\mathcal{A}}$ defines a uniformly continuous semigroup on $\mathbf{R}_+$.

The hypothesis whereby $T(t_0)\mathcal{A}$ is closed in $\mathcal{A}$ is fulfilled if $T(t_0)$ is invertible and $T(t_0)^{-1} \in \mathcal{L}(\mathcal{A})$, in which case $T(t_0)\mathcal{A} = \mathcal{A}$. The above lemma implies

Theorem 2.3.5 *If $T((t_0, t_1)) \subset \mathcal{M}(\mathcal{A})$ and if $T(t_0)$ is invertible in $\mathcal{L}(\mathcal{A})$, the conclusions of Theorem 2.3.3 hold.*

These results can be applied to the following situation.

Let $T : \mathbf{R}_+ \to \mathcal{L}(\mathcal{E})$ be a strongly continuous semigroup. If $\mathcal{F}$ is a complex Banach space, T defines a semigroup $\tilde{T} : \mathbf{R}_+ \to \mathcal{L}(\mathcal{L}(\mathcal{F}, \mathcal{E}))$ by

$$\tilde{T}(t)W = T(t) \circ W$$

for all $t \in \mathcal{R}_+$ and $W \in \mathcal{L}(\mathcal{F}, \mathcal{E})$.

The strong continuity of T implies that $\tilde{T}$ is continuous for the strong operator topology on $\mathcal{L}(\mathcal{F}, \mathcal{E})$, that is to say, for the locally convex topology defined on $\mathcal{L}(\mathcal{F}, \mathcal{E})$ by the family of seminorms

$$W \mapsto \|Wy\| \qquad (2.33)$$

indexed by $y \in \mathcal{F}$.

An infinitesimal generator $\tilde{X}$ of $\tilde{T}$ can be defined, [81], whose domain is

$$\mathcal{D}(\tilde{X}) = \{W \in \mathcal{L}(\mathcal{F}, \mathcal{E}) : Wy \in \mathcal{D}(X) \text{ for all } y \in \mathcal{F}\},$$

and on which $\tilde{X}$ is expressed by

$$\tilde{X}(W) = X \circ W.$$

If $\dim_{\mathbf{C}} \mathcal{F} < \infty$, the locally convex topology generated in $\mathcal{L}(\mathcal{F}, \mathcal{E})$ by the family of seminorms (2.33) is equivalent to the uniform topology, and thus the semigroup $\tilde{T}$ is strongly continuous on the Banach space $\mathcal{L}(\mathcal{F}, \mathcal{E})$. If $\mathcal{F} = \mathcal{E}$, Theorem 2.3.3 implies

Theorem 2.3.6 *Let $\dim_{\mathbf{C}} \mathcal{E} = \infty$. If the semigroup $\tilde{T} : \mathbf{R}_+ \to \mathcal{L}(\mathcal{L}(\mathcal{E}))$ is strongly continuous on the Banach space $\mathcal{L}(\mathcal{E})$, the semigroup $T : \mathcal{R}_+ \to \mathcal{L}(\mathcal{E})$ is uniformly continuous (and therefore also $\tilde{T}$ is uniformly continuous).*

We list some examples of Banach spaces $\mathcal{E}$ for which all strongly continuous semigroups $T : \mathbf{R}_+ \to \mathcal{L}(\mathcal{E})$ are necessarily uniformly continuous [2]:

---

[2]For proofs and bibliographical references, see: H.P. Lotz, *Semigroups on* L^∞ *and* H^∞ , in [1],pp. 54-59.

$\mathcal{E} = C(K)$ where K is a Stone space, i.e. a compact, Hausdorff space on which the closure of any open set is open.

$\mathcal{E} = L^\infty(\Xi, \mu)$, where (Ξ, μ) is a measure space.

The Banach space $\mathcal{E} = H^\infty(D)$ of all bounded holomorphic functions on a finitely connected domain $D \subset \mathbf{C}$.

The Banach space $\mathcal{E}$ of all bounded continuous solutions f of the Laplace equation $\Delta f = 0$ on an open set $D \subset \mathbf{R}^n$.

The Banach space $\mathcal{E}$ of all bounded continuous solutions of the equation

$$\sum_{i=1}^{n} \frac{\partial^2 f}{\partial x_i{}^2} = \frac{\partial f}{\partial x_{n+1}}$$

on an open set $D \subset \mathbf{R}^{n+1}$.

2.4 The spectrum of the infinitesimal generator

Let $T : \mathbf{R}_+ \to \mathcal{L}(\mathcal{E})$ be a strongly continuous semigroup. According to Theorem 2.1.1 there are two real constants a and $M \geq 1$ for which (2.6) holds.

Theorem 2.4.1 *The spectrum of the infinitesimal generator X of T is contained in the closed half-plane $\{\zeta \in \mathbf{C} : \Re\zeta \leq a\}$, and*

$$(\zeta I - X)^{-1}x = \int_0^{+\infty} e^{-t\zeta} T(t)x\,dt \qquad (2.34)$$

for all $x \in \mathcal{E}$ and all $\zeta \in \mathbf{C}$ with $\Re\zeta > a$.

Proof If $\Re\zeta > a$, then

$$|e^{-t\zeta} e^{at}| = e^{(a-\Re\zeta)t} < 1$$

whenever $t > 0$, and therefore, by (2.6),

$$\| \int_0^{+\infty} e^{-t\zeta} T(t)x\,dt \| \leq \int_0^{+\infty} \| e^{-t\zeta} T(t)x \|\,dt$$

$$\leq M \int_0^{+\infty} e^{(a-Re\zeta)t}\,dt \|x\|$$

$$= \frac{M}{\Re\zeta - a} \|x\|. \qquad (2.35)$$

Hence, the integral on the right hand side of (2.34) exists as a norm convergent improper integral. Let $\zeta \mapsto R(\zeta)$ be the function from the open half-plane $\{\zeta \in \mathbf{C} : \Re\zeta > a\}$ to $\mathcal{L}(\mathcal{E})$ defined by that integral, and, for $x \in \mathcal{E}$, let $y = R(\zeta)x$. Since, for $h > 0$,

$$
\begin{aligned}
T(h)\,y &= \int_0^{+\infty} e^{-t\zeta} T(t+h)x\,dt \\
&= \int_h^{+\infty} e^{-(t-h)\zeta} T(t)x\,dt \\
&= e^{h\zeta} \int_h^{+\infty} e^{-t\zeta} T(t)x\,dt \\
&= e^{h\zeta} \left\{ \int_0^{+\infty} e^{-t\zeta} T(t)x\,dt - \int_0^h e^{-t\zeta} T(t)x\,dt \right\},
\end{aligned}
$$

then

$$
\lim_{h\downarrow 0} \frac{1}{h}(T(h) - I)y =
$$

$$
= \lim_{h\downarrow 0} \left\{ -\frac{1}{h} e^{h\zeta} \int_0^h e^{-t\zeta} T(t)x\,dt + \frac{e^{h\zeta} - 1}{h} \int_0^{+\infty} e^{-t\zeta} T(t)x\,dt \right\}
$$

$$
- = x + \zeta R(\zeta)x = -x + \zeta y.
$$

Hence $y \in \mathcal{D}(X)$ and $Xy = -x + \zeta y$, i.e.

$$
(\zeta I - X)y = x \quad \forall\, x \in \mathcal{E}.
$$

Thus, $R(\zeta)(\mathcal{E}) \subset \mathcal{D}(X)$, and

$$
(\zeta I - X)\,R(\zeta) = I \quad \text{on } \mathcal{E}. \tag{2.36}
$$

As a consequence, the map $\zeta I - X : \mathcal{D}(X) \to \mathcal{E}$ is surjective. Since X is closed, in order to prove that $\zeta \in r(X)$, it suffices to show that $\zeta I - X$ is injective when $\Re\zeta > a$. Thus, let $x \in \mathcal{D}(X)$ be such that $(\zeta I - X)x = 0$. Letting, for $t \geq 0$, $x(t) = e^{\zeta t}x$, then:

$$
x(t) \in \mathcal{D}(X) \quad \forall\, t \geq 0,
$$

$$
x(0) = x,
$$

and

$$\dot{x}(t) = \zeta e^{\zeta t} = (e^{\zeta \bullet} X x)(t)$$
$$= X x(t) \ \forall t \geq 0.$$

By Theorem 2.2.4, $x(t) = T(t)x$ for all $t \geq 0$, so that

$$\|T(t)x\| = \|x(t)\| = e^{t\Re\zeta}\|x\|.$$

Hence, by (2.6),

$$e^{t(\Re\zeta - a)}\|x\| \leq M\|x\| \ \forall t \geq 0.$$

For $\Re\zeta > a$, letting $t \to +\infty$, we see that $x = 0$, thus establishing the injectivity of $\zeta I - X$ when $\Re\zeta > a$. Hence, by (2.36), $R(\zeta) = (\zeta I - X)^{-1}$ if $\Re\zeta > a$, and the theorem is proven. ∎

In view of (2.34) and (2.35),

$$\|(\zeta I - X)^{-1}\| \leq \frac{M}{\zeta - a}$$

whenever $\zeta \in (a, +\infty)$. Moreover, by (2.34) and Fubini's theorem, if $\Re\zeta > a$,

$$(\zeta I - X)^{-2}x = \int_0^{+\infty} e^{-t_2\zeta}T(t_2)\left(\int_0^{+\infty} e^{-t_1\zeta}T(t_1)x\,dt_1\right)dt_2$$
$$= \int_0^{+\infty}\int_0^{+\infty} e^{-(t_1+t_2)\zeta}T(t_1+t_2)x\,dt_1\,dt_2,$$

and (2.35) yields

$$\|(\zeta I - X)^{-2}\| \leq \frac{M}{(\zeta - a)^2} \quad \text{for all } \zeta \in (a, +\infty).$$

Iteration of this argument proves the following proposition.

Proposition 2.4.2 *If (2.6) holds, then* $(a, +\infty) \subset r(X)$ *and*

$$\|(\zeta I - X)^{-m}\| \leq \frac{M}{(\zeta - a)^m}$$

for all $\zeta \in (a, +\infty)$ *and* $m = 1, 2, \ldots$.

For $\zeta \in \mathbf{C}$ with $\Re\zeta > a$, and $t \geq 0$, let $Y_\zeta(t) \in \mathcal{L}(\mathcal{E})$ be defined by the integral

$$Y_\zeta(t)x = \int_0^t e^{(t-s)\zeta} T(s)x \, ds. \tag{2.37}$$

Let $R : \mathbf{R}_+ \to \mathcal{L}(\mathcal{E})$ be the strongly continuous semigroup

$$t \mapsto e^{-t\zeta} T(t)x \quad (x \in \mathcal{E}).$$

Since, for $t > 0$,

$$\frac{1}{t}(R(t) - I)x = \frac{e^{-t\zeta}}{t}(T(t) - I)x + \frac{e^{-t\zeta} - 1}{t}x,$$

the domain of the infinitesimal generator Z of R is $\mathcal{D}(Z) = \mathcal{D}(X)$, and $Z = X - \zeta I$. The integral (2.37) becomes

$$Y_\zeta(t)x = e^{t\zeta} \int_0^t R(s)x \, ds,$$

so that, by (2.16), for all $x \in \mathcal{E}$,

$$Z Y_\zeta(t)x = e^{t\zeta} Z \int_0^t R(s)x \, ds = e^{t\zeta}(R(t) - I)x$$
$$= T(t)x - e^{t\zeta}x,$$

i.e.,

$$X Y_\zeta(t)x = Z Y_\zeta(t)x + \zeta Y_\zeta(t)x$$
$$= T(t)x - e^{t\zeta}x + \zeta Y_\zeta(t)x.$$

Hence,

$$(\zeta I - X)Y_\zeta(t)x = e^{t\zeta}x - T(t)x \quad \forall x \in \mathcal{E}, \ \forall t \geq 0. \tag{2.38}$$

If $x \in \mathcal{D}(X)$ and $t \geq 0$, Lemma 2.2.2 yields

$$Y_\zeta(t)Xx = e^{t\zeta} \int_0^t e^{-s\zeta} T(s)Xx \, ds$$
$$= e^{t\zeta} \int_0^t R(s)(X - \zeta I)x \, ds + \zeta e^{t\zeta} \int_0^t R(s)x \, ds$$
$$= e^{t\zeta} \int_0^t R(s)Zx \, ds + \zeta Y_\zeta(t)x$$
$$= e^{t\zeta}(R(t) - I)x + \zeta Y_\zeta(t)x$$
$$= T(t)x - e^{t\zeta}x + \zeta Y_\zeta(t)x.$$

Hence,

$$Y_\zeta(t)(\zeta I - X)x = e^{t\zeta}x - T(t)x \; : \; \forall x \in \mathcal{D}(X) \; \forall t \geq 0. \qquad (2.39)$$

Furthermore, for $t, r \geq 0$,

$$Y_\zeta(t)T(r) = \int_0^t e^{(t-s)\zeta}T(s+r)ds = T(r)Y_\zeta(t). \qquad (2.40)$$

Theorem 2.4.3 *If X generates the strongly continuous semigroup T : $\mathbf{R}_+ \to \mathcal{L}(\mathcal{E})$, then*

$$e^{t\sigma(X)} \subset \sigma(T(t))\backslash\{0\}. \qquad (2.41)$$

Here

$$e^{t\sigma(X)} := \{e^{t\zeta} : \zeta \in \sigma(X)\}.$$

Proof We shall establish the theorem by showing that

$$r(T(t)) \cup \{0\} \subset e^{t\,r(X)} \cup \{0\}, \qquad (2.42)$$

where

$$e^{t\,r(X)} := \{e^{t\zeta} : \zeta \in r(X)\}.$$

Let $\zeta \in \mathbf{C}$ be such that $e^{t\zeta} \in r(T(t))$, and let $A = (e^{t\zeta}I - T(t))^{-1} \in \mathcal{L}(\mathcal{E})$. By (2.38) and (2.39)

$$(\zeta I - X)\,Y_\zeta(t)\,Ax = x \; \forall x \in \mathcal{E}, \; \forall t \in \mathbf{R}_+, \qquad (2.43)$$

$$A\,Y_\zeta(t)\,(\zeta I - X)x = x \; \forall x \in \mathcal{D}(X), \; \forall t \in \mathbf{R}_+. \qquad (2.44)$$

Since, by (2.40), $Y_\zeta(t)$ commutes with $(e^{t\zeta}I - T(t)$, and therefore also with) A, (2.44) can be written

$$Y_\zeta(t)\,A\,(\zeta I - X)x = x \; \forall x \in \mathcal{D}(X), \; \forall t \in \mathbf{R}_+,$$

and, together with (2.43) shows that

$$Y_\zeta(t)\,A = (\zeta I - X)^{-1},$$

and therefore $\zeta \in r(X)$. ∎

Let T be uniformly continuous. Then $X \in \mathcal{L}(\mathcal{E})$, and $T(t) = \exp tX$. Applying the spectral mapping theorem for unital Banach algebras to the holomorphic map $\zeta \mapsto \exp \zeta X$ of $\mathbf{C}$ into $\mathcal{L}(\mathcal{E})$, we obtain

$$\sigma(\exp \zeta X) = e^{\zeta \sigma(X)}$$

for all $\zeta \in \mathbf{C}$. This proves

Theorem 2.4.4 *If the semigroup $T : \mathbf{R}_+ \to \mathcal{L}(\mathcal{E})$ is uniformly continuous, (2.41) is an equality for all t.*

Later on in this chapter, we shall establish this theorem without appealing to the spectral mapping theorem, as a corollary of a more general statement.

Example Let $\mathcal{E}$ be the Banach space of all continuous functions $x : [0, 1] \to \mathbf{C}$ such that $x(0) = 0$, with the uniform norm. For $t \geq 0$ and $x \in \mathcal{E}$, let

$$(T(t)x)(r) \begin{cases} x(r - t) \text{ if } 0 \leq r - t \leq 1, \\ 0 \qquad \text{ if } t > r \text{ or } t < r - 1. \end{cases}$$

Then $T : \mathbf{R}_+ \to \mathcal{L}(\mathcal{E})$ is a strongly continuous semigroup, with $\|T(t)\| \leq 1$ for all $t \geq 0$, whose infinitesimal generator X is defined by $Xx = \dot{x}$ for all x in its domain

$$\mathcal{D}(X) = \{x \in \mathcal{E} : x \in C^1([0, 1]), \dot{x} \in \mathcal{E}\}.$$

Integrating the ordinary differential equation

$$\dot{x} - \zeta x = y$$

for $y \in \mathcal{E}$ and $\zeta \in \mathbf{C}$, we see that $r(X) = \mathbf{C}$ and, for all $\zeta \in \mathbf{C}$,

$$(\zeta I - X)^{-1} = -e^{t\zeta} \int_0^t e^{-s\zeta} y(s) ds.$$

Hence $\sigma(X) = \emptyset$. Since $\sigma(T(t)) \neq \emptyset$, this shows that the inclusion (2.41) may be proper.

The real number

$$s(X) := \sup\{\Re \zeta : \zeta \in \sigma(X)\}$$

is called the *spectral limit* of X or of the semigroup T. By (2.11) and Theorem 2.4.1,

$$s(X) \leq \omega(T), \tag{2.45}$$

where $\omega(T)$ is the growth bound of T.

Theorem 2.1.4 and Theorem 2.4.4 yield

Proposition 2.4.5 *If the semigroup T is uniformly continuous, then* $s(X) = \omega(T)$.

Under the hypothesis of this latter proposition, the compactness of $\sigma(X)$ implies that, if $\sigma(X) \subset \{\zeta \in \mathbf{C} : \Re\zeta < 0\}$, then $s(X) < 0$. Thus, by (2.11), there exist $M \geq 1$ and $a < 0$ for which (2.6) holds. That proves the following theorem.

Theorem 2.4.6 *If the semigroup T is uniformly continuous, and if, for every $\zeta \in \sigma(X)$, $\Re\zeta < 0$, then $\lim_{t \to +\infty} \|T(t)\| = 0$.*

Example Le X be a complex square matrix of order n, and consider the Cauchy problem

$$\dot{x}(t) = X\,x(t)$$
$$x(0) = x_o \in \mathbf{C}^n.$$

If all eigenvalues of X have negative real parts, for every initial condition $x_o \in \mathbf{C}^n$ the solution x is "stable", in the sense that $\lim_{t \to +\infty} x(t) = 0$.

The following example shows that (2.45) may be a strict inequality.

Let $l_n{}^2$ be the complex Hilbert space of dimension $n = 1, 2, \ldots$, and let $\mathcal{H}$ be the Hilbert space direct sum $\mathcal{H} = \oplus l_n{}^2$, whose elements are the sequences $x = (x_1, x_2, \ldots)$, with $x_n \in l_n{}^2$ and $\sum \|x_n\|^2 < \infty$. The inner product of $x, y \in \mathcal{H}$ is $(x|y) = \sum(x_n|y_n)$. Given $F_n \in \mathcal{L}(l_n{}^2)$ such that $\sup\{\|F_n\| : n = 1, 2, \ldots\} < \infty$, let $F \in \mathcal{L}(\mathcal{H})$ be the linear operator defined by

$$F(x) = (F_1(x_1), F_2(x_2), \ldots),$$

whose norm is $\|F\| \leq \sup\{\|F_n\| : n = 1, 2, \ldots\}$. Let $A_n \in \mathcal{L}(l_n{}^2)$ be the linear operator represented by a $n \times n$ matrix - that will be denoted by the same symbol A_n - all of whose entries are zero, except, when $n > 1$, for $A_{n;p,p+1} = 1$ as $p = 1, 2, \ldots, n-1$. Since $A_n{}^n = 0$, then $\sigma(A_n) = \{0\}$, so that, if I_n is the identity operator on $l_n{}^2$ and

$$X_n = i\, n I_n + A_n,$$

then $\sigma(X_n) = \{i\, n\}$. Being $\|A_n\| = 1$, then

$$\|\exp tX_n\| = \|e^{int}\exp tA_n\| = \|\exp tA_n\|$$

$$= \|I_n + tA_n + \frac{t^2}{2}A_n{}^2 + \cdots + \frac{t^{n-1}}{(n-1)!}A_n{}^{n-1}\|$$

$$\leq 1 + |t|\|A_n\| + \frac{|t|^2}{2!}\|A_n\|^2 + \cdots + \frac{|t|^{n-1}}{(n-1)!}\|A_n\|^{n-1}$$

$$\leq 1 + |t| + \frac{|t|^2}{2!} + \cdots + \frac{|t|^{n-1}}{(n-1)!}$$

$$\leq e^{|t|}$$

for all $t \in \mathbf{R}$. Thus, denoting by $T_n : \mathbf{R} \to \mathcal{L}(l_n{}^2)$ the group $t \mapsto \exp tX_n$, we have $\|T_n(t)\| \leq e^{|t|}$ for all $t \in \mathbf{R}$. We show now that

$$\limsup_{n \to +\infty} \|T_n(t)\| = e^{|t|} \tag{2.46}$$

for all $t \in \mathbf{R}$.

Choose $t \geq 0$ and let $u_n \in l_n{}^2$ be given by

$$u_n = \frac{1}{\sqrt{n}}\begin{pmatrix} 1 \\ \vdots \\ 1 \end{pmatrix}.$$

Then

$$(\exp tA_n)u_n = \frac{1}{\sqrt{n}}\begin{pmatrix} 1 + t + \frac{t^2}{2} + \cdots + \frac{t^{n-2}}{(n-2)!} + \frac{t^{n-1}}{(n-1)!} \\ 1 + t + \frac{t^2}{2} + \cdots + \frac{t^{n-2}}{(n-2)!} \\ \vdots \\ 1 + t \\ 1 \end{pmatrix},$$

152

and therefore

$$\|(\exp tA_n)u_n - e^t u_n\|^2 =$$
$$\frac{1}{n}[(1 + t + \frac{t^2}{2} + \cdots + \frac{t^{n-1}}{(n-1)!} - e^t)^2 + \cdots$$
$$+ (1 + t + \frac{t}{2} + \cdots + \frac{t^{n-2}}{(n-2)!} - e^t)^2 + \cdots$$
$$+ (1 + t - e^t)^2 + (1 - e^t)^2].$$

Since, for some $\theta \in (0,1)$,

$$e^t = 1 + t + \cdots + \frac{t^p}{p!} + \frac{t^{p+1}}{(p+1)!} \frac{d^{p+1}e^s}{ds^{p+1}}\Big|_{s=\theta t}$$
$$= 1 + t + \cdots + \frac{t^p}{p!} + \frac{t^{p+1}}{(p+1)!}e^{\theta t}$$
$$\leq 1 + t + \cdots + \frac{t^p}{p!} + \frac{t^{p+1}}{(p+1)!}e^t,$$

then, for $t > 0$,

$$\|(\exp tA_n)u_n - e^t u_n\|^2 \leq$$
$$\frac{1}{n}\left[\left(\frac{t^n}{n!}e^t\right)^2 + \left(\frac{t^{n-1}}{(n-1)!}e^t\right)^2 + \cdots + (te^t)^2\right]$$

$$= \frac{e^{2t}}{n}\left[t^2 + \frac{t^4}{(2!)^2} + \cdots + \frac{t^{2n}}{(n!)^2}\right]$$

$$< \frac{t^2 e^{2t}}{n}\left[1 + t^2 + \frac{t^4}{2!} + \cdots + \frac{t^{2n}}{n!} + \cdots\right]$$

$$= \frac{t^2 e^{2t+t^2}}{n},$$

whence

$$\lim_{n\to+\infty} \|(\exp tA_n)u_n - e^t u_n\| = 0.$$

If $\{n(1), n(2), \dots\}$ is an increasing sequence of positive integers such that $\lim_{p \to +\infty} e^{in(p)} = 1$, then

$$\lim_{p \to +\infty} \|(\exp tX_{n(p)})u_{n(p)} - e^t u_{n(p)}\| = 0.$$

That proves (2.46) when $t \geq 0$. A similar argument yields (2.46) for $t \leq 0$.

As a consequence, a group $T : \mathbf{R} \to \mathcal{L}(\mathcal{H})$ is defined by

$$T(t)x := (T_1(t)x_1, T_2(t)x_2, \dots)$$

for which

$$\|T(t)\| = e^{|t|}.$$

Again by (2.46)

$$\|T(t)x - x\|^2 = \sum_{1}^{+\infty} \|T_n(t)x_n - x_n\|^2 =$$

$$\sum_{1}^{N} \|T_n(t)x_n - x_n\|^2 + \sum_{N+1}^{+\infty} \|T_n(t)x_n - x_n\|^2$$

$$\leq \sum_{1}^{N} \|T_n(t)x_n - x_n\|^2 + \sum_{N+1}^{+\infty} \|T_n(t) - I_n\|^2 \|x_n\|^2$$

$$\leq \sum_{1}^{N} \|T_n(t)x_n - x_n\|^2 + \sum_{N+1}^{+\infty} (\|T_n(t)\| + 1)^2 \|x_n\|^2$$

$$\leq \sum_{1}^{N} \|T_n(t)x_n - x_n\|^2 + (e^{|t|} + 1)^2 \sum_{N+1}^{+\infty} \|x_n\|^2.$$

For any $\epsilon > 0$ there is a positive integer N for which $\sum_{N+1}^{+\infty} \|x_n\|^2 < \epsilon$. Let $\delta > 0$ be such that, whenever $t \in (-\delta, \delta)$, $\|T_n(t)x_n - x_n\|^2 < \epsilon$ for $n = 1, 2, \dots, N$. In conclusion, if $|t| < \delta$,

$$\|T(t)x - x\|^2 < N\epsilon + (e^\delta + 1)^2 \epsilon$$

and that proves that T is a strongly continuous group. The same symbol T will stand for its restriction to $\mathbf{R}_+$.

154

Let $\mathcal{D}$ be the algebraic direct sum of all l_n^2, that is, the vector space of all sequences $\{x_1, x_2, \dots\}$ with $x_n \in l_n^2$ such that $x_n \neq 0$ for only a finite number of $n's$. Let $X : \mathcal{D} \to \mathcal{D}$ be the linear operator defined by

$$X\,x = \{X_1 x_1, X_2 x_2, \dots\}.$$

If $x \in \mathcal{D}$, then

$$\lim_{t \downarrow 0} \frac{1}{t}(T(t)x - x) = X\,x.$$

Since $\mathcal{D}$ is dense in $\mathcal{H}$ and $T(t)\mathcal{D} \subset \mathcal{D}$ for all t, then $\mathcal{D}$ is a core for the infinitesimal generator Z of T, and $X = Z_{|\mathcal{D}}$. As a consequence, $\sigma(Z) = \sigma(X)$. On the other hand, if $\zeta \neq in$ for $n = 1, 2, \dots$, then $\zeta \in r(X_n)$, and

$$(\zeta\,I_n - X_n)^{-1} = ((\zeta - in)I_n - A_n)^{-1} =$$

$$\frac{1}{\zeta - in}\left(I_n + \frac{1}{\zeta - in}A_n + \dots + \frac{1}{(\zeta - in)^{n-1}}A_n^{\,n-1}\right).$$

If n is such that $|\zeta - in| \geq k > 1$, then

$$\|(\zeta\,I_n - X_n)^{-1}\| \leq \frac{1}{k}\frac{1}{1 - \frac{1}{k}} = \frac{1}{k - 1},$$

and therefore

$$\sup\{\|(\zeta\,I_n - X_n)^{-1}\| : n = 1, 2, \dots\} < \infty.$$

Let $(\zeta\,I - X)^{-1} \in \mathcal{L}(\mathcal{H})$ be the linear operator defined by

$$(\zeta\,I - X)^{-1} = ((\zeta\,I_1 - X_1)^{-1}x_1, (\zeta\,I_2 - X_2)^{-1}x_2, \dots). \qquad (2.47)$$

Since: $(\zeta\,I - X)^{-1}\mathcal{D} \subset \mathcal{D}$,

$$(\zeta\,I - X)(\zeta\,I - X)^{-1} = I$$

on $\mathcal{H}$, and

$$(\zeta\,I - X)^{-1}(\zeta\,I - X) = I$$

on $\mathcal{D}$, then $\mathbf{C}\backslash\{in: \ n = 1, 2, \dots\} \subset r(X)$. On the other hand, letting

$$v_n = \begin{pmatrix} 1 \\ 0 \\ \vdots \\ 0 \end{pmatrix}$$

then $X_n v_n = in v_n$, and therefore $v = (0, \dots, 0, v_n, 0, \dots)$ is an eigenvector of X with eigenvalue in. In conclusion,

$$\sigma(Z) = \sigma(X) = p\sigma(X) = \{in: \ n: 1, 2, \dots\}. \tag{2.48}$$

By (2.46) and (2.48),

$$\omega(T) = 1, \ \ s(Z) = 0.$$

Exercise Prove that $(\zeta I - X)^{-1}$ is a compact operator.

Note The above example is due to J.Zabczyc [82] and can be found also in [1] and [12] together with other examples showing that (2.45) may be a strict inequality.

Let now $T : \mathbf{R} \to \mathcal{L}(\mathcal{E})$ be a strongly continuous, uniformly bounded group, and let $X : \mathcal{D}(X) \subset \mathcal{E} \to \mathcal{E}$ be its infinitesimal generator. We shall show that $\sigma(X) \neq \emptyset$.

For $f \in L^1(\mathbf{R})$ (Lebesgue measure) we define $\hat{f}(T) \in \mathcal{L}(\mathcal{E})$ by

$$\hat{f}(T)(x) = \int_{-\infty}^{+\infty} f(t)T(t)x\,dt \ \ (x \in \mathcal{E}).$$

Lemma 2.4.7 *If* $f \in L^1(\mathbf{R})$ *and* $\hat{f} \in L^1(\mathbf{R})$, $\hat{f}(T)$ *is expressed by*

$$\hat{f}(T)(x) = \frac{1}{2\pi} \lim_{a \downarrow 0} \int_{-\infty}^{+\infty} \hat{f}(-t) \left(((a + it)I - X)^{-1} \right.$$
$$\left. - - ((-a + it)I - X)^{-1} \right) x\,dt.$$

Proof Let $a > 0$. Fubini's theorem and Theorem 2.4.1 yield

$$\int_{-\infty}^{+\infty} e^{-a|t|} f(t) T(t) x dt = \frac{1}{2\pi} \int_{-\infty}^{+\infty} e^{-a|t|} \left(\int_{-\infty}^{+\infty} e^{ist} \hat{f}(s) ds \right) T(t) x dt =$$

$$= \frac{1}{2\pi} \int_{-\infty}^{+\infty} \hat{f}(s) \left(\int_{-\infty}^{+\infty} e^{-a|t|} e^{ist} T(t) x dt \right) ds =$$

$$= \frac{1}{2\pi} \int_{-\infty}^{+\infty} \hat{f}(s) \left(\int_{0}^{+\infty} e^{-at} e^{ist} T(t) x dt - \int_{0}^{+\infty} e^{at} e^{ist} T(t) x dt \right) ds =$$

$$= \frac{1}{2\pi} \int_{-\infty}^{+\infty} \hat{f}(s) \left(\int_{0}^{+\infty} e^{-(a-is)t} T(t) x dt \right.$$

$$\left. = -\int_{0}^{+\infty} e^{-(-a-is)t} T(t) x dt \right) ds =$$

$$= \frac{1}{2\pi} \int_{-\infty}^{+\infty} \hat{f}(s) \left(((a-is)I - X)^{-1} - ((-a-is)I - X)^{-1} \right) x ds.$$

Letting $a \downarrow 0$, the dominated convergence theorem [3] yields the conclusion. ∎

Let $f \in L^1(\mathbf{R})$. If $\hat{f}$ is compactly supported and $\mathrm{Supp} f \cap i\sigma(X) = \emptyset$, then, by the dominated convergence theorem,

$$\lim_{a \downarrow 0} \int_{-\infty}^{+\infty} \hat{f}(-t) \left(((a+it)I - X)^{-1} - ((-a+it)I - X)^{-1} \right) x dt =$$

$$= \int_{-\infty}^{+\infty} \hat{f}(-t) \left((itI - X)^{-1} - (itI - X)^{-1} \right) x dt = 0.$$

That proves

Lemma 2.4.8 Let $f \in L^1(\mathbf{R})$. If $\hat{f}$ is compactly supported and vanishes in a neighbourhood of σX, then $\hat{f}(T) = 0$.

If $\sigma(X) = \emptyset$, this latter lemma holds for all $f \in L^1(\mathbf{R})$ whose Fourier transforms have compact support. Since the set of all these functions is dense in $L^1(\mathbf{R})$ (see, e.g., [46]), then $\hat{f}(T) = 0$ for all $f \in L^1(\mathbf{R})$.

---

[3] For the dominated convergence theorem in the case of vector-valued functions; see [40], Theorem 3.7.9, p. 83

Let now $f(t) = e^{-t}$ when $t \geq 0$, $f(t) = 0$ when $t < 0$. Then, by Theorem 2.4.1,

$$(I - X)^{-1} = \int_0^{+\infty} e^{-t} T(t) x \, dt$$

$$= \int_{-\infty}^{+\infty} f(t) T(t) x \, dt = \hat{f}(T) x = 0$$

for all $x \in \mathcal{E}$. This contradiction proves

Theorem 2.4.9 *If $T : \mathbf{R} \to \mathcal{L}(\mathcal{E})$ is a strongly continuous, uniformly bounded group, then $\sigma(X) \neq \emptyset$.*

Note Our presentation of Lemma 2.4.7, Lemma 2.4.8 and Theorem 2.4.9 follows the exposition given by J. van Neerven in [58] (pp. 46-48) in the more general context of Beurling algebras defined by non-quasi analytic weights.

2.5   The point spectrum

Let $T : \mathcal{R}_+ \to \mathcal{L}(\mathcal{E})$ be a strongly continuous semigroup, and let $X : \mathcal{D}(X) \subset \mathcal{E} \to \mathcal{E}$ be its infinitesimal generator.

If $x_o \in \mathcal{D}(X) \backslash \{0\}$ is an eigenvector of X, with eigenvalue ζ :

$$X x_o = \zeta x_o. \tag{2.49}$$

Setting $x(t) = e^{t\zeta} x_o$, for $t \geq 0$,

$$\dot{x}(t) = \zeta e^{t\zeta} x_o = e^{t\zeta} X x_o = X x(t).$$

Since $x(t) = T(t) x_o$, then $e^{t\zeta} x_o = T(t) x_o$, and the following lemma holds.

Lemma 2.5.1 *If $x_o \in \mathcal{D}(X) \backslash \{0\}$ is an eigenvector of X, with eigenvalue ζ, x_o is an eigenvector of $T(t)$, with eigenvalue $e^{t\zeta}$, for all $t \in \mathbf{R}_+$.*

Let now $\tau \neq 0$ be an eigenvalue of $T(t)$ for some $t > 0$. Once a value of $\log \tau = \log |\tau| + i \arg \tau$ has been chosen, all solutions μ_n of the equation $e^{t\mu} = \tau$ are given by $\mu_n = \frac{\log \tau}{t} + \frac{2\pi n}{t} i$, for $n \in \mathbf{Z}$. The space

$\mathcal{M} := \ker(\tau I - T(t))$ is a linear, non-zero, closed subspace of $\mathcal{E}$. For all $s \geq 0$ and all $x \in \mathcal{M}$,

$$(\tau I - T(t))\,T(s)x = T(s)\,(\tau I - T(t))x = 0,$$

showing that

$$T(s)\mathcal{M} \subset \mathcal{M} \quad \forall s \in \mathbf{R}_+.$$

Since, for $x \in \mathcal{M}$,

$$\lim_{s\downarrow 0} T(s)x = \lim_{s\downarrow 0} T(s)\frac{1}{\tau}T(t)x =$$

$$= \frac{1}{\tau}\lim_{s\downarrow 0} T(t+s)x = \frac{1}{\tau}T(t)x = x,$$

the restriction $s \mapsto T(s)_{|\mathcal{M}}$ is a strongly continuous semigroup. Thus, $s \mapsto S(s) := e^{-\mu_0 s}T(s)_{|\mathcal{M}}$ is a strongly continuous semigroup $S : \mathbf{R}_+ \to \mathcal{L}(\mathcal{M})$, for which we have

$$S(s+t) = e^{-\mu_0 s}T(s)_{|\mathcal{M}}\, e^{-\mu_0 t}T(t)_{|\mathcal{M}}$$

$$= e^{-\mu_0 s}T(s)_{|\mathcal{M}}\,\frac{1}{\tau}T(t)_{|\mathcal{M}}$$

$$= e^{-\mu_0 s}T(s)_{|\mathcal{M}} = S(s).$$

Hence S is a continuous periodic function with period t. By the theorem of Banach-Steinhaus, there is a constant $M \geq 1$ such that $\|S(s)\| \leq M$ for all $s \in \mathbf{R}_+$.

For $x \in \mathcal{M}$ and $n \in \mathbf{Z}$, the integral

$$J_n x := \frac{1}{t}\int_0^t e^{-\frac{2\pi n}{t}si}S(s)x\,ds$$

defines a linear operator J_n on $\mathcal{M}$, for which

$$\|J_n x\| \leq \frac{1}{t}\int_0^t \|S(s)x\|ds \leq M\|x\|,$$

whence $J_n \in \mathcal{L}(\mathcal{M})$ and $\|J_n\| \leq M$.

For $n, m \in \mathbf{Z}$ and $x \in \mathcal{M}$, Fubini's theorem yields

$$J_n \circ J_m x = \frac{1}{t} \int_0^t e^{-\frac{2\pi n}{t}si} S(s) J_m x \, ds$$

$$= \frac{1}{t^2} \int_0^t \int_0^t e^{-\frac{2\pi}{t}(sn+rm)i} S(s+r) x \, dr \, ds$$

$$= \frac{1}{t^2} \int_0^t \int_0^t e^{-\frac{2\pi}{t}(n-m)si} e^{-\frac{2\pi}{t}m(r+s)i} S(s+r) x \, dr \, ds$$

$$= \frac{1}{t^2} \int_0^t e^{-\frac{2\pi}{t}(n-m)si} \left(\int_0^t e^{-\frac{2\pi}{t}m(r+s)i} S(s+r) x \, dr \right) ds$$

$$= \frac{1}{t} \int_0^t e^{-\frac{2\pi}{t}(n-m)si} ds \, J_m x$$

$$= \int_0^1 e^{-2\pi(n-m)si} ds \, J_m x$$

$$= \delta_{n,m} J_m x,$$

that is,

$$J_n \circ J_m = \delta_{nm} J_m.$$

In other words, $\{J_n : n \in \mathbf{Z}\}$ is a family of mutually orthogonal projectors. Their ranges $\mathcal{M}_n := J_n \mathcal{M}$ are linear, closed, linear independent subspaces of $\mathcal{M}$. Furthermore, if $r \geq 0$ and $x \in \mathcal{M}$, then

$$T(r) J_n x = \frac{1}{t} \int_0^t e^{-\frac{2\pi n}{t}si} T(r) S(s) x \, ds$$

$$= \frac{1}{t} \int_0^t e^{-\frac{2\pi n}{t}si} S(s) T(r) x \, ds$$

$$= J_n T(r) x,$$

i. e.,

$$T(r) \circ J_n = J_n \circ T(r)_{|\mathcal{M}}$$

for all $n \in \mathbf{Z}$ and all $r \in \mathbf{R}_+$.

For $x \in \mathcal{M}$, $J_n x$ is the n-th Fourier coefficient of the continuous periodic function $s \mapsto S(s)x$. Hence, letting

$$S_n(s)x := \sum_{-n}^{n} e^{\frac{2\pi k}{t}si} J_k x,$$

the sequence $\{S_n(s)x\}$ converges $C.1$ to the function $s \mapsto S(s)x$. In other words, the Cesaro means $\Omega_N(s)x$, defined, for $N \in \mathbf{N}^*$, by

$$\Omega_N(s)x := \frac{1}{N}(S_0(s) + S_1(s) + \cdots + S_{N-1}(s))x,$$

converge to $S(s)x$, when $N \to \infty$, for all $x \in \mathcal{M}$, uniformly with respect to $s \in \mathbf{R}_+$. In particular, for $s = 0$,

$$S_n(0)x = \sum_{-n}^{n} J_k x,$$

and

$$x = S(0)x = C.1 - \sum_{-\infty}^{+\infty} J_n x \quad \forall x \in \mathcal{M}.$$

Thus, $\mathcal{M}$ is the closure of the linear submanifold spanned by all $\mathcal{M}_n$:

$$\mathcal{M} = \bigvee \mathcal{M}_n.$$

Since $\mathcal{M} \neq \{0\}$, then $\mathcal{M}_n \neq \{0\}$ for some $n \in \mathbf{Z}$. For $x \in \mathcal{M}_n$,

$$S(s)x = S(s)J_n x = J_n S(s)x$$

$$= \frac{1}{t}\int_0^t e^{-\frac{2\pi n}{t}ri}S(r+s)x dr$$

$$= \frac{1}{t}\int_s^{t+s} e^{-\frac{2\pi n}{t}(r-s)i}S(r)x dr$$

$$= e^{\frac{2\pi n s i}{t}}\frac{1}{t}\int_0^t e^{-\frac{2\pi n}{t}ri}S(r)x dr$$

$$= e^{\frac{2\pi n s i}{t}} J_n x = e^{\frac{2\pi n s i}{t}} x.$$

Hence, if $x \in \mathcal{M}_n$,

$$T(s)x = e^{\mu_0 s}S(s)x$$

$$= e^{(\mu_0 + \frac{2\pi n i}{t})s}x = e^{\mu_n s}x.$$

Consequently the limit $\lim_{s \downarrow 0} \frac{1}{s}(T(s) - I)x$ exists, and

$$Xx = \lim_{s \downarrow 0}\frac{1}{s}(T(s) - I)x$$

$$= \lim_{s \downarrow 0}\frac{e^{\mu_n s} - 1}{s}x = \mu_n x.$$

Thus,

$$\mathcal{M}_n \subset \ker(\mu_n I - X), \qquad (2.50)$$

and, since $\mathcal{M}_n \neq \{0\}$, then $\mu_n \in p\sigma(X)$.

In order to establish the opposite inclusion

$$\mathcal{M}_n \supset \ker(\mu_n I - X), \qquad (2.51)$$

choose $x \in \mathcal{D}(X)$ such that $Xx = \mu_n x$, and let $x(s) = T(s)x$. Since

$$\dot{x}(s) = XT(s)x = T(s)Xx$$
$$= \mu_n T(s)x = \mu_n x(s),$$

with $x(0) = x$, and since the C^1 function $s \mapsto y(s) = e^{\mu_n s}x$ satisfies this latter equation with the same initial condition $y(0) = x$, Theorem 2.2.4 implies that

$$T(s)x = e^{\mu_n s}x \quad \forall s \in \mathbf{R}_+,$$

i. e.

$$x \in \ker(e^{\mu_n s} I - T(s)) \quad \forall s \in \mathbf{R}_+.$$

In particular

$$x \in \ker(e^{\mu_n t} I - T(t)) = \ker(e^{\mu t} I - T(t)) = \mathcal{M}.$$

Furthermore,

$$S(s)x = e^{-\mu_0 s}T(s)x$$
$$= e^{(\mu_n - \mu_0)s}x = e^{\frac{2\pi n}{t} si}x,$$

and consequently

$$J_n x = \frac{1}{t}\int_0^t e^{-\frac{2\pi n}{t} si} S(s)x ds = \frac{1}{t}tx = x,$$

proving that $x \in \mathcal{M}_n$ and establishing thereby the inclusion (2.51), which, together with (2.50), yields

$$\mathcal{M}_n = \ker(\mu_n I - X).$$

In conclusion, the following theorem holds.

Theorem 2.5.2 *For all $t \in \mathbf{R}_+$*

$$\mathrm{e}^{t\,p\sigma(X)} \subset p\sigma\big(T(t)\big) \subset \mathrm{e}^{t\,p\sigma(X)} \cup \{0\}.$$

If v is an eigenvector of X with eigenvalue ζ, v is an eigenvector of $T(t)$ with eigenvalue $\mathrm{e}^{t\zeta}$ for all $t \geq 0$.

Viceversa, let $\tau \neq 0$ be an eigenvalue of $T(t)$ for some $t > 0$, and choose one of the values of $\log \tau$. Then there exists some $n \in \mathbf{Z}$ such that

$$\mu_n := \frac{\log \tau + 2\pi n i}{t} \in p\sigma(X),$$

and, if L is the set of all integers n for which this inclusion holds, then

$$\ker(\tau\,I - T(t)) = \bigvee_{n \in L} \ker(\mu_n\,I - X).$$

2.6 The mean ergodic theorem for semi-groups

With the same notations as before, suppose now that the semigroup T is uniformly bounded:

$$\|T(t)\| \leq M$$

for all $t \in \mathbf{R}_+$ and some finite $M \geq 1$.

Lemma 2.6.1 *For all $u \in \ker X$ and all $v \in \overline{\mathcal{R}(X)}$,*

$$\lim_{t \to +\infty} \frac{1}{t} \int_0^t T(s)(u + v)ds = u.$$

Proof Since $T(s)u = u$, then, if $t > 0$,

$$\frac{1}{t} \int_0^t T(s)u\, ds = u.$$

For $\epsilon > 0$, let $y \in \mathcal{R}(X)$ be such that $\|v - y\| < \epsilon$, and let $z \in \mathcal{D}(X)$ be such that $Xz = y$. Then,

$$T(s)y = T(s)Xz = XT(s)z.$$

Because $s \mapsto T(s)z$ is of class C^1 on $\mathbf{R}_+$, and

$$\frac{d}{ds}T(s)z = XT(s)z,$$

then

$$\frac{1}{t}\int_0^t T(s)yds = \frac{1}{t}\int_0^t XT(s)zds =$$

$$\frac{1}{t}\int_0^t \frac{d}{ds}T(s)zds = \frac{1}{t}(T(t)-I)z.$$

Hence

$$\frac{1}{t}\int_0^t T(s)vds = \frac{1}{t}\int_0^t T(s)(v-y)ds + \frac{1}{t}(T(t)-I)z,$$

and, as a consequence,

$$\left\|\frac{1}{t}\int_0^t T(s)vds\right\| \leq$$

$$\leq \frac{1}{t}\int_0^t \|T(s)\|\,\|v-y\|ds + \frac{1}{t}(\|T(t)\| + 1)\|z\|$$

$$< M\epsilon + \frac{M+1}{t}\|z\|.$$

In conclusion,

$$\lim_{t\to+\infty}\left\|\frac{1}{t}\int_0^t T(s)vds\right\| \leq M\epsilon,$$

and therefore

$$\lim_{t\to+\infty}\frac{1}{t}\int_0^t T(s)vds = 0,$$

because $\epsilon > 0$ is arbitrary. $\blacksquare$

Lemma 2.6.1 implies that

$$\ker X \cap \overline{\mathcal{R}(X)} = \{0\}. \tag{2.52}$$

Lemma 2.6.2 *The linear space $\mathcal{F}$ of all points $x \in \mathcal{E}$ at which*

$$\lim_{t \to +\infty} \frac{1}{t} \int_0^t T(s)x\,ds \qquad (2.53)$$

exists in $\mathcal{E}$, is closed.

Proof Let $\{x_\nu\}$ be a sequence in $\mathcal{F}$ converging to some $x \in \mathcal{E}$. For $t_1 > t_2 > 0$,

$$\frac{1}{t_1} \int_0^{t_1} T(s)x\,ds - \frac{1}{t_2} \int_0^{t_2} T(s)x\,ds =$$

$$\left(\frac{1}{t_1} \int_0^{t_1} - \frac{1}{t_2} \int_0^{t_2} \right) T(s)(x - x_\nu)\,ds +$$

$$+ \left(\frac{1}{t_1} \int_0^{t_1} - \frac{1}{t_2} \int_0^{t_2} \right) T(s)x_\nu\,ds.$$

For any $\epsilon > 0$, there exists ν_0 such that $\|x - x_{\nu_0}\| < \epsilon$. Let $t_0 > 0$ be such that, whenever $t_1 > t_0$ and $t_2 > t_0$, then

$$\left\| \left(\frac{1}{t_1} \int_0^{t_1} - \frac{1}{t_2} \int_0^{t_2} \right) T(s)x_{\nu_0}\,ds \right\| < \epsilon.$$

Hence, if $t_1 > t_0$ and $t_2 > t_0$, then

$$\left\| \left(\frac{1}{t_1} \int_0^{t_1} - \frac{1}{t_2} \int_0^{t_2} \right) T(s)x\,ds \right\| \leq$$

$$\left(\frac{1}{t_1} \int_0^{t_1} + \frac{1}{t_2} \int_0^{t_2} \right) \|T(s)\| \, \|x - x_{\nu_0}\| +$$

$$\left\| \left(\frac{1}{t_1} \int_0^{t_1} - \frac{1}{t_2} \int_0^{t_2} \right) T(s)x_{\nu_0}\,ds \right\|$$

$$\leq 2M\|x - x_{\nu_0}\| + \epsilon < (2M + 1)\epsilon.$$

That proves the existence of the limit (2.53). ∎

Setting

$$Px = \lim_{t \to +\infty} \frac{1}{t} \int_0^t T(s)x\,ds \qquad (2.54)$$

for all $x \in \mathcal{F}$, then $\|Px\| \le M\,\|x\|$, i. e., $P \in \mathcal{L}(\mathcal{F},\mathcal{E})$ and $\|P\| \le M$. For $r \ge 0$, $t > 0$ and $x \in \mathcal{E}$,

$$\frac{1}{t} \int_0^t T(s)T(r)x\,ds =$$

$$= \frac{1}{t} \int_r^{r+t} T(s)x\,ds =$$

$$= \frac{1}{r+t} \frac{r+t}{t} \int_0^{r+t} T(s)x\,ds - \frac{1}{t} \int_0^r T(s)x\,ds. \qquad (2.55)$$

Hence

$$\lim_{t \to +\infty} \frac{1}{t} \int_0^t T(s)T(r)x\,ds = \lim_{t \to +\infty} \frac{1}{r+t} \int_0^{r+t} T(s)x\,ds = Px,$$

showing that $T(r)\mathcal{F} \subset \mathcal{F}$ and $PT(r) = P$ on $\mathcal{F}$. Furthermore, if $x \in \mathcal{F}$, (2.55) yields

$$T(r)Px = \lim_{t \to +\infty} \frac{1}{t} \int_r^{r+t} T(s)x\,ds$$

$$= \lim_{t \to +\infty} \frac{1}{r+t} \int_0^{r+t} T(s)x\,ds = Px.$$

Therefore

$$P^2 x = \lim_{t \to +\infty} \frac{1}{t} \int_0^t T(s)Px\,ds = Px$$

for all $x \in \mathcal{F}$.

Summing up, P is a projector of $\mathcal{F}$ onto $\ker X$, for which $P(\overline{\mathcal{R}(X)}) = \{0\}$.

If the point x is in the closure of $\ker X \oplus \overline{\mathcal{R}(X)}$, there are sequences $\{u_\nu\}$ in $\ker X$ and $\{v_\nu\}$ in $\mathcal{R}(X)$ such that $\{u_\nu + v_\nu\}$ converges to x. Since $u_\nu = P(u_\nu + v_\nu)$, then $\{u_\nu\}$ is a Cauchy sequence, and therefore also $\{v_\nu\}$ is a Cauchy sequence. As such, they converge in $\ker X$ and in $\overline{\mathcal{R}(X)}$ respectively, and thus, $x \in \ker X \oplus \overline{\mathcal{R}(X)}$, showing that $\ker X \oplus \overline{\mathcal{R}(X)}$ is closed.

For any $b \in \mathbf{R}_+$ and $x \in \ker X \oplus \overline{\mathcal{R}(X)}$,

$$\lim_{t \to +\infty} \frac{1}{t} \int_b^t T(s)x\,ds =$$

$$\lim_{t \to +\infty} \frac{1}{t} \left(\int_0^t - \int_0^b \right) T(s)x\,ds$$

$$= \lim_{t \to +\infty} \frac{1}{t} \int_0^t T(s)x\,ds.$$

The results established so far are the contents of the *mean ergodic theorem*, [35], [36], which will now be stated in a slightly more general form.

Since, for $\zeta \in \mathbf{C}$, $X - \zeta I$ generates the strongly continuous semigroup $t \mapsto e^{-t\zeta}T(t)$, the following theorem holds.

Theorem 2.6.3 *Let $T : \mathbf{R}_+ \to \mathcal{L}(\mathcal{E})$ be a strongly continuous semigroup such that $\|T(t)\| \le M$ for some $M \ge 1$ and all $t \in \mathbf{R}_+$, and let $X : \mathcal{D}(X) \subset \mathcal{E} \to \mathcal{E}$ be its infinitesimal generator.*
For any $\theta \in \mathbf{R}$,

$$\ker(X - i\theta\,I) \cap \overline{\mathcal{R}(X - i\theta\,I)} = \{0\},$$

and, for any $b \in \mathbf{R}_+$, the limit

$$Px := \lim_{t \to +\infty} \frac{1}{t} \int_b^t e^{-is\theta}T(s)x\,ds \qquad (2.56)$$

exists in $\mathcal{E}$ on all points x of the closed subspace $\ker(X - i\theta\,I) \oplus \overline{\mathcal{R}(X - i\theta\,I)}$ of $\mathcal{E}$, defining a projector $P \in \mathcal{L}(\ker(X{-}i\theta I) \oplus \overline{\mathcal{R}(X{-}i\theta I)})$, with norm $\|P\| \le M$, whose range is $\ker(X - i\theta\,I)$.

The following lemma is a consequence of Lemma 2.10.6 which will be established at the end of this chapter.

Lemma 2.6.4 *The closed space* $\ker(X - i\theta\, I) \oplus \overline{\mathcal{R}(X - i\theta\, I)}$ *is the set of all points* $x \in \mathcal{E}$ *such that the limit defined by (2.56) exists in* $\mathcal{E}$.

Exercise Let $G : \mathcal{R} \to \mathcal{L}(\mathcal{E})$ be a strongly continuous group such that $\|G(t)\| \le 1$ for all $t \in \mathbf{R}$. Let $X : \mathcal{D}(X) \subset \mathcal{E} \to \mathcal{E}$ be its infinitesimal generator. Show that the projector P defined by (2.54) is also expressed by the limit

$$Px = \lim_{t \to +\infty} \frac{1}{2t} \int_{b-t}^{b+t} G(s)x\,ds$$

for all $x \in \ker X \oplus \overline{\mathcal{R}(X)}$ and for any fixed $b \in \mathbf{R}$.

It will be shown in Corollary 2.10.8 that, if $\mathcal{E}$ is reflexive, then

$$\ker(X - i\theta\, I) \oplus \overline{\mathcal{R}(X - i\theta\, I)} = \mathcal{E}$$

for every eigenvalue $i\theta$, with $\theta \in \mathbf{R}$, of X.

Proposition 2.6.5 *If* $\mathcal{E}$ *is reflexive, and if* $i\theta$ *and* $i\tau$ *are two distinct eigenvalues of* X, *with* $\theta, \tau \in \mathbf{R}$, *the corresponding projectors* P *and* Q *are orthogonal:*

$$PQ = QP = 0.$$

Proof The projectors P and Q are expressed by

$$Px = \lim_{a \to +\infty} \frac{1}{a} \int_0^a e^{-it\theta} T(t)x\,dt$$

$$Qx = \lim_{b \to +\infty} \frac{1}{b} \int_0^b e^{-is\tau} T(s)x\,ds$$

for all $x \in \mathcal{E}$. Then

$$P\,Qx = \lim_{a \to +\infty} \frac{1}{a} \int_0^a \mathrm{e}^{-it\theta} T(t) Qx\, dt$$

$$= \lim_{a \to +\infty} \frac{1}{a} \int_0^a \mathrm{e}^{-it\theta} T(t) \times$$

$$\left(\lim_{b \to +\infty} \frac{1}{b} \int_0^b \mathrm{e}^{-is\tau} T(s) x\, ds \right) dt$$

$$= \lim_{a \to +\infty} \frac{1}{a} \int_0^a \mathrm{e}^{-it\theta} \times$$

$$\left(\lim_{b \to +\infty} \frac{1}{b} \int_0^b \mathrm{e}^{-is\tau} T(t+s) x\, ds \right) dt$$

$$= \lim_{a \to +\infty} \frac{1}{a} \int_0^a \mathrm{e}^{-it\theta} \times$$

$$\left(\lim_{b \to +\infty} \frac{1}{b} \int_t^{t+b} \mathrm{e}^{-i(s-t)\tau} T(s) x\, ds \right) dt$$

$$= \lim_{a \to +\infty} \frac{1}{a} \int_0^a \mathrm{e}^{i(\tau-\theta)t} \times$$

$$\left(\lim_{b \to +\infty} \frac{1}{b} \int_t^{t+b} \mathrm{e}^{-is\tau} T(s) x\, ds \right) dt$$

$$= \lim_{a \to +\infty} \frac{1}{a} \int_0^a \mathrm{e}^{i(\tau-\theta)t} \times$$

$$\left(\lim_{b \to +\infty} \frac{1}{t+b} \frac{t+b}{b} \int_0^{t+b} \mathrm{e}^{-is\tau} T(s) x\, ds - \right.$$

$$\left. \lim_{b \to +\infty} \frac{1}{b} \int_0^t \mathrm{e}^{-is\tau} T(s) x\, ds \right) dt$$

$$= \lim_{a \to +\infty} \frac{1}{a} \int_0^a \mathrm{e}^{i(\tau-\theta)t} dt\, Qx$$

$$= \lim_{a \to +\infty} \frac{1}{a} \frac{\mathrm{e}^{i(\tau-\theta)a} - 1}{i(\tau - \theta)} Qx$$

$$= 0.$$

A similar computation yields $Q\,Px = 0$. ∎

If $\zeta \in r\sigma(X)$, then $\ker(X - \zeta I) = \{0\}$ and $\overline{\mathcal{R}(X - \zeta I)} \neq \mathcal{E}$ (and viceversa). Hence Theorem 2.6.3 yields

Proposition 2.6.6 *If the strongly continuous semigroup* $T : \mathbf{R}_+ \to \mathcal{E}$ *is uniformly bounded and* $\mathcal{E}$ *is reflexive, then* $r\sigma(X) \cap i\mathbf{R} = \emptyset$.[4]

Let T be uniformly bounded, and let ζ_o be an isolated point of $\sigma(X)$. As in Section 1.9, there is some $r_o > 0$ such that $\Delta(\zeta_o, r_o) \backslash \{\zeta_o\} \subset r(X)$, and, for every $\zeta \in \Delta(\zeta_o, r_o) \backslash \{\zeta_o\} \subset r(X)$, $(\zeta I - X)^{-1}$ is expressed by the Laurent series (1.65), whose coefficients $A_\nu \in \mathcal{L}(\mathcal{E})$ are given by the integrals (1.66). By (1.74), the projector $R = A_{-1}$ (that in Section 1.9 was denoted by P) is such that

$$A_{-2} = (X - \zeta_o I)R.$$

Since $R A_{-\nu} = A_{-\nu} R = A_{-\nu}$ for $\nu = 1, 2, \ldots$, then

$$(I - R)(X - \zeta_o I)Rx = (I - R)A_{-2}x$$
$$= A_{-2}x - RA_{-2}x = A_{-2}x - A_{-2}x = 0$$

for all $x \in \mathcal{E}$, whence

$$(X - \zeta_o I)\mathcal{R}(R) \subset \mathcal{R}(R).$$

By Lemma 1.14.1, $X - \zeta_o I \in \mathcal{L}(\mathcal{R}(R))$. Since

$$T(t)(X - \zeta I)x = (X - \zeta I)T(t)x \quad \forall t \geq 0, \zeta \in \mathbf{C}, x \in \mathcal{D}(X),$$

if $\zeta \in r(X)$ then

$$T(t)y = (X - \zeta I)T(t)(X - \zeta I)^{-1}y,$$

i. e.,

$$(X - \zeta I)^{-1}T(t)y = T(t)(X - \zeta I)^{-1}y \quad \forall t \geq 0, \zeta \in r(X), y \in \mathcal{E}.$$

Thus, if $t \geq 0$, $y \in \mathcal{E}$, and if γ is a circle with center ζ_o and radius r, with $0 < r < r_o$, oriented counterclockwise,

$$RT(t)y = \frac{1}{2\pi i}\int_\gamma (\zeta I - X)^{-1}T(t)yd\zeta = \frac{1}{2\pi i}\int_\gamma T(t)(\zeta I - X)^{-1}yd\zeta$$
$$= T(t)\frac{1}{2\pi i}\int_\gamma (\zeta I - X)^{-1}yd\zeta = T(t)\,Ry.$$

---

[4]This result generalizes Lemma 2 and Theorem 3 of ([78], pp. 309-310).

Hence
$$T(t)\mathcal{R}(R) \subset \mathcal{R}(R) \quad \forall t \geq 0.$$

The infinitesimal generator of the uniformly continuous semigroup $T_{|\mathcal{R}(R)} : t \mapsto T(t)_{|\mathcal{R}(R)}$ is $X_{|\mathcal{R}(R)} \in \mathcal{L}(\mathcal{R}(R))$ and $T(t)_{|\mathcal{R}(R)} = \exp(tX_{|\mathcal{R}(R)})$. Since, by Lemma 1.14.3,

$$\sigma\left((X - \zeta_o I)_{|\mathcal{R}(R)}\right) = \{0\},$$

i. e.,

$$\sigma\left(X_{|\mathcal{R}(R)}\right) = \{\zeta_o\},$$

then

$$\sigma\left(T(t)_{|\mathcal{R}(R)}\right) = e^{t\zeta_o}$$

for all $t \geq 0$.

Assume now that $\zeta_o \notin p\sigma(X)$, and therefore

$$p\sigma\left(X_{|\mathcal{R}(R)}\right) = \emptyset.$$

By Theorem 2.6.3

$$\lim_{t \to +\infty} \int_0^t e^{-is\theta} T(s)x \, ds = 0$$

for all $x \in \mathcal{R}(R)$ and all $\theta \in \mathbf{R}$. The same conclusion holds when the semigroup T is replaced by the semigroup $t \mapsto e^{-rt}T(t)$, for any fixed $r > 0$, whose infinitesimal generator is $X - rI$. Thus, choosing any $\lambda \in \mathcal{E}'$, the Fourier transform of the function $f \in L^1(\mathbf{R})$ defined by

$$f(t) = \begin{cases} e^{-rt}\langle T(t)x, \lambda \rangle & \text{if } t \geq 0 \\ 0 & \text{if } t < 0 \end{cases}$$

vanishes identically. Hence $f = 0$. This contradiction proves the following theorem, [58].

Theorem 2.6.7 *If the semigroup T is uniformly bounded, and ζ_o is an isolated point of $\sigma(X)$, then $\zeta_o \in p\sigma(X)$.*

Note on nilpotent generators

Given a linear operator $X : \mathcal{D}(X) \subset \mathcal{E} \to \mathcal{E}$, and $n > 1$, the *n-th iterate* of X is the linear operator X^n on $\mathcal{E}$, defined on

$$\mathcal{D}(X^n) := \{x \in \mathcal{D}(X) : Xx \in \mathcal{D}(X), X(Xx) \in \mathcal{D}(X), \dots ,$$
$$\underbrace{X(X(\dots(Xx)\dots))}_{n-1 \text{ times}} \in \mathcal{D}(X)\}$$

by $X^n x = Xx$.

The operator X is said to be *nilpotent* if $X^n = 0$ for some $n > 1$; *nilpotent of step* n, if n is the least positive integer for which $X^n = 0$, i.e., if $X^n = 0$ and $X^{n-1} \neq 0$.

If $\zeta \in \mathbf{C} \backslash \{0\}$, then

$$\zeta^n I = \zeta^n I - X^n = (\zeta I - X)(I + \zeta X + \dots + \zeta^{n-1} X^{n-1}$$
$$= (I + \zeta X + \dots + \zeta^{n-1} X^{n-1})(\zeta I - X)$$

on $\mathcal{D}(X)$. Thus, the following lemma holds.

Lemma 2.6.8 *If X is nilpotent, then $\mathbf{C} \backslash \{0\} \in r(X)$.*

Example [1] Let $C_o([0,1))$ be the Banach space of all compactly-supported, complex-valued functions on $[0,1)$, and let $T : \mathbf{R}_+ \to \mathcal{L}(C_o([0,1)))$ be the strongly continuous semigroup defined by

$$(T(t)x)(\xi) = \begin{cases} x(\xi + t) \text{ if } \xi + t < 1 \\ 0 \qquad \text{ if } \xi + t \geq 1. \end{cases}$$

Its infinitesimal generator is $X = \frac{d}{d\xi}$ with domain $\{x \in C^1([0,1)) : x(1) = 0\}$.

Since $T(t) = 0$ for $t \geq 1$, then the bounded operator $T(t)$ is nilpotent for all $t > 0$. By Lemma 2.6.8, $\sigma(T(t)) = \{0\}$ for all $t > 0$. Thus, Theorem 2.4.3 yields $\sigma(X) = \emptyset$, i.e., $r(X) = \mathbf{C}$ [5].

Lemma 2.6.9 *Any strongly continuous semigroup whose infinitesimal generator is nilpotent is not uniformly bounded.*

Proof If X is nilpotent of step $n > 1$, then $\mathcal{R}(X^{n-1}) \subset \ker X$. Since $\mathcal{R}(X^{n-1}) \subset \mathcal{R}(X)$, (2.52) yields $\mathcal{R}(X^{n-1}) = \{0\}$, contradicting the fact that X is nilpotent of step n. ∎

---

[5] See also the examples at the end of Section 1.6 of the previous chapter.

2.7 The residual spectrum

The following theorem characterizes the residual spectrum of the infinitesimal generator X of any strongly continuous semigroup $T : \mathbf{R}_+ \to \mathcal{L}(\mathcal{E})$.

Theorem 2.7.1 *Let $t > 0$. If $\zeta \in r\sigma(X)$ and if there is no $n \in \mathbf{Z}$ for which $\zeta_n := \zeta + \frac{2\pi n}{t}i \in p\sigma(X)$, then $e^{t\zeta} \in r\sigma(T(t))$.*

If $e^{t\zeta} \in r\sigma(T(t))$ for some $t > 0$, then $\zeta_n \notin p\sigma(X)$ for every $n \in \mathbf{Z}$, and $\zeta_m \in r\sigma(X)$ for some $m \in \mathbf{Z}$.

Proof If $\zeta \in r\sigma(X)$, there exists some continuous linear form $\lambda \in \mathcal{E}' \backslash \{0\}$ such that

$$< (\zeta I - X)x, \lambda >= 0$$

for all $x \in \mathcal{D}(X)$. By (2.38),

$$< (e^{t\zeta} I - T(t))x, \lambda >= 0$$

for all $x \in \mathcal{E}$, and therefore $\overline{\mathcal{R}(e^{t\zeta} I - T(t))} \neq \mathcal{E}$.

If $e^{t\zeta} I - T(t)$ is not injective, i.e., $e^{t\zeta} \in p\sigma(T(t))$, by Theorem 2.5.2 there is some $m \in \mathbf{Z}$ for which $\zeta_m \in p\sigma(X)$. Since, by hypothesis, $\zeta_n \notin p\sigma(X)$ for every $n \in \mathbf{Z}$, then $e^{t\zeta} I - T(t)$ must be injective, and therefore $e^{t\zeta} \in r\sigma(T(t))$.

Now, let $e^{t\zeta} \in r\sigma(T(t))$ for some $t > 0$. If there is $m \in \mathbf{Z}$ for which $\zeta_m \in p\sigma(X)$, then by Theorem 2.5.2, $e^{t\zeta_m} = e^{t\zeta} \in p\sigma(T(t))$. Thus, $\zeta_n \notin p\sigma(X)$ for all $n \in \mathbf{Z}$, and we are left to prove that there is some $m \in \mathbf{Z}$ for which $\zeta_m \in r\sigma(X)$, or - equivalently - that $\zeta_m \notin r(X) \cup c\sigma(X)$. Suppose then that $\zeta_n \in r(X) \cup c\sigma(X)$ for all $n \in \mathbf{Z}$. By (2.39),

$$\begin{aligned}
(e^{t\zeta} I - T(t))x &= (e^{t\zeta_n} I - T(t))x \\
&= Y_{\zeta_n}(t)(\zeta_n I - X)x \quad \forall x \in \mathcal{D}(X) \; \forall n \in \mathbf{Z}. \quad (2.57)
\end{aligned}$$

Since $e^{t\zeta} \in r\sigma(T(t))$, there is a linear, closed, proper subspace $\mathcal{F}$ of $\mathcal{E}$ such that

$$(e^{t\zeta} I - T(t))\mathcal{E} \subset \mathcal{F}.$$

Since $\zeta_n \in r(X) \cup c\sigma(X)$, $\mathcal{R}(\zeta_n I - X) = (\zeta_n I - X)\mathcal{D}(X)$ is dense in $\mathcal{E}$. By (2.57), $\mathcal{R}(Y_{\zeta_n}(t)) \subset \mathcal{F}$. Consider the continuous function

$[0, 1] \to \mathcal{E}$, defined, for $x \in \mathcal{E}$, by $s \mapsto e^{-s\zeta}T(s)x$. By (2.36), its n-th Fourier coefficient is given by the integral

$$\frac{1}{t}\int_0^t e^{-\frac{2\pi n}{t}is}e^{-s\zeta}T(s)x\,ds = \frac{1}{t}\int_0^t e^{-s\zeta_n}T(s)x\,ds$$

$$= \frac{e^{-t\zeta_n}}{t}\int_0^t e^{(t-s)\zeta_n}T(s)x\,ds$$

$$= \frac{e^{-t\zeta}}{t}Y_{\zeta_n}(t)x,$$

and the Fourier series

$$\sum_{-\infty}^{+\infty} Y_{\zeta_n}(t)x$$

converges C.1 to $e^{-s\zeta}T(s)x$ for $0 < s < t$. Since all its summands are contained in $\mathcal{F}$, then $e^{-s\zeta}T(s)x \in \mathcal{F}$ for every $x \in \mathcal{E}$. Hence

$$\lim_{s\downarrow 0} e^{-s\zeta}T(s)x \in \mathcal{F},$$

contradicting the fact that $\mathcal{F} = \overline{\mathcal{F}} \neq \mathcal{E}$. ∎

The information concerning $c\sigma(X)$ is less exhaustive.

Theorem 2.7.2 *Let $t > 0$. If $\zeta \in c\sigma(X)$ and if there is no $n \in \mathbf{Z}$ for which $\zeta_n = \zeta + \frac{2\pi n}{t}i \in p\sigma(X) \cup r\sigma(X)$, then $e^{t\zeta} \in c\sigma(T(t))$.*

Proof By Theorem 2.4.3, $e^{t\zeta} \in \sigma(T(t))$. If $e^{t\zeta} \in r\sigma(T(t))$, then, by Theorem 2.7.1, $\zeta_m \in r\sigma(X)$ for some $m \in \mathbf{Z}$. Similarly, Theorem 2.5.2 implies that, if $e^{t\zeta} \in p\sigma(T(t))$, then $\zeta_m \in p\sigma(X)$ for some $m \in \mathbf{Z}$. ∎

So far, we have investigated the behaviour of $\sigma(T(t))$ outside 0.

Let now $0 \in p\sigma(T(t))$ for some $t > 0$ and let $x \in \mathcal{E}\{0\}$ be such that $T(t)x = 0$. If $s > t$,

$$T(s)x = T(s - t + t)x = T(s - t)T(t)x = 0,$$

and therefore $0 \in p\sigma(T(s))$. It follows from

$$0 = T(t)x = T(\frac{t}{2} + \frac{t}{2})x = T(\frac{t}{2})T(\frac{t}{2})x$$

that $0 \in p\sigma(T(\frac{t}{2}))$, and thus $0 \in p\sigma T(\frac{t}{2^n}))$ for $n = 1, 2, \ldots$. In conclusion, $0 \in p\sigma(T(s))$ for all $s > 0$.

If $0 \in r\sigma(T(t))$ for some $t > 0$, there is a linear, closed, proper subspace $\mathcal{F}$ of $\mathcal{E}$ for which $T(t)\mathcal{E} \subset \mathcal{F}$. If $s > t$

$$T(s)\mathcal{E} = T(t + s - t)\mathcal{E} = T(t)T(s - t)\mathcal{E} \subset T(t)\mathcal{E} \subset \mathcal{F},$$

and therefore $0 \in r\sigma(T(s))$. Since

$$T(\frac{t}{2})T(\frac{t}{2})\mathcal{E} = T(t)\mathcal{E} \subset \mathcal{F},$$

if $\overline{\mathcal{R}(T(\frac{t}{2}))} = \mathcal{E}$, the continuity of $T(\frac{t}{2})$ and the fact that $\mathcal{F}$ is closed imply that $T(\frac{t}{2})\mathcal{E} \subset \mathcal{F}$. This contradiction shows that $T(\frac{t}{2})\mathcal{E}$ is not dense. On the other hand, $0 \notin p\sigma(T(\frac{t}{2}))$ because this would imply that $0 \in p\sigma(T(t))$. Hence $0 \in r\sigma(T(\frac{t}{2}))$, and iteration of this argument shows that $0 \in r\sigma(T(\frac{t}{2^n}))$ for $n = 1, 2, \ldots$. In conclusion, $0 \in r\sigma(T(s))$ for all $s > 0$.

If $0 \in r(T(t))$ for some $t > 0$, that is, if $T(t)$ is a continuous isomorphism of $\mathcal{E}$ onto $\mathcal{E}$, also $T(nt) = T(t)^n$ is a continuous isomorphism of $\mathcal{E}$ onto $\mathcal{E}$ for $n = 1, 2, \ldots$. If $r > 0$, $s > 0$ are such that $r + s = nt$, it follows from

$$T(r)T(s) = T(s)T(r) = T(nt)$$

that the continuous linear map $T(s)$ is injective and surjective, and therefore - by the Banach theorem - is a continuous isomorphism of $\mathcal{E}$ onto $\mathcal{E}$ whenever $s \in [0, nt]$. Thus, $0 \in r(T(s))$ for all $s > 0$.

All that proves the following

Theorem 2.7.3 *If, for some $t > 0$, 0 is contained in one of the disjoint sets $p\sigma(T(t))$, $r\sigma(T(t))$, $c\sigma(T(t))$, $r(T(t))$, then 0 is contained, respectively, in $p\sigma(T(s))$, $r\sigma(T(s))$, $c\sigma(T(s))$, $r(T(s))$ for all $s > 0$.*

Corollary 2.7.4 *If $T(t)$ is invertible, or continuously invertible for some $t > 0$, $T(s)$ is invertible or, respectively, continuously invertible for all $s > 0$.*

2.8 Eventually uniformly continuous semigroups

Let $T : \mathbf{R}_+ \to \mathcal{L}(\mathcal{E})$ be a strongly continuous semigroup, and let $\mathcal{A}$ be the maximal abelian subalgebra of $\mathcal{L}(\mathcal{E})$ containing $T(t)$ for all $t \in \mathbf{R}_+$. The existence of $\mathcal{A}$ follows from the Zorn lemma. The algebra $\mathcal{A}$ is a commutative Banach algebra for the norm topology.

Lemma 2.8.1 *The subalgebra $\mathcal{A}$ is closed in $\mathcal{L}(\mathcal{E})$ for the strong topology.*

Proof Let $C \in \mathcal{L}(\mathcal{E})$ be contained in the closure of $\mathcal{A}$ for the strong topology. This means that, for every choice of $\epsilon > 0$, of $N \in \mathbf{N}^*$ and of N points $x_1, x_2, \dots, x_N$ in $\mathcal{E}$, there exists some $A \in \mathcal{A}$ for which

$$\|Cx_n - Ax_n\| < \epsilon \text{ for } n = 1, 2, \dots, N. \qquad (2.58)$$

For all $L \in \mathcal{A}$, any $x \in \mathcal{E}$ and any $A \in \mathcal{A}$,

$$
\begin{aligned}
(CL - LC)x &= CLx - ALx + LAx - LCx \\
&= (C - A)Lx + L(A - C)x,
\end{aligned}
$$

and therefore

$$\|(CL - LC)x\| \le \|(C - A)Lx\| + \|L\| \, \|(A - C)x\|.$$

If $A \in \mathcal{A}$ satisfies (2.58) with $N = 2$, $x_1 = x$ and $x_2 = Lx$, then

$$\|(CL - LC)x\| \le (1 + \|L\|)\epsilon,$$

and - since $\epsilon > 0$ is arbitrary - that implies that $CLx = LCx$ for all $L \in \mathcal{A}$ and all $x \in \mathcal{E}$, i. e., $C \in \mathcal{A}$. ∎

Lemma 2.8.2 *For all $\zeta \in r(X)$, $(\zeta I - X)^{-1} \in \mathcal{A}$.*

Proof If $t \ge 0$ and $x \in \mathcal{D}(X)$,

$$(X - \zeta I)T(t)x = T(t)(X - \zeta I)x,$$

and therefore, setting $y = (X - \zeta I)x$,

$$T(t)(X - \zeta I)^{-1}y = (X - \zeta I)^{-1}T(t)y$$

for all $y \in \mathcal{E}$ because $(X - \zeta I)(\mathcal{D}(X)) = \mathcal{E}$. ∎

If $C \in \mathcal{A}$ is invertible in the Banach algebra $\mathcal{L}(\mathcal{E})$, then, for all $A \in \mathcal{A}$,

$$C(C^{-1}A - AC^{-1})C = AC - CA = 0,$$

i.e., $C^{-1} \in \mathcal{A}$. As a consequence, for all $A \in \mathcal{A}$, $\sigma(A)$ coincides with the spectrum of A in the Banach algebra $\mathcal{A}$.

Let $\Xi \subset \mathcal{A}'$ be the space of characters of $\mathcal{A}$ with the Gelfand topology. For any $A \in \mathcal{A}$,

$$\sigma(A) = \{\chi(A) : \chi \in \Xi\}.$$

The semigroup T will be said to be *eventually uniformly continuous* or also *eventually norm continuous* if there is some $b \in \mathbf{R}_+$ such that $T_{|[b,+\infty)} : [b,+\infty) \to \mathcal{L}(\mathcal{E})$ is a continuous function for the norm topology. If $b = 0$, T is a uniformly continuous semigroup.

We will show now that, if T is eventually uniformly continuous, the inclusion in (2.41) is an equality.

Theorem 2.8.3 *If T is eventually uniformly continuous, then*

$$\mathrm{e}^{t\sigma(X)} = \sigma(T(t))\backslash\{0\}$$

for all $t \in \mathbf{R}_+$.

Proof Let $b \geq 0$ be such that $T_{|[b,+\infty)}$ be norm continuous, and let $t > 0$. For any $\tau \in \sigma(T(t))\backslash\{0\}$ there is some $\chi \in \Xi$ such that $\chi(T(t)) = \tau$. For $s_1 \geq 0$ and $s_2 \geq 0$,

$$\chi(T(s_1 + s_2)) = \chi(T(s_1)\,T(s_2)) = \chi(T(s_1))\,\chi(T(s_2)),$$

and furthermore $(\chi \circ T)_{|[b,+\infty)}$ is a continuous function. Consequently, there is some $\kappa \in \mathbf{C}$ such that $\chi(T(s)) = \mathrm{e}^{s\kappa}$ for all $s \geq 0$ [6]

---

[6]Letting $\varphi(s) = \chi(T(s))$, then $\varphi(0)\,\varphi(t) = \varphi(0+t) = \varphi(t) = \tau \neq 0$, and therefore $\varphi(0) = 1$. Furthermore, $\varphi(nt) = \varphi(t)^n = \tau^n$ for $n = 1, 2, \ldots$. If $0 \leq s \leq nt$, then

Let $X : \mathcal{D}(X) \subset \mathcal{E} \to \mathcal{E}$ be the infinitesimal generator of T, and let $M \geq 1$ and $a \in \mathbf{R}$ be such that (2.6) holds. If $\Re\zeta > a$, $(\zeta I - X)^{-1}$ is espressed by (2.34). Hence

$$T(b)(\zeta I - X)^{-1} = \int_0^{+\infty} e^{-s\zeta} T(b+s) ds,$$

where, in view of our hypothesis, now the integral is norm-convergent. Thus, by Lemma 2.8.2, if furthermore $\Re\zeta > \Re\kappa$,

$$e^{b\kappa} \chi((\zeta I - X)^{-1}) = \chi(T(b))\, \chi((\zeta I - X)^{-1}) =$$
$$\chi((T(b)(\zeta I - X)^{-1}) = \int_0^{+\infty} e^{-s\zeta} \chi(T(b+s)) ds =$$
$$\int_0^{+\infty} e^{-s\zeta} e^{(b+s)\kappa} ds = e^{b\kappa} \int_0^{+\infty} e^{s(\kappa-\zeta)} ds = \frac{e^{b\kappa}}{\zeta - \kappa},$$

whence

$$\chi((\zeta I - X)^{-1}) = \frac{1}{\zeta - \kappa}.$$

Hence $\frac{1}{\zeta-\kappa} \in \sigma((\zeta I - X)^{-1})$. As a consequence, $\kappa \in \sigma(X)$ and $e^{t\kappa} = \tau$. ∎

Theorem 2.8.3 entails Theorem 2.4.4 as a corollary.

The following theorem provides a sufficient condition for the semigroup T to be eventually norm continuous.

Theorem 2.8.4 *If there is $b > 0$ for which $T(b)$ is a compact operator, then $T_{|[b,+\infty)}$ is norm continuous.*

Proof The image by $T(b)$ of the open unit disc of $\mathcal{E}$ has compact closure, K. Let $\epsilon > 0$. Since $T : \mathbf{R}_+ \times \mathcal{E} \to \mathcal{E}$ is continuous, there is some $\delta > 0$ such that, for all $x \in K$ and all $t \in [0, \delta)$,

$$\|T(t)x - x\| < \epsilon.$$

$\varphi(nt) = \varphi(nt - s)\,\varphi(s) \neq 0$, whence $\varphi(s) \neq 0$ for all $s \in \mathbf{R}_+$. Since

$$\varphi(s) = \frac{\varphi(b+s)}{\varphi(b)},$$

the scalar-valued function φ is continuous on $\mathbf{R}_+$. Then, as is well known, there is some $\kappa \in \mathbf{C}$ such that $\varphi(s) = e^{s\kappa}$ for all $s \in \mathbf{R}_+$.

If $c > b$, $t \in [c, c + \delta)$ and $\|x\| \le 1$, then

$$
\begin{aligned}
T(t)x - T(c)x &= T(t)x - T(c - b)T(b)x \\
&= T(t - c + c - b + b)x - T(c - b)T(b)x \\
&= T(c - b)T(t - c)T(b)x - T(c - b)T(b)x \\
&= T(c - b)(T(t - c) - I)T(b)x,
\end{aligned}
$$

and therefore

$$
\begin{aligned}
\|T(t)x - T(c)x\| &\le \|T(c - b)\|\, \|(T(t - c) - I)T(b)x\| \\
&\le \|T(c - b)\|\, \epsilon.
\end{aligned}
$$

Thus,

$$
\|T(t) - T(c)\| \le \|T(c - b)\|\, \epsilon
$$

whenever $t \in [c, c + \delta)$. ∎

Theorem 2.8.5 *If $T(b)$ is a compact operator for some $b > 0$, for every choice of $\alpha > 0$, $\mathcal{E}$ splits as a direct sum $\mathcal{E} = \mathcal{E}_0 \oplus \mathcal{E}_1$ of two closed subspaces $\mathcal{E}_0$ and $\mathcal{E}_1$, in such a way that:*
$T(t)\mathcal{E}_0 \subset \mathcal{E}_0$ and $T(t)\mathcal{E}_1 \subset \mathcal{E}_1$ for all $t \in \mathbf{R}_+$;

$$
\dim_{\mathbf{C}} \mathcal{E}_0 < \infty;
$$

$$
\lim_{t \to +\infty} e^{\alpha t} \|T(t)_{|\mathcal{E}_1}\| = 0.
$$

Furthermore, $\sigma(X) = p\sigma(X)$, and the eigenvalues of X are a (finite or) countable set, and have finite algebraic multiplicity. If $\dim_{\mathbf{C}} \mathcal{E}_1 = \infty$, then

$$
\sigma(T(t)) = \{0\} \cup e^{t\sigma(X)} \quad \forall t > 0. \tag{2.59}
$$

Proof If $\dim_{\mathbf{C}} \mathcal{E} = \infty$, the compactness of $T(b)$ implies that $0 \in \sigma(T(b))$ and therefore, by Theorem 2.7.3, $0 \in \sigma(T(t))$ for all $t > 0$. Theorems 2.8.3 and 2.8.4 yield (2.59).

Choose any $\alpha > 0$, and let $\mathcal{E}_1$ be the closed subspace of $\mathcal{E}$ spanned by the eigenspaces corresponding to the eigenvalues of $T(b)$ contained in the open disc $\Delta(0, e^{-b\alpha})$ with center 0 and radius $e^{-b\alpha}$ in $\mathbf{C}$; $\mathcal{E}_0$ will be the (finite dimensional) space spanned by the eigenspaces corresponding to the points of $\sigma(T(b))\backslash\Delta(0, \mathrm{e}^{-b\alpha})$.

Since $T(t)T(b) = T(b)T(t)$, $\mathcal{E}_0$ and $\mathcal{E}_1$ are invariant under the action of $T(t)$ for all $t \geq 0$. Let $S : \mathbf{R}_+ \to \mathcal{L}(\mathcal{E}_1)$ be the strongly continuous semigroup defined by $S(t) = T(t)_{|\mathcal{E}_1}$. By Theorem 2.1.4, there is a constant $\gamma \in \mathbf{R}$ such that

$$\rho(S(t)) \leq \mathrm{e}^{-t\gamma} \ \forall t \geq 0.$$

On the other hand, since $\sigma(S(b))$ is compact in $\Delta(0, \mathrm{e}^{-b\alpha})$, then

$$\rho(S(b)) \leq \mathrm{e}^{-b\alpha},$$

and therefore $-\gamma < -\alpha$, i. e., $\alpha < \gamma$. Hence, there exist real constants $N \geq 1$ and β, with $-\gamma < -\beta < -\alpha$ such that

$$\|S(t)\| \leq N\mathrm{e}^{-t\beta} \ \forall t \geq 0,$$

$$\lim_{t \to +\infty} \mathrm{e}^{t\alpha}\|S(t)\| \leq N \lim_{t \to +\infty} \mathrm{e}^{t(\alpha-\beta)} = 0.$$

The fact that $\sigma(X) = p\sigma(X)$ and that $p\sigma(X)$ consists of eigenvalues with finite algebraic multiplicity follows from Theorem 2.5.2 and from the compactness of $T(b)$. ∎

2.9 Characterization of the infinitesimal generator

In the first part of this section we shall characterize tha infinitesimal generator of a strongly continuous semigroup, proving the following *Feller-Miyadera-Phillips theorem.*

Theorem 2.9.1 *A linear, closed, densely defined operator $X : \mathcal{D}(X) \subset \mathcal{E} \to \mathcal{E}$ is the infinitesimal generator of a strongly continuous semigroup $\mathbf{R}_+ \to \mathcal{E}$ if, and only if, there exist two real constants a and $M \geq 1$ such that the open half line $(a, +\infty) = \{\zeta \in \mathbf{R} : \zeta > a\}$ is contained in $r(X)$ and*

$$\|(\zeta I - X)^{-n}\| \leq \frac{M}{(\zeta - a)^n} \ \forall \zeta \in (a, +\infty), \ \forall n = 1, 2, \dots. \quad (2.60)$$

If these conditions are satisfied, and if T is the semigroup generated by X, then (2.6) holds.

Viceversa, if (2.6) holds, then $(a, +\infty) \subset r(X)$ and (2.60) holds for all $\zeta \in (a, +\infty)$ and $n = 1, 2, \ldots$.

A part of the theorem is contained in Proposition 2.4.2. We are left to establish the existence of the semigroup. Thus, we will assume that the hypotheses on X are fulfilled and we shall construct the semigroup T by an approximation procedure.

If $\zeta \in (a, +\infty)$, the identity

$$X(\zeta I - X)^{-1} = -I + \zeta(\zeta I - X)^{-1} \tag{2.61}$$

implies that $X(\zeta I - X)^{-1} \in \mathcal{L}(\mathcal{E})$ whenever $\zeta \in r(X)$.

If $x \in \mathcal{D}(X)$, then

$$(\zeta I - X)^{-1} X x = -x + \zeta(\zeta I - X)^{-1} x,$$

that is,

$$(\zeta I - X)^{-1} X = -I + \zeta(\zeta I - X)^{-1} \text{ on } \mathcal{D}(X).$$

Hence,

$$X(\zeta I - X)^{-1} = (\zeta I - X)^{-1} X \text{ on } \mathcal{E}$$

for all $\zeta \in (a, +\infty)$.

We prove now that

$$\lim_{\zeta \to +\infty} X(\zeta I - X)^{-1} x = 0 \quad \forall x \in \mathcal{E}. \tag{2.62}$$

If $\zeta \in (a, +\infty)$ and $x \in \mathcal{D}(X)$,

$$\|X(\zeta I - X)^{-1} x\| = \|(\zeta I - X) X x\|$$
$$\leq \frac{M}{\zeta - a} \|X x\|,$$

and this inequality implies (2.62) when $x \in \mathcal{D}(X)$.

We still have to show that (2.62) holds on $\mathcal{E}$.

By (2.61)

$$\|X(\zeta I - X)^{-1}\| \leq \|\zeta(\zeta I - X)^{-1}\| + 1$$
$$\leq \frac{M|\zeta|}{\zeta - a} + 1,$$

and therefore $\|X(\zeta I - X)^{-1}\|$ is bounded by a finite constant $k > 0$ on $(a + 1, +\infty)$.

For $x \in \mathcal{E}$ and $y \in \mathcal{D}(X)$

$$X(\zeta I - X)^{-1}x = X(\zeta I - X)^{-1}(x - y) + X(\zeta I - X)^{-1}y.$$

Let $\epsilon > 0$. Since $\mathcal{D}(X)$ is dense in $\mathcal{E}$, for all $x \in \mathcal{E}$ there is some $y \in \mathcal{D}(X)$ with $\|x - y\| < \epsilon$. Let $b \in (a, +\infty)$ be such that

$$\|X(\zeta I - X)^{-1}y\| < \epsilon \quad \forall \zeta \in (b, +\infty).$$

If $\zeta > \max(a + 1, b)$,

$$\|X(\zeta I - X)^{-1}x\| < (k + 1)\epsilon.$$

Since $\epsilon > 0$ is arbitrary (2.62) is proved.

For $\zeta \in (a, +\infty)$, the linear operator

$$X_\zeta : \zeta X(\zeta I - X)^{-1} \in \mathcal{L}(\mathcal{E})$$

is called a *Yosida approximation* of X. A reason for this terminology is offered by the following equality.

$$\lim_{\zeta \to +\infty} X_\zeta x = Xx \quad \forall x \in \mathcal{D}(X). \tag{2.63}$$

In order to establish this equality, note first that, for all $\tau \in (a, +\infty)$, there is some $y \in \mathcal{E}$ for which $x = (\tau I - X)^{-1}y$, and therefore, if $\zeta \in (a, +\infty)$,

$$(X_\zeta - X)x = \zeta X(\zeta I - X)^{-1}(\tau I - X)^{-1}y - X(\tau I - X)^{-1}y.$$

Since

$$(\zeta I - X)^{-1} - (\tau I - X)^{-1} =$$
$$(\zeta I - X)^{-1}(\tau I - X - (\zeta I - X))(\tau I - X)^{-1}$$
$$= (\tau - \zeta)(\zeta I - X)^{-1}(\tau I - X)^{-1}$$

for all $\zeta, \tau \in r(X)$, then, if $\zeta, \tau \in (a, +\infty)$, and $\zeta \neq \tau$,

$$(X_\zeta - X)x = \frac{\zeta}{\tau - \zeta} X \left((\zeta I - X)^{-1} - (\tau I - X)^{-1} \right) y -$$
$$- X(\tau I - X)^{-1}y$$
$$= \frac{\zeta}{\tau - \zeta} X(\zeta I - X)^{-1}y - (\frac{\zeta}{\tau - \zeta} + 1)X(\tau I - X)^{-1}y$$
$$= \frac{\zeta}{\tau - \zeta} X(\zeta I - X)^{-1}y - \frac{\tau}{\tau - \zeta} X(\tau I - X)^{-1}y.$$

Hence

$$\|(X_\zeta - X)x\| \le \frac{|\zeta|}{|\tau - \zeta|}\|X(\zeta I - X)^{-1}y\|$$
$$+ \frac{|\tau|}{|\tau - \zeta|}\|X(\tau I - X)^{-1}y\|.$$

By (2.62), $\lim_{\zeta \to +\infty}\|(X_\zeta - X)x\| = 0$, and that proves (2.63).

Remark The equalities (2.62) and (2.63) hold for every linear, closed, densely defined operator X for which $(a, +\infty) \subset r(X)$ for some $a \in \mathbf{R}$ and for which (2.60) holds for $n = 1$ and for all $\zeta \in (a, +\infty)$.

Let $a \ge 0$. For $\zeta \in (a, +\infty)$ let T^ζ be the uniformly continuous semigroup generated by X_ζ. By (2.61), if $t \ge 0$, $T^\zeta(t)$ is expressed by

$$T^\zeta(t) = \sum_{n=0}^{+\infty} \frac{t^n}{n!}X_\zeta{}^n$$
$$= \sum_{n=0}^{+\infty} \frac{t^n}{n!}\left(-\zeta I + \zeta^2(\zeta I - X)^{-1}\right)^n$$
$$= \sum_{n=0}^{+\infty} \frac{(-t\zeta)^n}{n!} \sum_{n=0}^{+\infty} \frac{t^n\zeta^{2n}}{n!}(\zeta I - X)^{-n}$$
$$= e^{-t\zeta}\exp\left(t\zeta^2(\zeta I - X)^{-1}\right).$$

It follows that, if $\zeta \in (a, +\infty)$,

$$\|T^\zeta(t)\| \le e^{-t\zeta}Me^{\frac{t\zeta^2}{\zeta - a}} = Me^{\frac{at\zeta}{\zeta - a}}.$$

Therefore

$$\limsup_{\zeta \to +\infty}\|T^\zeta(t)\| \le Me^{at}, \tag{2.64}$$

and [7]

$$\|T^\zeta(t)\| \le Me^{2at} \ \forall \zeta \in (2a, +\infty) \ \forall t \in \mathbf{R}_+. \tag{2.65}$$

---

[7]The function $\zeta \mapsto \frac{\zeta}{\zeta - a}$ on $(a, +\infty)$ is constant if $a = 0$ and decreasing if $a > 0$. Its value for $\zeta = 2a > 0$ is 2.

Lemma 2.9.2 *For all $x \in \mathcal{E}$, the limit $\lim_{\zeta \to +\infty} T^\zeta(t)x$ exists in $\mathcal{E}$; the convergence is uniform when t varies on compact subsets of $\mathbf{R}_+$.*

Proof I. Let $x \in \mathcal{D}(X)$. If $\zeta, \tau \in (2a, +\infty)$, $t > 0$, $s \in (0, t]$, then

$$\frac{d}{ds} \left(T^\zeta(t-s) T^\tau(s) \right) x =$$

$$\left(\frac{d}{ds} T^\zeta(t-s) \right) T^\tau(s)x + T^\zeta(t-s) \left(\frac{d}{ds} T^\tau(s) \right) x$$
$$= -T^\zeta(t-s) X_\zeta T^\tau(s)x + T^\zeta(t-s) X_\tau T^\tau(s)x$$
$$= T^\zeta(t-s)(-X_\zeta + X_\tau) T^\tau(s)x$$
$$= T^\zeta(t-s) T^\tau(s)(-X_\zeta + X_\tau)x,$$

whence, by (2.65),

$$\left\| \frac{d}{ds} \left(T^\zeta(t-s) T^\tau(s) \right) x \right\| \leq M^2 e^{2a(t-s+s)} \|(-X_\zeta + X_\tau)x\|$$
$$= M^2 e^{2at} \|(-X_\zeta + X_\tau)x\|.$$

Integration from 0 to t yields

$$\left\| \left(T^\zeta(t) - T^\tau(t) \right) x \right\| = \left\| \int_0^t \frac{d}{ds} \left(T^\zeta(t-s) T^\tau(s) \right) x \, ds \right\|$$
$$\leq \int_0^t \left\| \frac{d}{ds} \left(T^\zeta(t-s) T^\tau(s) \right) x \right\| ds$$
$$\leq M^2 t \, e^{2at} \|(-X_\zeta + X_\tau)x\|.$$

Thus, (2.63) yields

$$\lim_{\zeta, \tau \to +\infty} \|T^\zeta(t)x - T^\tau(t)x\| = 0$$

for all $x \in \mathcal{D}(X)$, uniformly when t varies on compact subsets of $\mathbf{R}_+$.

II. Let $x \in \mathcal{E}$, and let J be a closed, bounded interval in $\mathbf{R}_+$. For any $\epsilon > 0$ there is some $y \in \mathcal{D}(X)$ for which $\|x - y\| < \epsilon$. In view of I., there exists $b > 2a$ such that, whenever $\zeta > b$ and $\tau > b$,

$$\|T^\zeta(t)y - T^\tau(t)y\| < \epsilon \tag{2.66}$$

for all $t \in J$. Since

$$\left\| \left(T^\zeta(t) - T^\tau(t)\right) x \right\| \le$$
$$\left\| \left(T^\zeta(t) - T^\tau(t)\right) (x - y) \right\| + \left\| \left(T^\zeta(t) - T^\tau(t)\right) y \right\|$$
$$\le \left(\|T^\zeta(t)\| + \|T^\tau(t)\| \right) \|x - y\| + \left\| \left(T^\zeta(t) - T^\tau(t)\right) y \right\|,$$

(2.65) and (2.66) yield

$$\left\| \left(T^\zeta(t) - T^\tau(t)\right) x \right\| \le 2Me^{2at} \|x - y\| + \epsilon$$
$$< (2M \max\{e^{2at} : t \in J\} + 1)\epsilon$$

for all $t \in J$, $\zeta, \tau \in (b, +\infty)$. ∎

Let $T(t)$ be the linear operator defined at $x \in \mathcal{E}$ by

$$T(t)x = \lim_{\zeta \to +\infty} T^\zeta(t)x.$$

By (2.64),

$$\|T(t)x\| \le \limsup_{\zeta \to +\infty} \|T^\zeta(t)x\|$$
$$\le Me^{at} \|x\|,$$

and that proves that $T(t) \in \mathcal{L}(\mathcal{E})$ and (2.6) holds for all $t \ge 0$. Since, moreover,

$$T(0) = I, \quad T(t + s) = T(t)T(s)$$

for all $t \ge 0$ and $s \ge 0$, $T : \mathbf{R}_+ \to \mathcal{L}(\mathcal{E})$ is a semigroup. To show that it is strongly continuous we need prove that (2.3) holds.

Note first that, for $x \in \mathcal{E}$,

$$\|(T(t) - I)x\| \le \|(T(t) - T^\zeta(t))x\| + \|(T^\zeta(t) - I)x\|.$$

By Lemma 2.9.2, for any $\epsilon > 0$ there is some $b > 0$ such that

$$\|(T(t) - T^\zeta(t))x\| < \epsilon$$

for all $\zeta \in (b, +\infty)$ and all $t \in [0, 1]$. For any $\zeta \in (b, +\infty)$, there exists some $\delta > 0$ such that

$$\|(T^\zeta(t) - I)x\| < \epsilon$$

whenever $0 \le t < \delta$. Thus, if $t \in [0, \delta)$,

$$\|(T(t) - I)x\| < 2\epsilon,$$

and (2.3) has been shown to hold.

Since

$$(T^\zeta(t) - I)x = \int_0^t T^\zeta(s) X_\zeta x \, ds$$

for all $x \in \mathcal{E}$. If $x \in \mathcal{D}(X)$,

$$\left\| T(t)x - x - \int_0^t T(s) X x \, ds \right\| \le$$

$$\|(T(t) - T^\zeta(t))x\| + \left\| \int_0^t (T(s)X - T^\zeta(s)X_\zeta)x \, ds \right\|$$

$$\le \|(T(t) - T^\zeta(t))x\| + \int_0^t \|(T(s)X - T^\zeta(s)X_\zeta)x\| ds,$$

and, by (2.65), if $\zeta \in [2a, +\infty)$ and $s \in [0, t]$,

$$\|(T(s)X - T^\zeta(s)X_\zeta)x\| =$$
$$\|(T^\zeta(s)X_\zeta - T^\zeta(s)X + T^\zeta(s)X - T(s)X)x\|$$
$$\le \|T^\zeta(s)(X_\zeta - X)x\| + \|(T^\zeta(s) - T(s))Xx\|$$
$$\le \|T^\zeta(s)\| \|X_\zeta - X)x\| + \|(T^\zeta(s) - T(s))Xx\|$$
$$\le Me^{2as}\|X_\zeta - X)x\| + \|(T^\zeta(s) - T(s))Xx\|$$
$$\le Me^{2at}\|X_\zeta - X)x\| + \|(T^\zeta(s) - T(s))Xx\|.$$

For all $\epsilon > 0$ there is some $b \in [2a, +\infty)$ such that, whenever $\zeta \in (b, +\infty)$, $\|(X_\zeta - X)x\| < \epsilon$, by (2.63), and

$$\sup\{\|(T^\zeta(s) - T(s))Xx\| : 0 \le s \le t\} < \epsilon,$$

$$\|(T(t) - T^\zeta(t))x\| < \epsilon,$$

by Lemma 2.9.2. Hence, if $x \in \mathcal{D}(X)$,

$$\left\| T(t)x - x - \int_0^t T(s) X x \, ds \right\| < (1 + t + e^{2at})\epsilon.$$

Since $\epsilon > 0$ is arbitrary, then

$$(T(t) - I)x = \int_0^t T^\zeta(s)Xx\,ds$$

for all $x \in \mathcal{D}(X)$, and therefore

$$\lim_{t \downarrow 0} \frac{1}{t}(T(t) - I)x = Xx \quad \forall x \in \mathcal{D}(X)$$

because the function $t \mapsto T(t)Xx$ is continuous.

Hence the infinitesimal generator Y of T is an extension of X. Since X is closed, and $(a, +\infty) \subset r(X)$, if Y were a proper extension of X, then $(a, +\infty) \subset p\sigma(Y) \subset \sigma(Y)$ contradicting Theorem 2.4.1. Thus $X = Y$, i.e., X is the infinitesimal generator of T.

The restriction $a \geq 0$ in the preceding argument can be removed by replacing T by the strongly continuous semigroup $t \mapsto e^{ct}T(t)$, for a suitable choice of the real constant c. In fact, first of all, the infinitesimal generator of this new semigroup is $X + cI$, with domain $\mathcal{D}(X)$, and (2.6) is replaced by

$$\|e^{ct}T(t)\| \leq Me^{(a+c)t} \quad \forall t \in \mathbf{R}_+.$$

Furthermore, if $b > a$, and if (2.60) holds for $\zeta \in (a, +\infty)$, then *a fortiori* it holds with a replaced by b and $\zeta \in (b, +\infty)$.

The proof of Theorem 2.9.1 is complete.

Theorem 2.2.7 and Theorem 2.9.1 yield

Theorem 2.9.3 *A closed, densely defined, linear operator X is the infinitesimal generator of a strongly continuous group $T : \mathbf{R} \to \mathcal{L}(\mathcal{E})$ if, and only if, there exist two finite real constants $M \geq 1$ and $a \geq 0$ such that $C : (-\infty, -a) \cup (a, +\infty) \subset r(X)$ and*

$$\|(\zeta I - X)^{-n}\| \leq \frac{M}{(|\zeta| - a)^n}$$

for all $\zeta \in \mathbf{R}$ and for $n = 1, 2, \ldots$. If these conditions are all satisfied, then

$$\|T(t)\| \leq Me^{a|t|} \quad \forall t \in \mathbf{R},$$

and

$$\sigma(X) \subset \{\zeta \in \mathbf{C} : |\Re\zeta| \leq a\}.$$

Let $T : \mathbf{R}_+ \to \mathcal{L}(\mathcal{E})$ be a strongly continuous semigroup of linear contractions (i.e.,

$$\|T(t)\| \leq 1 \ \forall t \in \mathbf{R}_+), \tag{2.67}$$

and let X be its infinitesimal generator. In view of (2.67), Proposition 2.4.2 implies that

$$\mathbf{R}_+^* \subset r(X) \tag{2.68}$$

and

$$\|(\zeta I - X)^{-1}\| \leq \frac{1}{\zeta} \ \forall \zeta \in \mathbf{R}_+^*. \tag{2.69}$$

Viceversa, if X is closed, densely defined, linear operator satisfying (2.68) and (2.69), this latter inequality implies that

$$\|(\zeta I - X)^{-n}\| = \|((\zeta I - X)^{-1})^n\|$$
$$\leq \|(\zeta I - X)^{-1}\|^n \leq \frac{1}{\zeta^n}$$

for all $\zeta \in \mathbf{R}_+^*$ and $n = 1, 2, \ldots$. By Theorem 2.9.1, X generates a strongly continuous semigroup $T : \mathbf{R}_+ \to \mathcal{L}(\mathcal{E})$, satisfying (2.6) with $M = 1$ and $a = 0$, i.e., (2.67). Thus the Feller-Miyadera-Phillips theorem implies the *Hille-Yosida theorem*:

Theorem 2.9.4 *A closed, densely defined, linear operator X is the infinitesimal generator of a strongly continuous semigroup of linear contractions if, and only if, $\mathbf{R}_+^* \subset r(X)$ and (2.69) holds.*

In the last part of this section, we shall indicate how to approximate a semigroup by means of its infinitesimal generator.

According to a well known fact of elementary calculus, if $\{\alpha_n\}$ is a sequence of real numbers α_n, all different from zero and such that $\lim_{n \to +\infty} |\alpha_n| = \infty$ and $1 + \frac{1}{\alpha_n} \neq 0$ for all $n = 1, 2, \ldots$, then

$$\lim_{n \to +\infty} \left(1 + \frac{t}{\alpha_n}\right)^{\alpha_n} = e^t$$

for all $t \in \mathbf{R}$.

The following theorem[8] may be viewed as an extension of this classical result to semigroups.

Theorem 2.9.5 *If* $T : \mathbf{R}_+ \to \mathcal{L}(\mathcal{E})$ *is a strongly continuous semigroup, and if* X *is its infinitesimal generator, then*

$$T(t)x = \lim_{n \to +\infty} \left(I - \frac{t}{n} X \right)^{-n} x$$

$$= \lim_{n \to +\infty} \left(\frac{n}{t} \left(\frac{n}{t} I - X \right)^{-1} \right)^{n} x$$

for all $t > 0$ *and all* $x \in \mathcal{E}$. *The limit is uniform when* t *varies on all compact subsets of* $\mathbf{R}_+^*$.

Proof According to Theorem 2.1.1 there exist $M \geq 1$ and $a \in \mathbf{R}$ for which (2.6) holds.

Throughout the following proof, ζ will be a real, non-negative number. Changing notations, for $\zeta > a$ we denote $(\zeta I - X)^{-1}$ by $R(\zeta, X)$, so that (2.34) becomes

$$R(\zeta, X)x = \int_0^{+\infty} e^{-s\zeta} T(s)x \, ds$$

for all $x \in \mathcal{E}$. The value $R'(\zeta, X)$ of the first derivative of $R(\zeta, X)$ with respect to ζ is expressed by

$$R'(\zeta, X)x = - \int_0^{+\infty} s e^{-s\zeta} T(s)x \, ds. \tag{2.70}$$

Setting $s = rt$, with $r \geq 0$ and $t > o$, we have, at $\zeta = \frac{1}{t}$,

$$R' \left(\frac{1}{t}, X \right) x = - \int_0^{+\infty} rt \, e^{-r} T(rt) x \, t \, dr$$

$$= -t^2 \int_0^{+\infty} (r e^{-r}) T(rt) x \, dr.$$

---

[8]whose proof follows the one given in [40], pp. 349-355; see also [63], pp. 33-35.

Differentiating once more with respect to ζ, (2.70) yields

$$R''(\zeta, X)x = \int_0^{+\infty} s^2 e^{-s\zeta} T(s)x \, ds.$$

Setting $s - rt$, with $r \geq 0$, $t > 0$, and $\zeta - \frac{2}{t}$, then

$$R'' \left(\frac{2}{t}, X \right) x = - \int_0^{+\infty} (rt)^2 e^{-2r} T(rt)x \, t \, dr$$

$$= t^3 \int_0^{+\infty} (re^{-r})^2 T(rt)x \, dr.$$

Iteration of this procedure yields

$$R^{(n)} \left(\frac{n}{t}, X \right) x = (-1)^n t^{n+1} \int_0^{+\infty} (re^{-r})^n T(tr)x \, dr \qquad (2.71)$$

for all $x \in \mathcal{E}$, $n = 0, 1, \ldots$ and $t > 0$. On the other hand,

$$R'(\zeta, X) = \frac{d}{d\zeta} (\zeta I - X)^{-1} = -R(\zeta, X)^2$$

$$\ldots$$

$$R^{(n)}(\zeta, X)(\zeta, X) = (-1)^n n! R(\zeta, X)^{n+1},$$

and therefore

$$R \left(\frac{n}{t}, X \right)^{n+1} x = \frac{t^{n+1}}{n!} \int_0^{+\infty} (re^{-r})^n T(tr)x \, dr,$$

i.e.,

$$\left(\frac{t}{n} R \left(\frac{n}{t}, X \right) \right)^{n+1} = \frac{n^{n+1}}{n!} \int_0^{+\infty} (re^{-r})^n T(tr)x \, dr$$

for all $x \in \mathcal{E}$, $t > 0$, $n = 1, 2, \ldots$.

Since[9]

$$\int_0^{+\infty} (re^{-r})^n dr = \frac{n!}{n^{n+1}}, \qquad (2.72)$$

$$\int_0^{+\infty} re^{-r} dr = - \int_0^{+\infty} r \, de^{-r}$$

$$= \int_0^{+\infty} e^{-r} dr = 1.$$

then

$$\left[\left(\frac{n}{t}R\left(\frac{n}{t},X\right)\right)^{n+1} - T(t)\right]x =$$

$$\frac{n^{n+1}}{n!}\int_0^{+\infty}(re^{-r})^n(T(tr) - T(t))x\,dr. \tag{2.73}$$

Since the function $s \mapsto \|T(s)x\|$ is continuous, given $\epsilon > 0$ and $t_o > 0$, there exist two real constants b and c, with $0 < b < 1 < c$, such that

$$\|T(tr)x - T(t)x\| < \epsilon \ \forall t \in [0, t_o], \ \forall r \in [b, c]. \tag{2.74}$$

The bounded continuous function $f : \mathbf{R}_+ \to \mathbf{R}_+$ defined by $f(r) = re^{-r}$ takes its absolute maximum for $r = 1$ with $f(1) = \frac{1}{e}$, is strictly increasing in $[0, 1]$ and strictly decreasing in $[1, +\infty)$. Hence, $be^{-b} = e^{-(1+\beta)}$, $ce^{-c} = e^{-(1+\gamma)}$ for some $\beta > 0$ and $\gamma > 0$. By the Stirling formula

$$\frac{n^{n+1}}{n!}(be^{-b})^n \asymp \frac{n^{n+1}}{\sqrt{2\pi}n^{n+\frac{1}{2}}}e^n(e^{-(1+\beta)})^n$$

$$= \frac{\sqrt{n}}{\sqrt{2\pi}e^{n\beta}} \to 0 \tag{2.75}$$

By induction on n, assume that (2.72) holds for some $n \geq 1$. Then, integrating by parts,

$$\int_0^{+\infty}(re^{-r})^{n+1}dr = -\frac{1}{n+1}\int_0^{+\infty}r^{n+1}d(e^{-(n+1)r})$$

$$= \int_0^{+\infty}r^n e^{-(n+1)}dr$$

$$= \frac{n^n}{(n+1)^n}\int_0^{+\infty}\left(\frac{n+1}{n}r\right)^n\left(e^{-\frac{n+1}{n}r}\right)^n dr$$

$$= \frac{n^n}{(n+1)^n}\frac{n}{n+1}\int_0^{+\infty}(re^{-r})^n dr$$

$$= \frac{n^{n+1}}{(n+1)^{n+1}}\frac{n!}{n^{n+1}} = \frac{n!}{(n+1)^{n+1}}$$

$$= \frac{(n+1)!}{(n+1)^{n+2}}.$$

as $n \to +\infty$. Similarly,

$$\frac{n^{n+1}}{n!}(ce^{-c})^n \asymp \frac{\sqrt{n}}{\sqrt{2\pi}e^{n\gamma}} \to 0 \qquad (2.76)$$

as $n \to +\infty$.

We split now the integral (2.73) as the sum of the three integrals:

$$I_1 = \frac{n^{n+1}}{n!} \int_0^b (re^{-r})^n (T(tr) - T(t))x \, dr,$$

$$I_2 = \frac{n^{n+1}}{n!} \int_b^c (re^{-r})^n (T(tr) - T(t))x \, dr,$$

$$I_3 = \frac{n^{n+1}}{n!} \int_c^{+\infty} (re^{-r})^n (T(tr) - T(t))x \, dr.$$

By (2.75),

$$\|I_1\| \leq \frac{n^{n+1}}{n!} \int_0^b (re^{-r})^n \|(T(tr) - T(t))x\| \, dr$$

$$\leq \frac{n^{n+1}}{n!}(be^{-b})^n \max\{\|(T(tr) - T(t))x\| : 0 \leq r \leq b\} \to 0$$

as $n \to +\infty$, uniformly with respect to $t \in [0, t_o]$.

By (2.6), if $n \geq n_o$,

$$\|I_3\| \leq \frac{n^{n+1}}{n!} \int_c^{+\infty} (re^{-r})^n \|(T(tr) - T(t))x\| \, dr$$

$$\leq \frac{n^{n+1}}{n!}(ce^{-c})^{n-n_o} M \int_c^{+\infty} (re^{-r})^{n_o}(e^{atr} + e^{at}) \, dr \, \|x\|.$$

Choosing the integer n_o such that $n_o > at$, the integral

$$\int_c^{+\infty} (re^{-r})^{n_o} e^{atr} \, dr = \int_c^{+\infty} r^{n_o} e^{-(n_o - at)r} \, dr$$

is finite. For every positive integer n_o, (2.72) yields

$$\int_c^{+\infty} (re^{-r})^{n_o} e^{at} \, dr \leq \int_0^{+\infty} (re^{-r})^{n_o} e^{at} \, dr$$

$$\leq \frac{n_o! \, e^{at}}{n_o^{n_o+1}},$$

and thus the integral is finite. Since, by (2.76),

$$\frac{n^{n+1}}{n!}(ce^{-c})^{n-n_o} = (ce^{-c})^{-n_o}\frac{n^{n+1}}{n!}(ce^{-c})^n \to 0$$

as $n \to +\infty$, then

$$\lim_{n \to 0}\|I_3\| = 0,$$

uniformly with respect to $t \in [0, t_o]$.

As for I_2, (2.72) and (2.74) yield

$$\|I_2\| \leq \frac{n^{n+1}}{n!}\int_b^c (re^{-r})^n\|(T(tr) - T(t))x\|\,dr$$

$$\leq \frac{n^{n+1}}{n!}\epsilon\int_b^c (re^{-r})^n dr$$

$$< \frac{n^{n+1}}{n!}\epsilon\int_0^{+\infty} (re^{-r})^n dr = \epsilon.$$

In conclusion, by (2.73)

$$\limsup_{n \to +\infty}\left\|\left[\left(\frac{t}{n}R\left(\frac{n}{t}, X\right)\right)^{n+1} - T(t)\right]x\right\| \leq \epsilon.$$

Since $\epsilon > 0$ is arbitrary, then

$$\lim_{n \to +\infty}\left\|\left(\frac{n}{t}R\left(\frac{n}{t}, X\right)\right)^{n+1}x - T(t)x\right\| = 0 \qquad (2.77)$$

for all $x \in \mathcal{E}$, uniformly with respect to $t \in [0, t_o]$.

By (2.61) and (2.62)

$$\lim_{\zeta \to +\infty}\zeta(\zeta I - X)^{-1}x = x,$$

and therefore

$$\lim_{n \to +\infty}\frac{n}{t}R\left(\frac{n}{t}, X\right)x = x$$

for all $x \in \mathcal{E}$, uniformly when $t \in [0, t_o]$. This latter equation, together with Proposition 2.4.2, yields

$$\left\| \left(\frac{n}{t} R \left(\frac{n}{t}, X \right) \right)^{n+1} x - \left(\frac{n}{t} R \left(\frac{n}{t}, X \right) \right)^n x \right\| \leq$$

$$\left\| \left(\frac{n}{t} R \left(\frac{n}{t}, X \right) \right)^n \right\| \left\| \left(\frac{n}{t} R \left(\frac{n}{t}, X \right) \right) x - x \right\|$$

$$\leq \frac{M}{\left(\frac{n}{t} - a \right)^n} \left(\frac{n}{t} \right)^n \left\| \left(\frac{n}{t} R \left(\frac{n}{t}, X \right) \right) x - x \right\|$$

$$\leq \frac{M}{(n - ta)^n} n^n \left\| \left(\frac{n}{t} R \left(\frac{n}{t}, X \right) \right) x - x \right\| \to 0$$

as $n \to +\infty$, for all $x \in \mathcal{E}$, uniformly when $t \in [0, t_o]$. Thus, being

$$\left(\frac{n}{t} R \left(\frac{n}{t}, X \right) \right)^n - T(t) =$$

$$\left(\frac{n}{t} R \left(\frac{n}{t}, X \right) \right)^n - \left(\frac{n}{t} R \left(\frac{n}{t}, X \right) \right)^{n+1}$$

$$+ \left(\frac{n}{t} R \left(\frac{n}{t}, X \right) \right)^{n+1} - T(t),$$

(2.77) completes the proof of the theorem.

2.10 Dual semigroups

Let $T : \mathbf{R}_+ \to \mathcal{L}(\mathcal{E})$ be a strongly continuous semigroup. The function $T' : \mathbf{R}_+ \to \mathcal{L}(\mathcal{E}')$ defined by $T'(t) = T(t)'$ is a semigroup. Let $X : \mathcal{D}(X) \subset \mathcal{E} \to \mathcal{E}$ the infinitesimal generator of T.

Theorem 2.10.1 *If the Banach space $\mathcal{E}$ is reflexive, T' is a strongly continuous semigroup whose infinitesimal generator is the dual X' of X.*

Proof For all $x \in \mathcal{E}$ and $\varphi \in \mathcal{E}'$,

$$\lim_{t \downarrow 0} < x, T(t)' \varphi > = \lim_{t \downarrow 0} < T(t)x, \varphi > = < x, \varphi > .$$

By Theorem 2.1.3, the semigroup T' is strongly continuous. If Z' is its infinitesimal generator, for all $x \in \mathcal{D}(X)$ and $\varphi \in \mathcal{D}(Z')$,

$$< Xx, \varphi > = \lim_{t \downarrow 0} < \frac{1}{t}(T(t)x - x), \varphi >$$

$$= \lim_{t \downarrow 0} < x, \frac{1}{t}(T(t)'\varphi - \varphi) > = < x, Z'\varphi > .$$

As a consequence, $\varphi \in \mathcal{D}(X')$ (i. e., $\mathcal{D}(Z') \subset \mathcal{D}(X')$) and

$$< x, X'\varphi > = < Xx, \varphi > = < x, Z'\varphi >$$

for all $x \in \mathcal{D}(X)$. Thus - $\mathcal{D}(X)$ being dense in $\mathcal{E}$ -

$$X'\varphi = Z'\varphi \ \forall \varphi \in \mathcal{D}(Z'),$$

that is to say, $Z' \subset X'$.

On the other hand, for all $x \in \mathcal{D}(X)$ and $\varphi \in \mathcal{D}(X')$,

$$\lim_{t \downarrow 0} < x, \frac{1}{t}(T(t)'\varphi - \varphi) > = \lim_{t \downarrow 0} < \frac{1}{t}(T(t)x - x), \varphi >$$

$$= < Xx, \varphi > = < x, X'\varphi > .$$

By Theorem 2.2.11 (applied to T'), $\varphi \in \mathcal{D}(Z')$ and $Z'\varphi = X'\varphi$, showing that $X' \subset Z'$. ∎

Corollary 2.10.2 *If $T : \mathbf{R}_+ \to \mathcal{L}(\mathcal{H})$ is a strongly continuous semigroup acting on a complex Hilbert space $\mathcal{H}$, the function $t \mapsto T(t)^*$, where $T(t)^*$ is the Hilbert space-adjoint of $T(t)$, defines a strongly continuous semigroup $T^* : \mathbf{R}_+ \to \mathcal{L}(\mathcal{H})$ whose infinitesimal generator is the Hilbert space-adjoint X^* of the infinitesimal generator X of T.*

Lemma 2.10.3 *If $\mathcal{E}$ is reflexive,*

$$\overline{\cup\{ \mathcal{R}(T(t)) : t > 0\}} = \mathcal{E}.$$

Proof Let $\lambda \in \mathcal{E}'$ be such that

$$< T(t)\xi, \lambda > = 0 \ \forall \xi \in \mathcal{E}, t > 0,$$

i.e.

$$, \xi, T'(t)\lambda >= 0 \ \forall \, \xi \in \mathcal{E}, t > 0.$$

Then $T'(t)\lambda = 0$ for all $t > 0$. Since the semigroup T' is strongly continuous, then

$$\lambda - \lim_{t \downarrow 0} T'(t)\lambda = 0.$$

∎

Example Let $\mathcal{E} = C_o(\mathbf{R})$ be the complex Banach space of all continuous functions $x : \mathbf{R} \to \mathbf{C}$ vanishing at infinity. Let $T : \mathbf{R} \to \mathcal{L}(\mathcal{E})$ be the strongly continuous group defined by the translations in $\mathbf{R}$: $(T(t)x)(r) = x(r - t)$. Let δ_r be the Dirac measure concentrated at the point $r \in \mathbf{R}$. Then, for $x \in \mathcal{E}$ and $t \in \mathbf{R}$,

$$< x, T(t)'\delta_r > = < T(t)x, \delta_r >$$
$$= (T(t)x)(r) = x(r - t) = < x, \delta_{r-t} >,$$

whence

$$T'(t)\delta_r = \delta_{r-t}.$$

If $t \neq 0$,

$$\|\delta_{r-t} - \delta_r\| = \sup\{| < x, \delta_{r-t} - \delta_r > | : x \in \mathcal{E}, \|x\| \leq 1\}$$
$$= \sup\{| x(r - t) - x(r)| : x \in \mathcal{E}, \|x\| \leq 1\} = 2.$$

Thus, if $t \neq 0$,

$$\|T'(t)\delta_r - \delta_r\| = 2,$$

and therefore T' is not strongly continuous.

We shall see now how Theorem 2.10.1 extends to any complex Banach space $\mathcal{E}$. As before let X' be the dual operator of X, let $\mathcal{F}'$ be the closure of $\mathcal{D}(X')$ in $\mathcal{E}'$, and let X^+ be the *part of X' in $\mathcal{F}'$*, that is, the restriction of X' to the linear space $\mathcal{D}(X^+) := \{\varphi \in \mathcal{D}(X') : X'\varphi \in \mathcal{F}'\}$.

Lemma 2.10.4 *The operator X^+, with domain $\mathcal{D}(X^+)$, is the infinitesimal generator of a strongly continuous semigroup $T^+ : \mathbf{R}_+ \to \mathcal{L}(\mathcal{F}')$.*

Proof I. Let $M \geq 1$ and $a \in \mathbf{R}$ be two constants such that (2.6) holds. By Theorem 2.9.1 and Lemma 1.17.3, $(a, +\infty) \subset r(X')$ and, whenever $\zeta \in (a, +\infty)$, $n = 1, 2, \ldots$,

$$\begin{aligned}
\|(\zeta I - X')^{-n}\| &= \|(\zeta I - X')^{-1} \cdots (\zeta I - X')^{-1}\| \\
&= \|(\zeta I - X)^{-1} \cdots (\zeta I - X)^{-1}\| \\
&= \|(\zeta I - X)^{-n}\| \leq \frac{M}{(\zeta - a)^n}.
\end{aligned}$$

Let $\zeta \in (a, +\infty)$ and let $J(\zeta) = ((\zeta I - X')^{-1})_{|\mathcal{F}'} \in \mathcal{L}(\mathcal{F}')$. Since, for $\varphi \in \mathcal{E}'$, $(\zeta I - X')^{-1} \varphi \in \mathcal{D}(X') \subset \mathcal{F}'$, for any $\varphi \in \mathcal{E}'$, $J(\zeta)\varphi \in \mathcal{F}'$. Furthermore

$$\|J(\zeta)^n\| \leq \frac{M}{(\zeta - a)^n}, \tag{2.78}$$

for $n = 1, 2, \ldots$, and

$$J(\zeta) - J(\tau) = (\tau - \zeta)J(\zeta)J(\tau) \quad \forall \zeta, \tau \in (a, +\infty). \tag{2.79}$$

II. If $\varphi \in \mathcal{D}(X')$,

$$\|(\zeta I - X')^{-1}X'\varphi\| \leq \frac{M}{\zeta - a}\|X'\varphi\| \to 0$$

as $\zeta \to +\infty$. Hence, being

$$\begin{aligned}
\|\zeta(\zeta I - X')^{-1}\varphi - \varphi\| &= \\
\|(\zeta I - X')^{-1}(\zeta I - X')\varphi &+ (\zeta I - X')^{-1}X'\varphi - \varphi\| \\
&= \|(\zeta I - X')^{-1}X'\varphi\|,
\end{aligned}$$

then

$$\lim_{\zeta \to +\infty} \zeta(\zeta I - X')^{-1}\varphi = \varphi \quad \forall \varphi \in \mathcal{D}(X'). \tag{2.80}$$

Furthermore, by (2.78),

$$\begin{aligned}
\|\zeta J(\zeta)\| &\leq \|(\zeta - a)J(\zeta)\| + \|aJ(\zeta)\| \\
&\leq M + \frac{|a|M}{\zeta - a} < 2M \quad \text{if } \zeta \geq a + |a|. \tag{2.81}
\end{aligned}$$

For every $\psi \in \mathcal{F}'$ and all $\epsilon > 0$ there is some $\varphi \in \mathcal{D}(X')$ for which $\|\psi - \varphi\| < \epsilon$. Since,

$$\|\zeta J(\zeta)\psi - \psi\| \leq \|\zeta J(\zeta)(\psi - \varphi) - (\psi - \varphi)\| +$$
$$\|\zeta J(\zeta)\varphi - \varphi\|$$
$$\leq (\|\zeta J(\zeta)\| + 1)\|\psi - \varphi\| + \|\zeta J(\zeta)\varphi - \varphi\|,$$

(2.80) and (2.81) yield

$$\limsup_{\zeta \to +\infty} \|\zeta J(\zeta)\psi - \psi\| \leq \epsilon$$

for all $\epsilon > 0$, and therefore

$$\lim_{\zeta \to +\infty} \zeta J(\zeta)\psi = \psi \quad \forall \psi \in \mathcal{F}'.$$

This equation shows that the range of $J(\zeta)$ - which, by (2.79), is a pseudoresolvent - is dense in $\mathcal{F}'$. By Theorem 1.13.3 there exists a closed, densely defined, linear operator $X^+ : \mathcal{D}(X^+) \subset \mathcal{F}' \to \mathcal{F}'$ such that $(a, +\infty) \subset r(X^+)$ and

$$J(\zeta) = (\zeta I - X^+)^{-1} \quad \forall \zeta \in (a, +\infty).$$

Theorem 2.9.1 and (2.78) imply that X^+ is the infinitesimal generator of a strongly continuous semigroup T^+. $\blacksquare$

By Theorem 2.9.5, for any $x \in \mathcal{E}$, any $\varphi \in \mathcal{F}'$ and any $t \geq 0$,

$$< x, T(t)'\varphi > = < T(t)x, \varphi > = \lim_{n \to +\infty} < (I - \frac{t}{n}X)^{-n}x, \varphi >$$
$$= \lim_{n \to +\infty} < x, (I - \frac{t}{n}X')^{-n}\varphi >$$
$$= \lim_{n \to +\infty} < x, (I - \frac{t}{n}X^+)^{-n}\varphi >$$
$$= < x, T^+(t)\varphi > .$$

Hence, $T^+(t)\varphi = T(t)'\varphi$ for all $\varphi \in \mathcal{F}'$, i.e.,

$$T^+(t) = T(t)'_{|\mathcal{F}'} \quad \forall t \in \mathbf{R}_+.$$

To prove that X^+ is the part of X' in $\mathcal{F}'$, note first that

$$\mathcal{D}(X^+) = \mathcal{R}(J(\zeta)) = \mathcal{R}(((\zeta I - X')^{-1})_{|\mathcal{F}'})$$
$$= \{\varphi \in \mathcal{D}(X') : X'\varphi \in \mathcal{F}'\}.$$

Let $\varphi \in \mathcal{D}(X^+)$, that is to say, let $\varphi \in \mathcal{D}(X')$ be such that $X'\varphi \in \mathcal{F}'$. Then

$$(\zeta I - X^+)^{-1}(\zeta I - X')\varphi = (\zeta I - X')^{-1}(\zeta I - X')\varphi = \varphi.$$

Multiplication on the left by $\zeta I - X^+$ yields $(\zeta I - X')\varphi = (\zeta I - X^+)\varphi$, i. e., $X^+\varphi = X'\varphi$.

In conclusion, the following theorem has been proven, which, by Proposition 1.17.7, generalizes Theorem 2.10.1.

Theorem 2.10.5 *The function $T^+ : \mathbf{R}_+ \to \mathcal{L}(\mathcal{F}')$ defined by $T^+(t) = (T(t)')_{|\mathcal{F}'}$ for all $t \in \mathbf{R}_+$, is a strongly continuous semigroup whose infinitesimal generator X^+ is the part of the dual X' of X in $\mathcal{F}'$.*

Going back to Theorem 2.6.3 and Proposition 2.6.5, let $T : \mathbf{R}_+ \to \mathcal{L}(\mathcal{E})$ be a uniformly bounded, strongly continuous semigroup and let X be its infinitesimal generator. With the same notations as in section 2.6, let $x_o \in \mathcal{F} \backslash (\ker X \oplus \overline{\mathcal{R}(X)})$. By the Hahn-Banach theorem, there is some $\lambda \in \mathcal{E}'$ such that $< x, \lambda > \neq 0$ and $\lambda = 0$ on $\ker X \cap \overline{\mathcal{R}(X)}$. Since $< Xx, \lambda >= 0$ for all $x \in \mathcal{D}(X)$, then $\lambda \in \mathcal{D}(X')$ and $X'\lambda = 0$. As a consequence, $\lambda \in \mathcal{D}(X^+)$ and $X^+\lambda = 0$. Hence $T^+(t)\lambda = \lambda$, and therefore also

$$< T^+(t)x, \lambda >=< x, \lambda > \quad \forall t \in \mathbf{R}_+.$$

Since $x_o \in \mathcal{F}$, then

$$Px_o = \lim_{t \to +\infty} \frac{1}{t} \int_0^t T(s)x_o ds, \tag{2.82}$$

whence

$$< Px_o, \lambda >= \lim_{t \to +\infty} \frac{1}{t} \int_0^t < T(s)x_o, \lambda > ds =< x_o, \lambda > .$$

But, being $Px_o \in \ker X$, then $< Px_o, \lambda >= 0$. This contradiction proves

Lemma 2.10.6 *The closed space $\mathcal{F}$ is the set of all $x_o \in \mathcal{E}$ for which the limit (2.82) exists in $\mathcal{E}$.*

Lemma 2.10.7 *[35]If $\mathcal{E}$ is reflexive,*

$$\ker X \oplus \overline{\mathcal{R}(X)} = \mathcal{E}.$$

Proof Let $\overline{\mathcal{R}(X)}^{\perp}$ be the annihilator of $\overline{\mathcal{R}(X)}$:

$$\overline{\mathcal{R}(X)}^{\perp} = \{\lambda \in \mathcal{E}' :< Xx, \lambda >= 0 \ \forall x \in \mathcal{D}(X)\}$$
$$= \{\lambda \in \mathcal{E}' :< x, X'\lambda >= 0 \ \forall x \in \mathcal{D}(X)\}.$$

Identifying canonically $\mathcal{E}''$ with $\mathcal{E}$ and X'' with X, then

$$\overline{\mathcal{R}(X)}^{\perp} = \ker X'.$$

Similarly

$$\overline{\mathcal{R}(X')}^{\perp} = \ker X,$$

whence

$$\overline{\mathcal{R}(X')} = \ker X^{\perp}.$$

Since

$$(\ker X \oplus \overline{\mathcal{R}(X)})^{\perp} \subset \ker X^{\perp} \cap \overline{\mathcal{R}(X)}^{\perp}$$
$$= \overline{\mathcal{R}(X')} \cap \ker X',$$

if $\ker X \oplus \overline{\mathcal{R}(X)} \neq \mathcal{E}$, then $\ker X' \cap \overline{\mathcal{R}(X')} \neq \{0\}$, contrary to Theorem 2.6.3 applied to X'. ∎

Corollary 2.10.8 *If $\mathcal{E}$ is reflexive,*

$$\ker(X - i\theta I) \oplus \overline{\mathcal{R}(X - i\theta I)} = \mathcal{E}$$

for all $\theta \in \mathbf{R}$ for which $i\theta \in p\sigma(X)$.

If 0 is contained in the compression spectrum of X, there exists $\lambda \in \mathcal{E}' \setminus \{0\}$ such that $\langle Xx, \lambda \rangle = 0$ for all $x \in \mathcal{D}(X)$. Hence, $\lambda \in \mathcal{D}(X')$ and $X'\lambda = 0$. As a consequence, $\lambda \in \mathcal{D}(X^{+})$ and $X^{+}\lambda = 0$, i. e., $0 \in p\sigma(X^{+})$.

Viceversa, if $0 \in p\sigma(X^{+})$, there is some $\lambda \in \ker X \setminus \{0\}$. Hence $\langle x, X^{+}\lambda \rangle = 0$, that is, $\langle Xx, \lambda \rangle = 0$ for all $x \in \mathcal{D}(X)$, implying that 0 is contained in the compression spectrum of X.

In conclusion, the following lemma holds.

Lemma 2.10.9 *The compression spectrum $k\sigma(X)$ of X coincides with $p\sigma(X^+)$.*

Thus, by Lemma 1.17.6,

$$p\sigma(X') = p\sigma(X^+). \tag{2.83}$$

We prove now that

$$p\sigma(T(t)') = p\sigma(T^+(t)) \ \forall \, t \in \mathbf{R}_+. \tag{2.84}$$

Since $T^+(t)$ is the restriction of $T(t)'$ to $\mathcal{E}^+$, then

$$p\sigma(T(t)') \supset p\sigma(T^+(t)).$$

In order to establish the opposite inclusion, let $\zeta \in p\sigma(T(t)')$, and let $\lambda \in \mathcal{E}'\backslash\{0\}$ be such that

$$T(t)'\lambda = \zeta\lambda.$$

If $\tau \in r(X) \subset r(X')$, then $(\tau I - X')^{-1}\lambda \in \mathcal{D}(X') \subset \mathcal{E}^+$, and moreover

$$T(t)'(\tau I - X')^{-1}\lambda = (\tau I - X')^{-1}T(t)'\lambda$$

because, for all $x \in \mathcal{E}$,

$$\langle x, T(t)'(\tau I - X')^{-1}\lambda \rangle = \langle (\tau I - X)^{-1}T(t)x, \lambda \rangle$$
$$\langle T(t)(\tau I - X)^{-1}x, \lambda \rangle = \langle x, (\tau I - X')^{-1}T(t)'\lambda \rangle.$$

Hence,

$$T^+(t)(\tau I - X')^{-1}\lambda = T(t)'(\tau I - X')^{-1}\lambda$$
$$(\tau I - X')^{-1}T(t)'\lambda = \zeta(\tau I - X')^{-1}\lambda.$$

By Lemma 1.17.6 and (2.84),

$$k\sigma(T(t)) = p\sigma(T(t)') = p\sigma(T^+(t)).$$

Thus, Theorem 2.5.2 and (2.83) yield

$$k\sigma(T(t))\backslash\{0\} = e^{t\,k\sigma(X)}.$$

According to Lemma 1.17.3, if the linear operator X is closed and densely defined, then $\sigma(X') \subset \sigma(X)$. We will show now that, if X is the infinitesimal generator of a strongly continuous, uniformly bounded semigroup T, then $p\sigma(X) \cap i\mathbf{R} \subset p\sigma(X') \cap i\mathbf{R}$.

Following the proof given in [58], we introduce first the notion of *invariant mean* in the Banach space $BUC(\mathbf{R}_+)$ of all complex-valued, uniformly continuous functions on $\mathbf{R}_+$. A invariant mean on $BUC(\mathbf{R}_+)$ is a positive continuous linear form ϕ on $BUC(\mathbf{R}_+)$, *i. e.* ,

$$0 \leq \phi \in (BUC(\mathbf{R}_+))', \tag{2.85}$$

such that

$$\langle 1, \phi \rangle = 1 \tag{2.86}$$

(where, on the left hand side, 1 stands for the constant function equal to 1 on $\mathbf{R}_+$), and, for all $s \in \mathbf{R}_+$,

$$\langle f, \phi \rangle = \langle f_s, \phi \rangle, \tag{2.87}$$

where $f_s = f(\bullet + s)$. To show that invariant means do exist, choose any ϕ_o such that $0 \leq \phi_o \in BUC(\mathbf{R}_+)'$, and $\langle 1, \phi_o \rangle = 1$, and, for $n = 1, 2, \dots$, define $\phi_n \in (BUC(\mathbf{R}_+))'$ by

$$\langle f, \phi_n \rangle = \frac{1}{n} \int_0^n \langle f_t, \phi_o \rangle dt \quad (f \in BUC(\mathbf{R}_+)).$$

Then, for all $n = 1, 2, \dots$, $\langle 1, \phi_n \rangle = 1$ and

$$|\langle f, \phi_n \rangle| \leq \|f\| \, \|\phi_n\|,$$

so that $\|\phi_n\| \leq 1$. By the weak-star compactness of the closed unit ball of $(BUC(\mathbf{R}_+))'$, the sequence $\{\phi_n\}$ has at least one limit value ϕ, which satisfies both (2.85) and (2.86). For any $f \in BUC(\mathbf{R}_+)$ and any $s \geq 0$ there is a subsequence $\{n_j\}$ of $\{n\}$ such that

$$\langle f, \phi \rangle = \lim_{j \to +\infty} \langle f, \phi_{n_j} \rangle$$

and

$$\langle f_s, \phi \rangle = \lim_{j \to +\infty} \langle f_s, \phi_{n_j} \rangle.$$

Thus,

$$\langle f_s, \phi \rangle = \lim_{j \to +\infty} \frac{1}{n_j} \int_0^{n_j} \langle f_{s+t}, \phi \rangle dt$$

$$= \lim_{j \to +\infty} \frac{1}{n_j} \int_s^{n_j+s} \langle f_t, \phi \rangle dt$$

$$= \lim_{j \to +\infty} \frac{n_j + s}{n_j} \frac{1}{n_j + s} \int_0^{n_j+s} \langle f_t, \phi \rangle dt -$$

$$\lim_{j \to +\infty} \frac{1}{n_j} \int_0^s \langle f_t, \phi \rangle dt$$

$$= \lim_{j \to +\infty} \frac{1}{n_j} \int_0^{n_j} \langle f_t, \phi \rangle dt = \langle f, \phi \rangle.$$

Hence, (2.87) holds for all $s \geq 0$, showing that ϕ is an invariant mean.

Theorem 2.10.10 *If the strongly continuous semigroup is uniformly bounded, then $p\sigma(X) \cap i\mathbf{R} \subset p\sigma(X') \cap i\mathbf{R}$.*

Proof Let $i\eta \in p\sigma(X) \cap i\mathbf{R}$. Replacing T by the uniformly bounded semigroup $\mathbf{R}_+ \ni e^{-i\eta}T(t)$, we may assume $\eta = 0$. Let $y \in \mathcal{D}(X) \backslash \{0\}$ be such that $Xy = 0$, and therefore $T(t)y = y$ for all $t \geq 0$. Let $\lambda \in \mathcal{E}'$ be such that $\langle y, \lambda \rangle = 1$. If ϕ is any invariant mean on $\mathrm{BUC}(\mathbf{R}_+)$, the map

$$\mathcal{E} \ni x \mapsto \langle \langle T(\bullet)x, \lambda \rangle, \phi \rangle$$

is a continuous linear form φ on $\mathcal{E}$:

$$\langle x, \varphi \rangle = \langle \langle T(\bullet)x, \lambda \rangle, \phi \rangle \quad \forall x \in \mathcal{E}.$$

Since

$$\langle x, T(t)'\varphi \rangle = \langle T(t)x, \varphi \rangle = \langle \langle T(\bullet)T(t)x, \lambda \rangle, \phi \rangle$$
$$= \langle \langle T(\bullet + t)x, \lambda \rangle, \phi \rangle = \langle \langle T(\bullet)x, \lambda \rangle, \phi \rangle = \langle x, \varphi \rangle,$$

for all $x \in \mathcal{E}$, then

$$T(t)'\varphi = \varphi \quad \forall t \geq 0.$$

Hence, $\varphi \in \mathcal{F}'$ and

$$\lim_{t \downarrow 0} \frac{1}{t}(T(t)'\varphi - \varphi) = 0,$$

so that $\varphi \in \mathcal{D}(X^+)$ and $X^+\varphi = 0$. Since, furthermore,

$$\langle y, \varphi \rangle = \langle \langle T(\bullet)y, \lambda \rangle, \phi \rangle = \langle \langle y, \lambda \rangle 1, \phi \rangle$$
$$= \langle 1, \phi \rangle = 1,$$

then $\varphi \neq 0$, showing that $0 \in p\sigma(X^+) = p\sigma(X')$. ∎

Chapter 3

Dissipative operators

To illustrate the basic questions that will be raised and discussed in this chapter, we begin with some heuristics, borrowed essentially from [4].

Let X be the infinitesimal generator of a strongly continuous semigroup $\mathbf{R}_+ \to \mathcal{L}(\mathcal{H})$ acting on a complex Hilbert space $\mathcal{H}$ with inner product $(\,|\,)$. Let X be its infinitesimal generator, and let

$$\begin{cases} \dot{x}(t) = X\,x(t) \ \text{ for } t \geq 0 \\ x(0) = x_o \in \mathcal{D}(X) \end{cases}$$

be the Cauchy problem with initial condition x_o, where the vector $x(t)$ represents the state of a physical system at the time t, for which $\|x(t)\|^2$ is some form of energy. Assume that

$$\Re(Xy|y) \leq \alpha\|y\|^2$$

for some constant $\alpha \geq 0$ and all $y \in \mathcal{D}(X)$. If $t \geq 0$,

$$\begin{aligned} \frac{d}{dt}\|x(t)\|^2 &= (\dot{x}(t)|x(t)) + (x(t)|\dot{x}(t)) \\ &= 2\Re(\dot{x}(t)|x(t)) = 2\Re(Xx(t)|x(t)) \\ &\leq 2\alpha\|x(t)\|^2. \end{aligned}$$

Since $X(0) = x_o$, by Gronwall's lemma [1],

$$\|x(t)\|^2 \leq e^{2\alpha t}\|x_o\|^2$$

---

[1] Let $a > 0$ and let $f : [0, a] \to \mathbf{R}$ be a continuous function for which there exist

for all $t \in \mathbf{R}_+$. If $\alpha = 0$, then $\|x(t)\| \leq \|x_o\|$, that is to say, the energy is *dissipated*, whereas, if $\alpha > 0$, the energy cannot be generated at a larger rate than 2α.

3.1 Semi inner products

Let $\mathcal{E}$ be a complex normed space, and let $\mathcal{E}'$ be its topological dual. By the Hahn-Banach theorem, for every $x \in \mathcal{E}$, the set

$$\Lambda(x) := \{\lambda \in \mathcal{E}' :< x, \lambda >= \|x\|, \|\lambda\| = 1\}$$

is non-empty. If $x = 0$, $\Lambda(0) = \{\lambda \in \mathcal{E}'; \|\lambda\| = 1\}$

For example if $\mathcal{E}$ is a pre-Hilbert space with inner product $(\,|\,)$, and if $x \neq 0$, $\Lambda(x)$ consists of the unique element $\lambda = \frac{1}{\|x\|}(\bullet|x)$. But there are other spaces for which $\mathrm{Card}(\Lambda(x)) = 1$ for all $x \neq 0$.

Lemma 3.1.1 *If $x \neq 0$, $\Lambda(x)$ is convex.*

Proof For $\lambda_0, \lambda_1 \in \Lambda(x)$ and $t \in [0, 1]$, let $\lambda_t = t\lambda_1 + (1-t)\lambda_0$. Since

$$< x, \lambda_t >= t < x, \lambda_1 > +(1-t) < x, \lambda_0 >$$
$$= t\|x\| + (1-t)\|x\| = \|x\|, \tag{3.1}$$

then

$$\|x\| = < x, \lambda_t > \leq \|x\| \, \|\lambda_t\|$$
$$\leq \|x\| \, (t\|\lambda_1\| + (1-t)\|\lambda_0\|)$$
$$= \|x\|(t + 1 - t) = \|x\|,$$

and therefore

$$\|x\| \, \|\lambda_t\| = \|x\|.$$

positive constants a and b such that

$$f(t) \leq a + b \int_0^t f(s)ds \quad \forall t \in [0, a].$$

Then

$$f(t) \leq ae^{bt} \quad \forall t \in [0, a].$$

For a proof, see [4] or [64].

Thus $\|\lambda_t\| = 1$ and, by (3.1), $\lambda_t \in \Lambda(x)$ for all $t \in [0,1]$. ∎

Let B and B' be the open unit discs of $\mathcal{E}$ and of $\mathcal{E}'$ respectively.

Corollary 3.1.2 *If B' is strictly convex (i.e., every $\lambda \in \partial B'$ is a real extreme point of the closure $\overline{B'}$ of B'), for every $x \in \mathcal{E}\backslash\{0\}$, $\Lambda(x)$ consists of a unique element.*

When $\mathrm{Card}(\Lambda(x)) = 1$, we will denote $\{\Lambda(x)\}$ by $\lambda(x)$. In this case,

$$< tx, \lambda(x) >= t < x, \lambda(x) >= t\|x\| = \|tx\|,$$

and therefore

$$\lambda(tx) = \lambda(x) \ \forall t > 0, \ \forall x \in \mathcal{E}\backslash\{0\}. \tag{3.2}$$

Recall that $\mathcal{E}$ is said to be *uniformly convex* if, given any $\epsilon > 0$, there is some $\delta > 0$ such that, for all $x, y \in \mathcal{E}$ for which $\|x\| = \|y\| = 1$ and $\|x - y\| \geq \epsilon$, one has $\|\frac{1}{2}(x + y)\| \leq 1 - \delta$.

For example, a pre-Hilbert space is uniformly convex.

Recall also from n. 1.3 that, if $\mathcal{E}$ is uniformly convex, then it is strictly convex.

Proposition 3.1.3 *If $\mathcal{E}'$ is uniformly convex, the map $x \mapsto \lambda(x)$ from $\mathcal{E}\backslash\{0\}$ to $\partial B'$ is continuous.*

Proof Since the map $x \mapsto \frac{1}{\|x\|}x$ of $\mathcal{E}\backslash\{0\}$ onto ∂B is continuous, by (3.2) it suffices to prove that the map $x \mapsto \lambda(x)$ of ∂B into $\partial B'$ is continuous.

Suppose that, on the contrary, there exist $\epsilon > 0$ and a sequence $\{x_\nu\}$ in ∂B, converging to some $x_o \in \partial B$, such that

$$\|\lambda(x_o) - \lambda(x_\nu)\| \geq \epsilon$$

for $\nu = 1, 2, \ldots$. Since $\mathcal{E}'$ is uniformly convex, there is $\delta > 0$ such that

$$\left\|\frac{1}{2}(\lambda(x_0) + \lambda(x_\nu))\right\| \leq 1 - \delta$$

for $\nu = 1, 2, \ldots$. Being

$$< x_\nu, \lambda(x_o) + \lambda(x_\nu) > = < x_\nu, \lambda(x_o) > + < x_\nu, \lambda(x_\nu) >$$
$$= < x_\nu, \lambda(x_o) > +1$$
$$= < x_\nu - x_o, \lambda(x_o) > +2$$

and

$$| < x_\nu, \lambda(x_o + x_\nu) > | \leq \|x_\nu\| \, \|\lambda(x_o + x_\nu)\|$$
$$= \|\lambda(x_o + x_\nu)\|$$
$$| < x_\nu - x_o, \lambda(x_o) > | \leq \|x_\nu - x_o\| \, \|\lambda(x_o)\|$$
$$= \|x_\nu - x_o\|,$$

then

$$1 - \delta \geq \left| \frac{1}{2} < x_\nu, \lambda(x_o) + \lambda(x_\nu) > \right|$$
$$\geq 1 - \frac{1}{2}| < x_\nu - x_o, \lambda(x_o) > |$$
$$\geq 1 - \frac{1}{2}\|x_\nu - x_o\|,$$

i. e.,

$$\delta \leq \frac{1}{2}\|x_\nu - x_o\|,$$

contradicting the fact that $\lim_{\nu \to +\infty} \|x_\nu - x_o\| = 0$. ∎

Exercise Show that the function $x \mapsto \lambda(x)$ is uniformly continuous on all bounded subsets of $\mathcal{E}\backslash\{0\}$.

The complex normed space $\mathcal{E}$ is said to be *smooth* [18] if, for every $x \in \partial B$, there is a unique $\lambda \in \partial B'$ for which $< x, \lambda >= 1$.

Exercise Prove that, if $\mathcal{E}$ is smooth, $\mathrm{Card}(\Lambda(x)) = 1$.

Proposition 3.1.4 *If the Banach space $\mathcal{E}$ is smooth, the map $\lambda : x \mapsto \lambda(x)$ of $\mathcal{E}\backslash\{0\}$ into $\partial B'$ is continuous for the weak star topology.*

Proof As in the proof of Proposition 3.1.3, in view of (3.2) it is enough to show that $\lambda_{|\partial B}$ is continuous for the weak star topology.

Let $\{x_\nu\}$ be a sequence in ∂B, converging to some $x_o \in \partial B$, and note first that

$$| < x_o, \lambda(x_\nu) > -1| = | < x_o, \lambda(x_\nu - x_o) > |$$
$$\leq |\lambda(x_\nu - x_o)| \leq \|x_\nu - x_o\|. \qquad (3.3)$$

If the sequence $\{\lambda(x_\nu)\}$ does not converge to $\lambda(x_o)$ for the weak star topology, there exist a weak star neighbourhood U' of $\lambda(x_o)$ in $\mathcal{E}'$ and a subsequence $\{y_\nu\}$ of $\{x_\nu\}$ such that

$$\lambda(y_\nu) \notin U' \ \forall \nu = 1, 2, \ldots . \qquad (3.4)$$

On the other hand, the compactness of $\overline{B'}$ for the weak star topology implies that there is subsequence $\{z_\nu\}$ of $\{y_\nu\}$ such that $\{\lambda(z_\nu)\}$ converges, for the weak star topology, to some μ_o, with $\|\mu_o\| \leq 1$. Thus, for any $\epsilon > 0$, there is an index ν_o such that, for all $\nu \geq \nu_o$, $\lambda(z_\nu)$ is contained in the weak star neighbourhood of μ_o:

$$\{\mu \in \mathcal{E}' : | < x_o, \mu_o - \mu > | < \epsilon\};$$

i. e.,

$$| < x_o, \mu_o - \lambda(z_\nu) > | < \epsilon \ \forall \nu \geq \nu_o.$$

Hence, by (3.3),

$$| < x_o, \mu_o > -1| \leq | < x_o, \mu_o - \lambda(z_\nu) > | + | < x_o, \lambda(z_\nu) > -|$$
$$< \epsilon\|z_\nu - z_o\| \ \forall \nu \geq \nu_o,$$

showing that $< x_o, \mu_o >= 1$, and therefore ($\|\mu_o\| = 1$ and) $\mu_o \in \Lambda(x_o)$. Since $\mathcal{E}$ is smooth, then $\mu_o = \lambda(x_o)$. But the fact that $\{\lambda(z_\nu)\}$ converges to $\lambda(x_o)$ for the weak star topology contradicts (3.4). ∎

Let $\mathcal{E}$ be a complex vector space. A *semi inner product*[2] is a function $\langle \, , \, \rangle : \mathcal{E} \times \mathcal{E} \to \mathbf{C}$ mapping any pair $x, y \in \mathcal{E}$ to $\langle x, y \rangle \in \mathbf{C}$ in such a

---

[2]Semi inner products have been introduced by G.Lumer in [50]; see also [51]

way that, for all $x, y, z \in \mathcal{E}$ and all $\zeta \in \mathbf{C}$,

$$\langle x + y, z \rangle = \langle x, z \rangle + \langle y, z \rangle,$$
$$\langle \zeta x, y \rangle = \zeta \langle x, y \rangle,$$
$$\langle x, x \rangle \in \mathbf{R}_+^*, \text{ if } x \neq 0$$
$$|\langle x, y \rangle|^2 \leq \langle x, x \rangle \langle y, y \rangle.$$

Lemma 3.1.5 *The function* $\| \ \| : x \mapsto \|x\| := \sqrt{\langle x, x \rangle}$ *is a norm on* $\mathcal{E}$.

Proof The chain of inequalities

$$\begin{aligned}
\|x + y\|^2 &= \langle x + y, x + y \rangle \\
&= \langle x, x + y \rangle + \langle y, x + y \rangle \\
&= |\langle x, x + y \rangle + \langle y, x + y \rangle| \\
&\leq |\langle x, x + y \rangle| + |\langle y, x + y \rangle| \\
&\leq \sqrt{\langle x, x \rangle} \sqrt{\langle x + y, x + y \rangle} + \sqrt{\langle y, y \rangle} \sqrt{\langle x + y, x + y \rangle} \\
&= \|x\| \|x + y\| + \|y\| \|x + y\|,
\end{aligned}$$

implies that

$$\|x + y\| \leq \|x\| + \|y\|$$

for all $x, y \in \mathcal{E}$. Furthermore

$$\begin{aligned}
\|\zeta x\|^2 &= \langle \zeta x, \zeta x \rangle \\
&= \zeta \langle x, \zeta x \rangle \\
&= |\zeta \langle x, \zeta x \rangle| \\
&\leq |\zeta| \|x\| \|\zeta x\|,
\end{aligned}$$

and therefore

$$\|\zeta x\| \leq |\zeta| \|x\|.$$

On the other hand, if $\zeta \neq 0$,

$$\begin{aligned}
\|x\| &= \left\| \frac{1}{\zeta} \zeta x \right\| \\
&\leq \frac{1}{|\zeta|} \|\zeta x\|,
\end{aligned}$$

i. e., $|\zeta|\,\|x\| \le \|\zeta x\|$, and, in conclusion, $|\zeta|\,\|x\| = \|\zeta x\|$. ∎

Now, let $\mathcal{E}$ be normed. For $x \in \mathcal{E}$ and $\lambda \in \Lambda(x)$, we denote by $\tilde{\lambda}$ the continuous linear form on $\mathcal{E}$ defined by $\tilde{\lambda} = \lambda\|x\|$. Then $< x, \tilde{\lambda} >= \|x\|^2 = \|\tilde{\lambda}\|^2$. Viceversa, given $x \in \mathcal{E}\setminus\{0\}$ and any continuous linear form $\tilde{\lambda}$ on $\mathcal{E}$ satisfying these latter equations, the linear form $\lambda := \frac{1}{\|x\|}\tilde{\lambda}$ is contained in $\Lambda(x)$. A map $x \mapsto \tilde{\lambda}(x)$ of $\mathcal{E}$ into $\mathcal{E}'$, where

$$< x, \tilde{\lambda}(x) >= \|x\|^2 = \|\tilde{\lambda}(x)\|^2,$$

is called a *duality section* of $\mathcal{E}$. By the Hahn-Banach theorem, duality sections exist always.

Choose a duality section $\tilde{\lambda}$ and, for $x, y \in \mathcal{E}$, set

$$\langle x, y \rangle =< x, \tilde{\lambda}(y) > .$$

Since

$$\begin{aligned}
|\langle x, y \rangle|^2 &=| < x, \tilde{\lambda}(y) > |^2 \\
&\le \|x\|^2\|\tilde{\lambda}(x)\|^2 \\
&=< x, \tilde{\lambda}(x) > < y, \tilde{\lambda}(y) > \\
&= \langle x, x \rangle \, \langle y, y \rangle,
\end{aligned}$$

then $\langle\,,\,\rangle$ is a semi inner product for which we have, furthermore,

$$\begin{aligned}
\|x\|^2 &= \langle x, x \rangle =< x, \tilde{\lambda}(x) > \\
&= \|x\|^2 \ \forall x \in \mathcal{E}.
\end{aligned}$$

Viceversa, let $\langle\,,\,\rangle$ be a semi inner product on $\mathcal{E}$, whose associated norm $\|\ \|$ is $\|\ \|$. Since

$$|\langle y, x \rangle| \le \|y\|\,\|x\|,$$

the linear form $\tilde{\lambda} : y \mapsto \langle y, x \rangle$ is continuous, and

$$< x, \tilde{\lambda} >= \langle x, x \rangle = \|x\|^2.$$

That proves the following proposition.

Proposition 3.1.6 *If $\tilde{\lambda}$ is a duality section in a normed vector space $\mathcal{E}$, with norm $\| \ \|$, the function $x, y \mapsto <x, \tilde{\lambda}(y)>$ of $\mathcal{E} \times \mathcal{E}$ onto $\mathbf{C}$ is a semi inner product whose associated norm is $\| \ \|$.*

Viceversa, for every semi inner product $\langle \ , \ \rangle$ on $\mathcal{E}$, whose associated norm is $\| \ \|$, there exists a duality section $\tilde{\lambda}$ for which $\langle x, y \rangle = <x, \tilde{\lambda}(y)>$ for all $x, y \in \mathcal{E}$.

Corollary 3.1.7 *If the normed space $\mathcal{E}$ is smooth, there exists a unique semi inner product whose associated norm is $\| \ \|$.*

Corollary 3.1.8 *If $\mathcal{E}$ is a pre-Hilbert space with inner product $(\ | \ )$, this one is the unique semi inner product whose associated norm is $\| \ \|$.*

Exercise In a vector space $\mathcal{E}$, a semi inner product is an inner product if, and only if, the associated norm satisfies the parallelogram identity.

3.2 Dissipative operators

Let $\mathcal{E}$ be a complex Banach space, and let $X : \mathcal{D}(X) \subset \mathcal{E} \to \mathcal{E}$ be a linear operator. If, for every $x \in \mathcal{E}$, there is $\lambda \in \Lambda(x)$ such that

$$\Re <Xx, \lambda > \le 0, \tag{3.5}$$

X is said to be a *dissipative* operator. This notion can be rephrased in terms of semi inner products:

Lemma 3.2.1 *The operator X is dissipative if, and only if, there is a semi inner product $\langle \ , \ \rangle$ in $\mathcal{E}$ - whose associated norm is the norm in $\mathcal{E}$ - such that*

$$\Re \langle Xx, x \rangle \le 0 \ \ \forall x \in \mathcal{D}(X).$$

If X is dissipative, for all $\zeta \in \mathbf{R}_+^*$, $x \in \mathcal{D}(X)$ and $\lambda \in \Lambda(x)$ such that $\Re <Xx, \lambda > \le 0$, we have

$$\Re <(\zeta I - X)x, \lambda > = \zeta \Re <x, \lambda> - \Re <Xx, \lambda>$$
$$\ge \zeta \Re <x, \lambda> = \zeta \|x\|,$$

and, on the other hand,

$$\Re < (\zeta I - X)x, \lambda > \leq | < (\zeta I - X)x, \lambda > |$$
$$\leq \|(\zeta I - X)x\| \, \|\lambda\| \leq \|(\zeta I - X)x\|.$$

Thus

$$\|(\zeta I - X)x\| \geq \zeta \|x\| \quad \forall x \in \mathcal{D}(X) \quad \forall \zeta \in \mathbf{R}_+^*. \qquad (3.6)$$

Assuming now this latter condition to be satisfied, we shall show that X is a dissipative. By (3.6), if $x \in \mathcal{D}(X)$ and $\mu_\zeta \in \Lambda((\zeta I - X)x)$,

$$\zeta \|x\| \leq \|(\zeta I - X)x\| = < (\zeta I - X)x, \mu_\zeta >$$
$$= \Re < (\zeta I - X)x, \mu_\zeta >= \zeta < x, \mu_\zeta > -\Re < Xx, \mu_\zeta >$$
$$\leq \zeta \|x\| \, \|\mu_\zeta\| - \Re < Xx, \mu_\zeta >= \zeta \|x\| - \Re < Xx, \mu_\zeta >,$$

and therefore

$$\Re < Xx, \mu_\zeta >\leq 0.$$

Furthermore

$$\Re < x, \mu_\zeta > \geq \|x\| + \frac{1}{\zeta} \Re < Xx, \mu_\zeta >$$

$$\geq \|x\| - \frac{1}{\zeta} | < Xx, \mu_\zeta > |$$

$$\geq \|x\| - \frac{1}{\zeta} | \, \|Xx\| \quad \forall x \in \mathcal{D}(X) \quad \forall \zeta \in \mathbf{R}_+^*.$$

For $\zeta = n = 1, 2, \ldots,$

$$\Re < Xx, \mu_n >\leq 0, \quad \Re < x, \mu_n >\geq \|x\| - \frac{1}{n}\|Xx\|. \qquad (3.7)$$

Since the closed unit disc of $\mathcal{E}'$ is compact for the weak star topology, and since $\|\mu_n\| = 1$ for $n = 1, 2, \ldots$, the sequence $\{\mu_n\}$ has at least one limit value μ, with $\|\mu\| \leq 1$. By (3.7),

$$\Re < Xx, \mu >\leq 0, \quad \Re < x, \mu >\geq \|x\|, \qquad (3.8)$$

and this latter inequality implies

$$\|x\| \leq \Re < x, \mu >\leq | < x, \mu > | \leq \|x\|,$$

whence

$$< x, \mu >= \|x\|.$$

Thus, ($\|\mu\| = 1$, and therefore) $\mu \in \Lambda(x)$. The first inequality in (3.8) shows then that X is dissipative.

In conclusion, the following proposition holds.

Proposition 3.2.2 *The linear operator* $X : \mathcal{D}(X) \subset \mathcal{E} \to \mathcal{E}$ *is dissipative if, and only if, (3.6) holds for all* $x \in \mathcal{D}(X)$ *and all* $\zeta \in \mathbf{R}_+^*$.

Corollary 3.2.3 *The linear operator* X *is dissipative if, and only if, for all* $\zeta \in \mathbf{R}_+^*$, $I - \zeta X$ *is injective,* $(I - \zeta X)^{-1} \in \mathcal{L}(\mathcal{R}(I - \zeta X))$ *and* $\|(I - \zeta X)^{-1}\| \le 1$.

Corollary 3.2.4 *If the dissipative operator* X *is closable, its closure* $\tilde{X}$ *is dissipative.*

Lemma 3.2.5 *If the dissipative operator* X *is densely defined,* X *is closable.*

Proof Let $\{x_\nu\}$ be a sequence in $\mathcal{D}(X)$ converging to 0 and such that the sequence $\{X x_\nu\}$ converges to some $y \in \mathcal{E}$. We have to show that $y = 0$. Condition (3.6), which can be written:

$$\|(I - \zeta X)z\| \ge \|z\| \quad \forall z \in \mathcal{D}(X), \ \forall \zeta \in \mathbf{R}_+^*,$$

implies

$$\|(x_\nu + \zeta z) - \zeta X(x_\nu + \zeta z)\| \ge \|x_\nu + \zeta z\|.$$

Passing to the limit, for $\nu \to +\infty$, we obtain

$$\|\zeta z - \zeta y - \zeta^2 X z\| \ge \|\zeta z\|$$

i. e.,

$$\|z - y - \zeta X z\| \ge \|z\|$$

for all $\zeta \in \mathbf{R}_+^*$ and all $z \in \mathcal{D}(X)$. Letting $\zeta \downarrow 0$ yields

$$\|z\| \le \|z - y\| \quad \forall z \in \mathcal{D}(X).$$

Since $\mathcal{D}(X)$ is dense in $\mathcal{E}$, there exists a sequence $\{z_\nu\}$ in $\mathcal{D}(X)$ converging to y. Thus,

$$\|y\| \le \|y - y\| = 0,$$

and therefore $y = 0$. ∎

By (3.6), $\zeta I - X$ is injective and $(\zeta I - X)^{-1} \in \mathcal{L}(\mathcal{R}(\zeta I - X))$ for all $\zeta \in \mathbf{R}_+^*$. Thus, the following lemma holds.

Lemma 3.2.6 *If X is dissipative, $\zeta \in r(X) \cup r\sigma(X)$ for all $\zeta \in \mathbf{R}_+^*$.*

Lemma 3.2.7 *If the dissipative operator is closed, $\mathcal{R}(\zeta I - X)$ is closed for all $\zeta \in \mathbf{R}_+^*$.*
If X is dissipative and $\mathcal{R}(\zeta I - X)$ is closed for some $\zeta \in \mathbf{R}_+^*$, X is closed.

Proof Let X be closed. Choose any $\zeta \in \mathbf{R}_+^*$, and let $\{x_\nu\}$ be a sequence in $\mathcal{D}(X)$ such that the sequence $\{(X - \zeta I)x_\nu\}$ converges to some $y \in \mathcal{E}$. By (3.6), $\{x_\nu\}$ converges. If $x := \lim_{\nu \to +\infty} x_\nu$, X being closed implies that $x \in \mathcal{D}(X)$ and $y = (X - \zeta I)x \in \mathcal{R}(X - \zeta I)$, showing that $\mathcal{R}(X - \zeta I)$ is closed.

Viceversa, let $\mathcal{R}(X - \zeta I)$ be closed for some $\zeta \in \mathbf{R}_+^*$, and let us show that the dissipative operator X is closed. Let the sequence $\{x_\nu\} \subset \mathcal{D}(X)$ converge to $x \in \mathcal{E}$ and be such that $\{Xx_\nu\}$ converges to some $y \in \mathcal{E}$. The sequence $\{w_\nu\}$, where $w_\nu = x_\nu - \zeta Xx_\nu$, converges to $w = x - \zeta y$. Since $\mathcal{R}(X - \zeta I)$ is closed, there exists $z \subset \mathcal{D}(X)$ for which $w = z - \zeta Xz$. Being, by (3.6),

$$\|x_\nu - z\| \leq \|x_\nu - z - \zeta X(x_\nu - z)\|,$$

letting $\nu \to +\infty$ yields

$$\|x - z\| \leq \|x - z - \zeta y + \zeta Xz\|$$
$$= \|w - w\| = 0.$$

Thus $x = z \in \mathcal{D}(X)$ and $x - \zeta y = w = x - \zeta Xx$, i. e., $y = Xx$. ∎

Suppose now that the dissipative operator X is closed, and that the range of $\zeta I - X$ is $\mathcal{E}$ for some $\zeta \in \mathbf{R}_+^*$. Then $\zeta \in r(X)$. The intersection $r(X) \cap \mathbf{R}_+^*$ is open. We will show that it is closed. Let ζ_ν be a sequence in $\mathbf{R}_+^* \cap r(X)$ converging to some $\zeta \in \mathbf{R}_+^*$. For every $y \in \mathcal{E}$ there exists a sequence $\{x_\nu\}$ in $\mathcal{D}(X)$ for which

$$(\zeta_\nu I - X)x_\nu = y \quad \forall \nu = 1, 2, \dots . \tag{3.9}$$

By (3.6),

$$\|y\| = \|(\zeta_\nu I - X)x_\nu\| \geq \zeta_\nu \|x_\nu\|,$$

and - since the set $\{\zeta_\nu\}$ is bounded from below by a positive constant - there is some $k > 0$ for which

$$\|x_\nu\| \leq \frac{1}{\zeta_\nu}\|y\| \leq k.$$

If p and q are positive integers,

$$
\begin{aligned}
\zeta_p(x_p - x_q) - X(x_p - x_q) &= y - \zeta_p x_q + X x_q \\
&= y - (\zeta_p - \zeta_q)x_q - \zeta_q x_q + X x_q \\
&= y - (\zeta_p - \zeta_q)x_q - y \\
&= -(\zeta_p - \zeta_q)x_q.
\end{aligned}
$$

Since, by (3.6),

$$\zeta_p \|x_p - x_q\| \leq \|\zeta_p(x_p - x_q) - X(x_p - x_q)\|,$$

then

$$\zeta_p \|x_p - x_q\| \leq |\zeta_p - \zeta_q|\|x_q\| \leq k\,|\zeta_p - \zeta_q|,$$

and therefore $\{x_\nu\}$ is a Cauchy sequence. Letting $x = \lim_{\nu \to +\infty} x_\nu$, by (3.9) $\{X x_\nu\}$ converges to $\zeta x - y$. Since X is closed, then $x \in \mathcal{D}(X)$ and $Xx = \zeta x - y$, i. e., $y = (\zeta I - X)x$, showing that $\mathcal{R}(\zeta I - X) = \mathcal{E}$, and therefore $\zeta \in r(X) \cap \mathbf{R}_+^*$.

Thus $r(X) \cap \mathbf{R}_+^*$ is open and closed, and, because $\zeta_o \in r(X) \cap \mathbf{R}_+^*$, then $\mathbf{R}_+^* \subset r(X)$. That proves

Proposition 3.2.8 *If the linear operator X is dissipative and closed, then $r(X) \cap \mathbf{R}_+^* \neq \emptyset$ if, and only if, $\mathbf{R}_+^* \subset r(X)$.*

The linear, dissipative operators X which are closed and for which $r(X) \cap \mathbf{R}_+^* \neq \emptyset$ are called *m-dissipative*. The above proposition can be rephrased by saying that *a linear, closed, dissipative operator X is m-dissipative if, and only if, $\mathbf{R}_+^* \subset r(X)$*.

For any $\eta \in \mathbf{R}$,

$$
\begin{aligned}
\sigma(X + i\eta\, I) &= \sigma(X) + i\eta \\
&= \{\zeta + i\eta : \zeta \in \sigma(X)\};
\end{aligned}
$$

furthermore, if $x \in \mathcal{D}(X) \backslash \{0\}$ and $\lambda \in \Lambda(x)$,

$$\Re < (X + i\eta\, I)x, \lambda >= \Re < Xx, \lambda > .$$

Proposition 3.2.8 implies then

Theorem 3.2.9 *If X is an m-dissipative operator, then*

$$\{\zeta \in \mathbf{C} : \Re\zeta > 0\} \subset r(X).$$

If X is closed and dissipative, and if

$$\{\zeta \in \mathbf{C} : \Re\zeta > 0\} \cap r(X) \neq \emptyset,$$

X is m-dissipative.

A dissipative operator X is said to be *maximal dissipative* if every dissipative operator Y for which $X \subset Y$ coincides with X. In view of the Zorn lemma, every dissipative operator is contained in a maximal dissipative operator.

Let X be m-dissipative, and let Y be a dissipative operator for which $X \subset Y$. If $\zeta \in \mathbf{R}_+^*$, then $\zeta \in r(X)$, and therefore - by Theorem 1.9.7 - if $X \neq Y$, then $\zeta \in p\sigma(Y)$, contradicting Lemma 3.2.6. That proves

Proposition 3.2.10 *Every m-dissipative operator is maximal dissipative.*

Exercise [35] Let $\mathcal{E}$ be the complex Banach space

$$\mathcal{E} = \{x \in C([0,1]) : x(0) = x(1) = 0\}$$

with the uniform norm, and let $X = \frac{d}{dr}$ with the maximal domain

$$\mathcal{D}(X) = \{x \in \mathcal{E} : \frac{d}{dr}x \in \mathcal{E}\}.$$

Prove that

$$\bigcap_{\zeta \in \mathbf{R}_+^*} \mathcal{R}(I - \zeta\, X) = \{0\}$$

and that the densely defined and maximal dissipative operator X is not m-dissipative.

Let $X : \mathcal{D}(X) \subset \mathcal{E} \to \mathcal{E}$ be a linear operator. If $\zeta \neq 0$ is such that $\frac{1}{\zeta} \in r(X)$, then, for all $x \in \mathcal{D}(X)$, $(I - \zeta X)^{-1}x \in \mathcal{D}(X)$, and

$$(I - \zeta X)^{-1}x - x = \zeta(I - \zeta X)^{-1}Xx.$$

If X is m-dissipative and if $h > 0$, then $h \in r(X)$ and, by Corollary 3.2.3, $\|(I - hX)^{-1}\| \leq 1$. Hence

$$\|(I - hX)^{-1}x - x\| \leq h\|Xx\|, \tag{3.10}$$

and therefore

$$\lim_{h \downarrow 0}(I - hX)^{-1}x = x \quad \forall x \in \mathcal{D}(X). \tag{3.11}$$

Furthermore,

$$X(I - hX)^{-1}x = (I - hX)^{-1}Xx = \frac{1}{h}((I - hX)^{-1} - I)x,$$

whence, by (3.10),

$$\|X(I - hX)^{-1}x\| \leq \|Xx\| \quad \forall x \in \mathcal{D}(X), \forall h > 0. \tag{3.12}$$

Proposition 3.2.11 *If $\mathcal{E}$ is reflexive, and if X is m-dissipative, then $\mathcal{D}(X)$ is dense in $\mathcal{E}$.*

Proof Since $\mathcal{E}$ is reflexive, (3.12) implies that, for every $x \in \mathcal{D}(X)$ there exist $z \in \mathcal{E}$ and a sequence $\{h_n\}$ in $\mathbf{R}_+^*$ such that $\lim_{n \to +\infty} h_n = 0$ and

$$\lim_{n \to +\infty} < X(I - h_nX)^{-1}x, \lambda >=< z, \lambda > \quad \forall \lambda \in \mathcal{E}'. \tag{3.13}$$

The linear operator X is closed, and therefore also weakly closed. Thus, by (3.11) and (3.13),

$$\lim_{n \to +\infty} < X(I - h_nX)^{-1}x, \lambda > = < Xx, \lambda >$$
$$= < z, \lambda > \quad \forall \lambda \in \mathcal{E}', \tag{3.14}$$

and therefore $Xx = z$. Let

$$< x, \lambda >= 0 \quad \forall x \in \mathcal{D}(X). \tag{3.15}$$

Since

$$X(I - h_n X)^{-1}x = (I - h_n X)^{-1}Xx \in \mathcal{D}(X),$$

(3.11) and (3.13) imply that also $< Xx, \lambda >= 0$, and therefore

$$< Xx - x, \lambda >= 0$$

for all $x \in \mathcal{D}(X)$; that is to say, λ vanishes on $\mathcal{R}(I - X)$. Since X is m-dissipative, and therefore $\mathcal{R}(I - X) = \mathcal{E}$, $\lambda = 0$. ∎

Let X be a linear dissipative operator in a complex Hilbert space $\mathcal{H}$, i. e., such that

$$\Re(Xx|x) \leq 0 \ \ \forall x \in \mathcal{D}(X).$$

Theorem 3.2.12 *A linear, densely defined, dissipative operator in a complex Hilbert space is maximal dissipative if, and only if, it is m-dissipative.*

By Proposition 3.2.10, what is left to prove is that, if the operator is maximal dissipative, then it is m-dissipative. This fact is an immediate consequence of the following lemma.

Lemma 3.2.13 *Every densely defined, linear, closed, dissipative operator X on the Hilbert space $\mathcal{H}$ has an m-dissipative extension.*

Proof Let $X : \mathcal{D}(X) \subset \mathcal{H} \to \mathcal{H}$ be a linear, closed, densely defined. dissipative operator. By Lemma 3.2.7, $\mathcal{R}(\zeta I - X)$ is closed for all $\zeta \in \mathbf{R}_+^*$. Suppose that X is not m-dissipative. For all $\zeta \in \mathbf{R}_+^*$, $\mathcal{R}(\zeta I - X) \neq \mathcal{H}$. For any fixed $\zeta \in \mathbf{R}_+^*$,

$$\mathcal{N} := (\mathcal{R}(\zeta I - X))^{\perp} \neq \{0\}. \tag{3.16}$$

If $u \in \mathcal{N}$ and $x \in \mathcal{D}(X)$, then $(Xx|u) = \zeta(x|u)$, and therefore the linear form $x \mapsto (Xx|u)$ is continuous on $\mathcal{D}(X)$. As a consequence, $u \in \mathcal{D}(X^*)$, where X^* is the adjoint of X. Therefore $\mathcal{N} \subset \mathcal{D}(X^*)$ and $(x|(X^* - \zeta I)u) = 0$ for all $x \in \mathcal{D}(X)$, and thus

$$X^*u = \zeta u \ \ \forall u \in \mathcal{N}.$$

Now we define a linear operator $\tilde{X}$, with domain $\mathcal{D}(\tilde{X}) = \mathcal{D}(X) + \mathcal{N}$, by setting

$$\tilde{X}(x+u) = Xx - \zeta u \quad (x \in \mathcal{D}(X),\ u \in \mathcal{N}). \tag{3.17}$$

For the definition to make sense, it suffices to show that, if $x+u=0$, then $x = 0$. Indeed, if $x = -u \in \mathcal{N}$, then $u \in \mathcal{D}(X) \cap \mathcal{D}(X^*)$, and therefore

$$(Xu|u) = (u|X^*u) = \zeta\|u\|^2.$$

Since $\zeta > 0$ and $\Re(Xu|u) \leq 0$, then $u = 0$, and therefore $x = 0$.

If $x \in \mathcal{D}(X)$ and $u \in \mathcal{N}$,

$$\begin{aligned}
(\tilde{X}(x+u)|x+u) &= (Xx - \zeta u|x+u) = \\
(Xx|x) &+ (Xx|u) - \zeta(u|x) - \zeta\|u\|^2 \\
&= (Xx|x) + (x|X^*u) - \zeta(u|x) - \zeta\|u\|^2 \\
&= (Xx|x) + \zeta(x|u) - \zeta(u|x) - \zeta\|u\|^2 \\
&= (Xx|x) + 2\zeta i\Im(x|u) - \zeta\|u\|^2,
\end{aligned}$$

and therefore

$$\Re(\tilde{X}(x+u)|x+u) = \Re(Xx|x) - \zeta\|u\|^2 \leq 0,$$

that is, $\tilde{X}$ is a dissipative operator extending X. We prove now that $\tilde{X}$ is maximal dissipative, showing, first of all, that

$$\mathcal{R}(\tilde{X} - \zeta I) = \mathcal{H}. \tag{3.18}$$

Let $y \in \mathcal{H}$, and let $z \in \mathcal{R}(X - \zeta I)$ and $w \in \mathcal{N}$ be such that $y = z + w$. Let $x \in \mathcal{D}(X)$ be such that

$$Xx - \zeta x = z = y - w, \tag{3.19}$$

Setting $u = -\frac{1}{2\zeta}w$, (3.17) and (3.19) yield

$$\begin{aligned}
\tilde{X}(x+u) - \zeta(x+u) &= Xx - \zeta u - \zeta x - \zeta u \\
&= Xx + w - \zeta x = y.
\end{aligned}$$

Hence (3.18) holds and, by Lemma 3.2.7, $\tilde{X}$ is closed. Since $\zeta \in r(X) \cap \mathbf{R}_+^*$, $\tilde{X}$ is m-dissipative, hence maximal dissipative by Proposition 3.2.10 [3]. ∎

The proof of Theorem 3.2.12 follows [48]. For further results on linear dissipative operators in Hilbert spaces, cf. e. g., [48] and [4].

Example Let

$$\mathcal{E} = l^1 = \{x = (x_0, x_1, \dots) : \|x\| = \sum_{n=0}^{+\infty} |x_n| < \infty\},$$

and let X be the linear operator defined by the matrix

$$X = \begin{pmatrix} 0 & 0 & 0 & \cdots \\ 0 & -1 & 0 & \cdots \\ 0 & 0 & -2 & \cdots \\ & \cdot & \cdot & \cdot & \cdots \\ & \cdot & \cdot & \cdot & \cdots \\ & \cdot & \cdot & \cdot & \cdots \end{pmatrix} \tag{3.20}$$

acting on $\mathcal{E}$, with maximal domain

$$\mathcal{D}(X) = \{x \in \mathcal{E} : \sum_{n=1}^{+\infty} n|x_n| < \infty\}. \tag{3.21}$$

Since the set of all $x = (x_0, x_1, \dots)$ containing only a finite number of $x_n \neq 0$ is contained in $\mathcal{D}(X)$ and is dense in $\mathcal{E}$, then $\overline{\mathcal{D}(X)} = \mathcal{E}$. Let $x \in \mathcal{E}$ with $\|x\| = \sum_{n=0}^{+\infty} |x_n| = 1$; if $x_n \neq 0$, let $\theta_n \in [0, 2\pi)$ be the argument of x_n and $y_n = e^{-i\theta_n}$; let $y_n = 0$ if $x_n = 0$. The linear form λ on $\mathcal{E}$ whose value at $z = (z_0, z_1, \dots)$ is

$$< z, \lambda > = \sum_{0}^{+\infty} z_n y_n,$$

---

[3]Here is a direct proof. If that were not the case, there would exist a proper dissipative extension Z of X and some $v \in \mathcal{D}(Z)\backslash\{0\}$, such that $(Z - \zeta I)v = 0$. Hence $\mathcal{R}(Zv|v) = \zeta\|v\|^2 > 0$, contradicting the fact that Z is dissipative.

is continuous on $\mathcal{E}$ with norm $\|\lambda\| = 1$. Since

$$< x, \lambda > = \sum_{0}^{+\infty} x_n y_n = \sum_{0}^{+\infty} |x_n| e^{i\theta_n} e^{-i\theta_n}$$

$$= \sum_{n=0}^{+\infty} |x_n| = 1$$

and

$$< Xx, \lambda > = -\sum_{0}^{+\infty} n x_n y_n = -\sum_{0}^{+\infty} n |x_n| e^{i\theta_n} e^{-i\theta_n}$$

$$= -\sum_{n=0}^{+\infty} n|x_n| \leq 0,$$

X is dissipative. To determine the spectrum $\sigma(X)$ (or, equivalently, $r(X)$), one has to find the values of $\zeta \in \mathbf{C}$ for which the equation

$$(\zeta I - X)x = y \tag{3.22}$$

is solvable with respect to $x \in \mathcal{D}(X)$ for any given $y \in \mathcal{E}$. Since the left hand side reads

$$(\zeta I - X)x = (\zeta x_0, (\zeta + 1)x_1, (\zeta + 2)x_2, \dots),$$

setting $y = 0$ we see that $x \in \ker(\zeta I - X)\backslash\{0\}$ if, and only if, there is $n_o \in \mathbf{N}$ such that $x_n = 0$ whenever $n \neq n_o$ and $\zeta = -n_o$. Thus,

$$p\sigma(X) = -\mathbf{N} = \{-n : n \in \mathbf{N}\},$$

and every eigenspace has complex dimension one.

For

$$\zeta \notin p\sigma(X) \tag{3.23}$$

and any $y \in \mathcal{E}$ we solve (3.22), obtaining $x_n = \frac{1}{\zeta+n} y_n$ for $n \in \mathbf{N}$. Since $\lim_{n\to+\infty} \frac{n}{\zeta+n} = 1$, then $\sum_{n=0}^{+\infty} n|x_n| < \infty$, and therefore $x \in \mathcal{D}(X)$. Hence $x \mapsto (\zeta I - X)x$ maps $\mathcal{D}(X)$ bijectively onto $\mathcal{E}$. We show now that $(\zeta I - X)^{-1} \in \mathcal{L}(\mathcal{E})$.

In view of (3.23), $\zeta + n \neq 0$ for all $n \in \mathbf{N}$, and there is some positive integer n_o such that $|\zeta + n| \geq 1$ whenever $n > n_o$. As a consequence,

$$\sum_{n=0}^{+\infty} |x_n| \leq \frac{|y_0|}{|\zeta|} + \frac{|y_1|}{|\zeta + 1|} + \cdots + \frac{|y_{n_o}|}{|\zeta + n_o|} +$$

$$|y_{n_o+1}| + |y_{n_o+2}| + \cdots$$

$$\leq \frac{1}{\min\{|\zeta|, |\zeta + 1|, \ldots, |\zeta + n_o|\}} \sum_{n=0}^{+\infty} |y_n|,$$

whence $(\zeta I - X)^{-1} \in \mathcal{L}(\mathcal{E})$ and

$$\|(\zeta I - X)^{-1}\| \leq \frac{1}{\min\{|\zeta|, |\zeta + 1|, \ldots, |\zeta + n_o|\}}. \tag{3.24}$$

That proves that

$$\sigma(X) = p\sigma(X) = -\mathbf{N}.$$

In particular the open half plane $\{\zeta \in \mathbf{C} : \Re\zeta > 0\}$ is contained in $r(X)$.

Since $(\zeta I - X)^{-1}$ is continuous, $\zeta I - X$ is closed, and therefore X is closed.

In conclusion, the following lemma has been proven.

Lemma 3.2.14 *The linear operator X acting on l^1, defined by (3.20) is m-dissipative.*

For this example, see [4].

Lemma 3.2.15 *If the closed, densely defined operator X is dissipative, and also its dual X' is dissipative, then both X and X' are m-dissipative.*

Proof If $\zeta \in \mathbf{R}_+^*$, then $\zeta \in r(X) \cup r\sigma(X)$. Hence, by Lemma 1.17.3 and Lemma 1.17.4, $\zeta \in r(X') \cup p\sigma(X')$. If $\zeta \in r\sigma(X)$, then $\zeta \in p\sigma(X')$, contradicting the hypothesis whereby X' is dissipative. Thus, $\zeta \in r(X)$ and, consequently, $\zeta \in r(X')$. ∎

3.3 Semigroups of linear contractions

Let X be a densely defined, m-dissipative linear operator acting on $\mathcal{E}$. By Theorem 3.2.8, $\mathbf{R}_+^* \subset r(X)$ and thus, by (3.6) and Theorem 2.9.4, X is the infinitesimal generator of a strongly continuous semigroup T of linear contractions of $\mathcal{E}$.

Viceversa, given any strongly continuous semigroup T of linear contractions of $\mathcal{E}$, by Theorem 2.9.4 its infinitesimal generator X is closed, densely defined, such that $\mathbf{R}_+^* \subset r(X)$ and that (3.6) holds. At this point, Theorem 3.2.8, implies that X is m-dissipative. However, the fact that X is dissipative is also a consequence of the following considerations, which yield a more precise information on X. Let $x \in \mathcal{D}(X)$ and let λ be *any* continuous linear form on $\mathcal{E}$ such that $<x,\lambda>=\|x\|, \|\lambda\|=1$. Then

$$\Re<Xx,\lambda>=\lim_{t\downarrow 0}\frac{1}{t}(\Re<T(t)x,\lambda>-\|x\|)$$

$$\leq \lim_{t\downarrow 0}\frac{1}{t}(\|T(t)x\|-\|x\|)\leq 0.$$

In conclusion, the Lumer-Phillips theorem holds:

Theorem 3.3.1 *Any linear, densely defined, m-dissipative operator* $X : \mathcal{D}(X) \subset \mathcal{E} \to \mathcal{E}$ *is the infinitesimal generator of a strongly continuous semigroup* $T : \mathbf{R}_+ \to \mathcal{L}(\mathcal{E})$ *of linear contractions of $\mathcal{E}$. Viceversa, the infinitesimal generator of any such semigroup satisfies the inequality* $\Re < Xx,\lambda >\leq 0$ *for all* $x \in \mathcal{D}(X)$ *and all* $\lambda \in \mathcal{E}'$ *for which* $<x,\lambda>=\|x\|$ *and* $\|\lambda\|=1$.

We have seen that Theorem 2.9.1 implies Theorem 3.3.1. We will show that, viceversa, this latter theorem implies the first one.

Let $T : \mathbf{R}_+ \to \mathcal{L}(\mathcal{E})$ be a strongly continuous semigroup that is *uniformly bounded, i.e.* such that

$$\|T(t)\| \leq M \ \forall t \in \mathbf{R}_+^* \tag{3.25}$$

and for some $M \geq 1$. Setting

$$\|x\| = \sup\{\|T(t)x\| : t \in \mathbf{R}_+\},$$

$\| \; \|$ is a norm on $\mathcal{E}$ for which

$$\|x\| \le \|x\| \le M \, \|x\| \quad \forall x \in \mathcal{E}, \tag{3.26}$$

and therefore is equivalent to the norm $\| \, \|$. Furthermore,

$$\|T(t)x\| = \sup\{\|T(s)T(t)x\| : s \in \mathbf{R}_+^*\}$$
$$\le \sup\{\|T(s)x\| : s \in \mathbf{R}_+^*\} = \|x\|,$$

proving thereby that $T(t)$ is a contraction for the norm $\| \; \|$. Since moreover, by (3.26),

$$\|T(t)x - x\| \le M \, \|T(t)x - x\|,$$

the following lemma holds.

Lemma 3.3.2 *If the strongly continuous semigroup T is uniformly bounded, there is an equivalent norm on $\mathcal{E}$ with respect to which T is a strongly continuous semigroup of linear contractions[4].*

The infinitesimal generator remains unchanged, together with its being closed and densely defined.

Remark If the strongly continuous semigroup T is such that

$$\sup\{| < T(t)x, \lambda > | : t \in \mathbf{R}_+\} < \infty \quad \forall x \in \mathcal{E} \; \forall \lambda \in \mathcal{E}', \tag{3.27}$$

by the Banach-Steinhaus theorem

$$\sup\{\|T(t)x\| : t \in \mathbf{R}_+\} < \infty \quad \forall x \in \mathcal{E}.$$

Again by the Banach-Steinhaus theorem,

$$\sup\{\|T(t)\| : t \in \mathbf{R}_+\} < \infty.$$

Thus, (3.27) is equivalent to requiring that the strongly continuous semigroup T be uniformly bounded.

---

[4]Answering a question raised by J.A.Goldstein, P.R.Chernoff [11] has constructed an example of a strongly continuous semigroup $T : \mathbf{R}_+ \to \mathcal{L}(\mathcal{H})$ in a complex Hilbert space $\mathcal{H}$, for which there is no Hilbert space norm which is equivalent to the given norm in $\mathcal{H}$, and there is no $\alpha \in \mathbf{R}$ such that $t \mapsto e^{-\alpha t} T(t)$ becomes a semigroup of contractions.

By Theorem 3.3.1, $\mathbf{R}_+^* \subset r(X)$ and

$$\|(\zeta I - X)^{-1}\| \leq \frac{1}{\zeta}\|x\| \ \ \forall x \in \mathcal{D}(X), \ \forall \zeta \in \mathbf{R}_+^*.$$

Hence

$$\|(\zeta I - X)^{-n}\| \leq \|(\zeta I - X)^{-n}\|$$
$$\leq \frac{1}{\zeta^n}\|x\| \leq \frac{M}{\zeta^n}\|x\|,$$

and therefore

$$\|(\zeta I - X)^{-n}\| \leq \frac{M}{\zeta^n} \ \ \forall \zeta \in \mathbf{R}_+^*, \ \ \forall n = 1, 2, \dots . \tag{3.28}$$

Assume now that T satisfies (2.6). The strongly continuous semigroup $S : t \mapsto e^{-at}T(t)$ is generated by $Y := X - aI$ with domain $\mathcal{D}(Y) = \mathcal{D}(X)$. Since $\|S(t)\| \leq M$ for all $t \in \mathbf{R}_+$, (3.28) yields the inequality

$$\|(\tau I - Y)^{-n}\| \leq \frac{M}{\tau^n} \ \ \forall \tau \in \mathbf{R}_+^*, \ \ \forall n = 1, 2, \dots , \tag{3.29}$$

which - being $\tau I - X = (\tau - a)I - Y$ - becomes (2.60).

Let now X be a closed, densely defined, linear operator such that

$$\mathbf{R}_+^* \subset r(X) \tag{3.30}$$

and that (3.28) holds. We will show that X is the infinitesimal generator of a strongly continuous semigroup $T : \mathbf{R}_+ \to \mathcal{L}(\mathcal{E})$ satisfying (3.25).

Lemma 3.3.3 *Let $X : \mathcal{D}(X) \subset \mathcal{E} \to \mathcal{E}$ be a closed linear operator for which (3.30), and also (3.28) for some $M \geq 1$, hold. There exists in $\mathcal{E}$ an equivalent norm $\|\ \|$ to $\|\ \|$ that satisfies (3.26) and is such that*

$$\|\tau(\tau I - X)^{-1}\| \leq 1 \ \ \forall \tau \in \mathbf{R}_+^*. \tag{3.31}$$

Proof For $\mu > 0$ and $x \in \mathcal{E}$ let

$$\|x\|_\mu := \sup\{\|\mu^n(\mu I - X)^{-n}x\| : n = 0, 1, \dots \}.$$

Because

$$\|x\| \le \|x\|_\mu \le M \, \|x\|, \tag{3.32}$$

$\| \; \|_\mu$ is an equivalent norm to $\| \; \|$, for which

$$
\begin{aligned}
\|\mu(\mu I - X)^{-1}x\|_\mu &= \sup\{\|\mu^n(\mu I - X)^{-n}(\mu(\mu I - X)^{-1})x\| : n = 0, 1, \ldots\} \\
&= \sup\{\|\mu^n(\mu I - X)^{-n}x\| : n = 1, 2, \ldots\} \\
&\le \sup\{\|\mu^n(\mu I - X)^{-n}x\| : n = 0, 1, 2, \ldots\} \\
&= \|x\|_\mu
\end{aligned}
$$

for all $x \in \mathcal{E}$. Therefore

$$\|\mu(\mu I - X)^{-1}x\|_\mu \le 1 \ \ \forall \mu \in \mathbf{R}_+^*. \tag{3.33}$$

We show now that

$$\|\tau(\tau I - X)^{-1}\|_\mu \le 1 \ \ \forall \tau \in (0, \mu]. \tag{3.34}$$

For $x \in \mathcal{E}$, let $y = (\tau I - X)^{-1}x$. The identity

$$(\tau I - X)^{-1} - (\mu I - X)^{-1} = (\mu - \tau)(\mu I - X)^{-1}(\tau I - X)^{-1},$$

that can also be written,

$$(\tau I - X)^{-1} = (\mu I - X)^{-1})(I + (\mu - \tau)(\tau I - X)^{-1}),$$

yields

$$y = (\mu I - X)^{-1}(x + (\mu - \tau)y).$$

Hence, by (3.33),

$$
\begin{aligned}
\|y\|_\mu &\le \|(\mu I - X)^{-1}x\|_\mu + (\mu - \tau)\|(\mu I - X)^{-1}y\|_\mu \\
&\le \frac{1}{\mu}\|x\|_\mu + (\mu - \tau)\frac{1}{\mu}\|y\|_\mu \\
&= \frac{1}{\mu}\|x\|_\mu + (1 - \frac{\tau}{\mu})\|y\|_\mu,
\end{aligned}
$$

whence

$$\tau\|y\|_\mu \le \|x\|_\mu.$$

This inequality can also be written

$$\|\tau(\tau I - X)^{-1}x\|_\mu \leq \|x\|_\mu \ \forall x \in \mathcal{E},$$

and is equivalent to (3.34). By this latter inequality and by (3.32),

$$\|\tau^n(\tau I - X)^{-n}x\| \leq \|\tau^n(\tau I - X)^{-n}x\|_\mu$$
$$\leq \|x\|_\mu \ \forall n \in \mathbf{N}, \tag{3.35}$$

and that proves that

$$\|x\|_\tau \leq \|x\|_\mu \ \forall \tau \in (0, \mu],$$

i. e., that $\mu \mapsto \|x\|_\mu$ is a non-decreasing function of $\mu \in \mathbf{R}_+^*$. Setting

$$\|x\| = \lim_{\mu \to +\infty} \|x\|_\mu,$$

(3.32) yields (3.26).

Choosing $n = 1$ in (3.35), and letting $\mu \to +\infty$ we obtain (3.31).

■

In view of Lemma 3.3.3 and of Theorem 3.3.1, any linear, closed, densely defined operator X which satisfies (3.30) and also (3.28) for some $M \geq 1$, generates a strongly continuous semigroup $T : \mathbf{R}_+ \to \mathcal{L}(\mathcal{E})$ for which $\|T(t)\| \leq 1$, i. e., $\|T(t)\| \leq M$ for all $t \geq 0$.

Finally, if the linear operator X is closed, densely defined, and if there exist $M \geq 1$ and $a \in \mathbf{R}$ for which $(a, +\infty) \subset r(X)$ and (2.60) holds, the operator $Y = X - aI$ is such that (3.30) and (3.29) hold. Hence Y generates a strongly continuous semigroup $S : \mathbf{R}_+ \to \mathcal{L}(\mathcal{E})$ such that $\|S(t)\| \leq M$ for all $t \in \mathbf{R}_+$. The strongly continuos semigroup $T : \mathbf{R}_+ \to \mathcal{L}(\mathcal{E})$ defined by $T(t) = e^{at}S(t)$ (2.6) and is generated by $Y - aI + aI = X$.

Let $T : \mathbf{R} \to \mathcal{L}(\mathcal{E})$ be a strongly continuous group of linear contractions of $\mathcal{E}$. For all $x \in \mathcal{E}$ and all $t \in \mathbf{R}$,

$$\|x\| = \|T(t)^{-1}T(t)x\|$$
$$\leq \|T(t)^{-1}\| \ \|T(t)x\| \leq |\ T(t)x\| \leq \|x\|.$$

Thus $\|T(t)x\| = \|x\|$ for all $x \in \mathcal{E}$ and all $t \in \mathbf{R}$; that is to say, $T(t)$ is a linear surjective isometry of $\mathcal{E}$ for all $t \in \mathbf{R}$, and the following theorem holds.

Theorem 3.3.4 *If* $T : \mathbf{R} \to \mathcal{L}(\mathcal{E})$ *is a strongly continuous group of linear contractions of a complex Banach space* $\mathcal{E}$, *both its infinitesimal generator* X *and* $-X$ *are m-dissipative. Viceversa, a closed, densely defined, linear operator* X, *which is m-dissipative together with* $-X$, *is the infinitesimal generator of a strongly continuous group of surjective linear isometries of* $\mathcal{E}$.

We will now describe a uniformly continuous semigroup of contractions of $\mathcal{E}$, which will be useful later on.

Let $C \in \mathcal{L}(\mathcal{E})$. For $t \in \mathbf{R}_+$,

$$\exp t(C - I) = \mathrm{e}^{-t} \exp tC$$

$$= \mathrm{e}^{-t} \left(I + tC + \frac{t^2}{2!}C^2 + \cdots \right),$$

and therefore

$$\| \exp t(C - I) \| \leq \mathrm{e}^{-t} \sum_{n=0}^{+\infty} \frac{t^n}{n!} \|C\|^n$$

$$= \mathrm{e}^{-t} \mathrm{e}^{t\|C\|} = \mathrm{e}^{t(\|C\|-1)}.$$

Hence , if $\|C\| \leq 1$, then $\| \exp t(C - I) \| \leq 1$ for all $t \in \mathbf{R}_+$. Let $\|C\| \leq 1$. For $x \in \mathcal{E}$ and $n \in \mathbf{N}^*$,

$$(\exp n(C - I) - C^n)x = \mathrm{e}^{-n}(\exp nC)x - C^n x =$$

$$\mathrm{e}^{-n} \left\{ \left(I + nC + \frac{n^2}{2!}C^2 + \cdots \right) x - \right.$$

$$\left. - \left(1 + n + \frac{n^2}{2!} + \cdots \right) C^n x \right\}$$

$$= \mathrm{e}^{-n} \left\{ (I - C^n) + n(C - C^n) + \frac{n^2}{2!}(C^2 - C^n) + \cdots \right.$$

$$\left. + \frac{n^p}{p!}(C^p - C^n) + \cdots \right\} x.$$

Therefore

$$\|(\exp n(C - I) - C^n)x\| \leq \mathrm{e}^{-n} \sum_{p=0}^{+\infty} \frac{n^p}{p!} \|(C^p - C^n)x\|$$

$$\leq \mathrm{e}^{-n} \sum_{p=0}^{+\infty} \frac{n^p}{p!} \|(C^{|p-n|}x - x\|.$$

But, since, for $q = 1, 2, \dots,$

$$\|C^q x - x\| = C^q x - C^{q-1} x + C^{q-1} x - C^{q-2} x + \cdots +$$
$$C^2 x - Cx + Cx - x$$
$$\leq \|C^q x - C^{q-1} x\| + \|C^{q-1} x - C^{q-2} x\| + \cdots +$$
$$\|C^2 x - Cx\| + \|Cx - x\|$$
$$\leq (\|C^{q-1}\| + \|C^{q-2}\| + \cdots + 1)\|Cx - x\|$$
$$\leq q\|Cx - x\|,$$

then

$$\|(\exp n(C - I) - C^n)x\| \leq e^{-n} \sum_{p=0}^{+\infty} \frac{n^p}{p!} |p - n| \, \|Cx - x\|.$$

By the Schwarz inequality,

$$e^{-n} \sum_{p=0}^{+\infty} \frac{n^p}{p!} |p - n| \leq \left(e^{-n} \sum_{p=0}^{+\infty} \frac{n^p}{p!} (p - n)^2 \right)^{\frac{1}{2}}$$
$$= \left(e^{-n} \sum_{p=0}^{+\infty} \frac{n^p}{p!} (n^2 - 2np + p^2) \right)^{\frac{1}{2}}.$$

Being

$$\sum_{p=0}^{+\infty} \frac{n^p}{p!} n^2 = n^2 e^n,$$

$$\sum_{p=0}^{+\infty} \frac{n^p}{p!} np = \sum_{p=1}^{+\infty} \frac{n^{p+1}}{(p-1)!}$$
$$= \sum_{p=0}^{+\infty} \frac{n^{p+2}}{p!} = n^2 e^n,$$

$$\sum_{p=0}^{+\infty} \frac{n^p}{p!} p^2 = \sum_{p=1}^{+\infty} \frac{n^p}{(p-1)!} p$$

$$= \sum_{p=1}^{+\infty} \frac{n^p}{(p-1)!}(p-1+1)$$

$$= \sum_{p=2}^{+\infty} \frac{n^p}{(p-2)!} + \sum_{p=1}^{+\infty} \frac{n^p}{(p-1)!}$$

$$= \sum_{p=0}^{+\infty} \frac{n^{p+2}}{p!} + \sum_{p=0}^{+\infty} \frac{n^{p+1}}{p!} n^2$$

$$= (n^2 + n)e^n,$$

then

$$\sum_{p=0}^{+\infty} \frac{n^p}{p!}(n^2 - 2np + p^2) = (n^2 - 2n^2 + n^2 + n)e^n = ne^n,$$

and, in conclusion,

$$\|(\exp n(C - I) - C^n)x\| \leq \sqrt{n}\,\|Cx - x\|. \qquad (3.36)$$

That proves

Lemma 3.3.5 *If $C \in \mathcal{L}(\mathcal{E})$ is a contraction, then $\mathbf{R}_+ \ni t \to \exp t(C - I)$ defines a uniformly continuous semigroup of linear contractions of $\mathcal{E}$. Moreover, (3.36) holds for all $x \in \mathcal{E}$ and for all positive integers n.*

3.4 Linear contractions of a Hilbert space

Let $T : \mathbf{R}_+ \to \mathcal{L}(\mathcal{H})$ be a semigroup of linear contractions of a complex Hilbert space $\mathcal{H}$, and let

$$\mathcal{K} = \{x \in \mathcal{H} : \|T(t)x\| = \|x\| \ \forall t \in \mathbf{R}_+\}.$$

If $x, y \in \mathcal{K}$, then

$$
\begin{aligned}
\|x\|^2 + \|y\|^2 &= \|T(t)x\|^2 + \|T(t)y\|^2 \\
&= \frac{1}{2} \left(\|T(t)(x+y)\|^2 + \|T(t)(x-y)\|^2 \right) \\
&\leq \frac{1}{2} \left(\|x+y\|^2 + \|x-y\|^2 \right) \\
&= \|x\|^2 + \|y\|^2
\end{aligned}
$$

for all $t \geq 0$. Hence $\|T(t)(x+y)\| = \|x+y\|$. i. e ., $x + y \in \mathcal{K}$. Since, furthermore, $\zeta \in \mathbf{C}$, $x \in \mathcal{K} \Rightarrow \zeta x \in \mathcal{K}$, then $\mathcal{K}$ is a linear subspace of $\mathcal{H}$.

If a sequence $\{x_\nu\}$ in $\mathcal{K}$ converges to some $x \in \mathcal{H}$, for $\nu \gg 0$ the chain of inequalities

$$
\begin{aligned}
\|x\| \geq \|T(t)x\| &= \|T(t)(x - x_\nu) + T(t)x_\nu\| \\
&\geq \|T(t)x_\nu\| - \|T(t)(x - x_\nu)\| \\
&\geq \|x_\nu\| - \|x - x_\nu\|
\end{aligned}
$$

implies that $\|T(t)x\| = \|x\|$ for all $t \geq 0$, showing, in conclusion, that $\mathcal{K}$ is a closed linear subspace of $\mathcal{H}$.

If $x \in \mathcal{K}$ and $t \in \mathbf{R}_+$,

$$
\|T(t)x\| = \|x\| = \|T(s+t)x\| = \|T(s)\,T(t)x\|
$$

for all $s \geq 0$. Thus $\mathcal{K}$ is invariant under $T(s)$, which acts on $\mathcal{K}$ as an isometry. As a consequence, also the closed linear space

$$
\mathcal{H}_o := \bigcap_{t \in \mathbf{R}_+} T(t)\mathcal{K}
$$

is invariant under $T(s)$, and $T(s)_{|\mathcal{H}_o}$ is an isometry.

If $x \in \mathcal{H}_o$ and $t \geq 0$, for every $s \geq 0$ there exist a unique $y_s \in \mathcal{K}$ and a unique $y_{t+s} \in \mathcal{K}$ for which

$$
x = T(s)y_s = T(s+t)y_{s+t} = T(s)\,T(t)y_{s+t}.
$$

Since $T(s)$ - being an isometry - is injective, then

$$
y_s = T(t)y_{s+t} \in T(t)\mathcal{K} \quad \forall t \in \mathbf{R}_+,
$$

and therefore $y_s \in \mathcal{H}_o$. That proves that $T(t)_{|\mathcal{H}_o}$ is a surjective isometry. As a consequence of the polarization formula we have then

$$(T(t)x_1|T(t)x_2) = (x_1|x_2)$$

for all $x_1, x_2 \in \mathcal{H}_o$, and therefore

$$(T(t)^*T(t))_{|\mathcal{H}_o} = I_{\mathcal{H}_o} \quad \forall t \in \mathbf{R}_+. \tag{3.37}$$

If $x \in \mathcal{H}_o$ and $t \geq 0$, there is a unique $u_t \in \mathcal{H}_o$ for which $x = T(t)u_t$. Since, by (3.37),

$$T(t)^*x = T(t)^*T(t)u_t = u_t \in \mathcal{H}_o,$$

$\mathcal{H}_o$ is invariant also under $T(t)^*$ for all $t \geq 0$.

Let

$$T(t) = \begin{pmatrix} A(t) & B(t) \\ C(t) & D(t) \end{pmatrix}$$

be the matrix representation of $T(t)$ with respect to the orthogonal decomposition

$$\mathcal{H} = \mathcal{H}_o \oplus \mathcal{H}_o^\perp, \tag{3.38}$$

with $A(t) \in \mathcal{L}(\mathcal{H}_o)$, $B(t) \in \mathcal{L}(\mathcal{H}_o^\perp, \mathcal{H}_o)$, $C(t) \in \mathcal{L}(\mathcal{H}_o, \mathcal{H}_o^\perp)$, $D(t) \in \mathcal{L}(\mathcal{H}_o^\perp)$. Since $T(t)\mathcal{H}_o \subset \mathcal{H}_o$, then $C(t) = 0$. Being

$$T(t)^* = \begin{pmatrix} A(t)^* & 0 \\ B(t)^* & D(t)^* \end{pmatrix},$$

the inclusion $T(t)^*\mathcal{H}_o \subset \mathcal{H}_o$ implies that $B(t)^* = 0$, and therefore $B(t) = 0$. Hence

$$T(t)^* = \begin{pmatrix} A(t)^* & 0 \\ 0 & D(t)^* \end{pmatrix},$$

and, as a consequence, $T(t)(\mathcal{H}_o^\perp) \subset \mathcal{H}_o^\perp$ for all $t \geq 0$. Furthermore, (3.38) yields

$$T(t)^*T(t) = \begin{pmatrix} A(t)^*A(t) & 0 \\ 0 & D(t)^*D(t) \end{pmatrix}$$
$$= \begin{pmatrix} I_{\mathcal{H}_o} & 0 \\ 0 & D(t)^*D(t) \end{pmatrix},$$

with $\|D(t)\| \leq 1$.

If $\mathcal{F}$ is a closed invariant subspace for the semigroup of contractions $T : \mathbf{R}_+ \to \mathcal{L}(\mathcal{H})$, in the sense that $T(t)\mathcal{F} \subset \mathcal{F}$ for all $t \geq 0$, and if $T(t)_{|\mathcal{F}}$ is a surjective isometry for all $t \geq 0$, we say that T is *unitary on $\mathcal{F}$* or that $\mathcal{F}$ is a *unitary subspace* for the semigroup T. If there are no non-trivial unitary subspaces for T, we say that T is *completely non-unitary*.

If $\mathcal{F}$ is a unitary subspace for the semigroup T, then $T(t)\mathcal{F} = \mathcal{F}$ for all $t \geq 0$, and therefore $\mathcal{F} \subset \mathcal{H}_o$. Hence the following theorem holds.

Theorem 3.4.1 *For every semigroup $T : \mathbf{R}_+ \to \mathcal{L}(\mathcal{H})$ of linear contractions of the complex Hilbert space $\mathcal{H}$ there is a unique orthogonal decomposition*

$$\mathcal{H} = \mathcal{H}_0 \oplus \mathcal{H}_1$$

of $\mathcal{H}$, where $\mathcal{H}_0$ and $\mathcal{H}_1$ are closed, invariant subspaces for T, and T is unitary on $\mathcal{H}_0$ and completely non-unitary on $\mathcal{H}_1$.

Remark The proof requires no hypothesis on the continuity of the semigroup T. Theorem 3.4.1 has been proved by B.Sz.Nagy and C.Foias [55] for the iterates of a single operator, and by E.B.Davies [15] in the general case.

Let now $Z : \mathcal{D}(Z) \subset \mathcal{H} \to \mathcal{H}$ be a linear, densely defined operator. Since Z^* is closed, if Z is symmetric, i. e.[5], if $Z \subset Z^*$, then Z is closable, $Z^{**} = (Z^*)^*$ exists and $Z^{**} = \overline{Z}$ (the closure of Z).

If Z is symmetric, then $\sigma(Z) \neq \emptyset$, and one of the following situations necessarily arises:

$$\sigma(Z) \subset \mathbf{R}$$
$$\sigma(Z) = \{\zeta \in \mathbf{C} : \Im\zeta \geq 0\},$$
$$\sigma(Z) = \{\zeta \in \mathbf{C} : \Im\zeta \leq 0\},$$
$$\sigma(Z) = \mathbf{C}.$$

Furthermore, if $\zeta \in \sigma(Z)$ and $\Im\zeta \neq 0$, then $\zeta \in r\sigma(Z)$. As a consequence, if $\zeta \in p\sigma(Z) \cup c\sigma(Z)$, then $\Im\zeta = 0$. Eigenspaces corresponding to different eigenvalues are orthogonal to each other.

---

[5]The basic facts of the geometry of Hilbert spaces that are need here can be found, e.g., in [67], [72], [40], [3].

If Z is self-adjoint, i. e., $Z^* = Z$, then Z is closed, $\sigma(Z) \subset \mathbf{R}$ and $r\sigma(Z) = \emptyset$. Viceversa, if Z is symmetric and if $\sigma(Z) \subset \mathbf{R}$, then $\overline{Z}$ is self-adjoint (in which case, Z is said to be essentially self-adjoint). Thus, if Z is closed and symmetric, and if $\sigma(Z) \subset \mathbf{R}$, Z is self-adjoint.

Let $X : \mathcal{D}(X) \subset \mathcal{H} \to \mathcal{H}$ be a densely defined, linear operator such that iX is symmetric, i. e.,

$$\mathcal{D}(X) \subset \mathcal{D}(X^*), \ X + X^* = 0 \text{ on } \mathcal{D}(X). \tag{3.39}$$

The spectrum of X is non-empty, and satisfies one of the following relations:

$$\sigma(X) \subset i\mathbf{R}$$
$$\sigma(X) = \{\zeta \in \mathbf{C} : \Re\zeta \leq 0\},$$
$$\sigma(X) = \{\zeta \in \mathbf{C} : \Re\zeta \geq 0\},$$
$$\sigma(X) = \mathbf{C}.$$

The first one characterizes the case in which iX is essentially self-adjoint (or self-adjoint, if X is closed).

By (3.39), if iX is symmetric, for all $x \in \mathcal{D}(X)$

$$\Re(Xx|x) = \frac{1}{2}\left((Xx|x) + (x|Xx)\right)$$
$$= \frac{1}{2}\left((Xx|x) + (X^*x|x)\right)$$
$$= 0,$$

and therefore X is dissipative. What we know about the spectrum of a symmetric operator implies that, if iX is symmetric, then X is m-dissipative if, and only if, either $\sigma(X) \subset i\mathbf{R}$ or $\sigma(X) = \{\zeta \in \mathbf{C} : \Re\zeta \leq 0\}$. Theorem 3.2.12 implies then

Proposition 3.4.2 *If iX is a symmetric operator, X is maximal dissipative if, and only if, either $\sigma(X) \subset i\mathbf{R}$ or $\sigma(X) = \{\zeta \in \mathbf{C} : \Re\zeta \leq 0\}$.*

Proposition 3.4.3 *If X is densely defined in $\mathcal{H}$, iX is self-adjoint if, and only if, both X and $-X$ are maximal dissipative.*

Theorem 3.3.4 yields then the *Stone theorem*:

Theorem 3.4.4 *If iX is self-adjoint, X is the infinitesimal generator of a strongly continuous group of unitary operators in $\mathcal{H}$. Viceversa, every strongly continuous group of unitary operators in $\mathcal{H}$ is generated by an operator X such that iX is self-adjoint.*

Now, let iX be symmetric, let X be maximal dissipative but let $-X$ be not maximal dissipative. In other words, let iX be closed, symmetric and

$$\sigma(X) = \{\zeta \in \mathbf{C} : \Re\zeta \leq 0\}.$$

Let $T : \mathbf{R}_+ \to \mathcal{L}(\mathcal{H})$ be the strongly continuous semigroup generated by X. If $x \in \mathcal{D}(X)$, then $T(t)x \in \mathcal{D}(X)$, and therefore

$$\begin{aligned}
\frac{d}{dt}\|T(t)x\|^2 &= \frac{d}{dt}(T(t)x|T(t)x) \\
&= (XT(t)x|T(t)x) + (T(t)x|XT(t)x) \\
&= (XT(t)x|T(t)x) + (X^*T(t)x|T(t)x) \\
&= ((X + X^*)T(t)x|T(t)x) = 0
\end{aligned}$$

for all $t \geq 0$. Thus the function $t \mapsto \|T(t)x\|$ is constant, and consequently

$$\|T(t)x\| = \|T(0)x\| = \|x\|$$

for all $t \geq 0$ and all ($x \in \mathcal{D}(X)$, hence for all) $x \in \mathcal{H}$.

Viceversa, assume that, for all $t \geq 0$, $T(t)$ is a linear isometry of $\mathcal{H}$, i. e.,

$$(T(t)x|T(t)y) = (x|y)$$

for all $x, y \in \mathcal{H}$. This equality is equivalent to

$$((T(t) - I)x|(T(t) - I)y) + ((T(t) - I)x|y) + (x|(T(t) - I)y) = 0.$$

Thus, if $x, y \in \mathcal{D}(X)$,

$$(Xx|y) + (x|Xy) = 0. \tag{3.40}$$

The linear form $x \mapsto (Xx|y)$ is continuous on $\mathcal{D}(X)$, and therefore $y \in \mathcal{D}(X^*)$, i. e., $\mathcal{D}(X) \subset \mathcal{D}(X^*)$; (3.40) reads now

$$((X + X^*)x|y) = 0 \ \forall x, y \in \mathcal{D}(X),$$

whence $X + X^* = 0$ on $\mathcal{D}(X)$. Hence the following theorem holds [48].

Theorem 3.4.5 *If iX is symmetric and X is maximal dissipative, X is the infinitesimal generator of a strongly continuous semigroup of linear isometries. Viceversa, every such semigroup is generated by a maximal dissipative operator X such that iX is symmetric.*

3.5 Conservative operators

Following [78], some of the results on semigroups of linear isometries of Hilbert spaces will now be extended to general Banach spaces.

If both the linear operators $X : \mathcal{D}(X) \subset \mathcal{E} \to \mathcal{E}$ and $-X$ are dissipative, i.e., if $\Re < Xx, \lambda \geq 0$ for all $x \in \mathcal{D}(X)$ and some $\lambda \in \Lambda$, X is called a *conservative operator*.

Lemma 3.5.1 *If $u \in \mathcal{D}(X)$ and $\|u\| = 1$, then*

$$\liminf_{t \to 0} \frac{\|u + t\,Xu\| - 1}{|t|} \geq 0.$$

If $v \in \mathcal{D}(X^2) = \{x \in \mathcal{D}(X) : Xx \in \mathcal{D}(X)\}$ and $\|v\| = 1$, then

$$\limsup_{t \to 0} \frac{\|v + t\,Xv\| - 1}{|t|} \leq 0.$$

Proof Let $\lambda \in \Lambda(u)$ be such that $\Re < Xu, \lambda \geq 0$. Since

$$\|u + t\,Xu\|^2 \geq |< u + t\,Xu, \lambda > |^2 = |1 + it\Im < Xu, \lambda > |^2$$
$$= 1 + (t\Im < Xu, \lambda >)^2 \geq 1,$$

the first part of the lemma is proved.

Let now $y = v + t\,Xv \in \mathcal{D}(X)$, and let $\mu \in \Lambda(y)$ be such that $\Re < Xy, \mu \geq 0$. Then

$$\|y - t\,Xy\| \geq |< y - t\,Xy, \mu > | = |\|y\| - t < Xy, \mu > |$$
$$= ((\|y\| - t\Re < Xy, \mu >)^2 + (t\Im < Xy, \mu >)^2)^{\frac{1}{2}}$$
$$= (\|y\|^2 + t^2(\Im < Xx, \mu >)^2)^{\frac{1}{2}} \geq \|y\|.$$

Since, on the other hand,

$$\|y - t\,Xy\| = \|v + tXv - tXv - t^2X^2v\|$$
$$= \|v - t^2X^2v\| = 1 + o(t)$$

as $t \to 0$, then, if $t \neq 0$,

$$\frac{\|v + t\,Xv\| - 1}{|t|} = \frac{\|y\| - 1}{|t|} \leq \frac{\|y - t\,Xy\| - 1}{|t|} = o(1)$$

as $t \to 0$, and the second part of the lemma follows. ∎

Corollary 3.5.2 *If X is a linear conservative operator, then*

$$\lim_{t \to 0} \frac{\|v + t\,Xv\| - \|v\|}{t} = 0$$

for all $v \in \mathcal{D}(X^2)$.

Lemma 3.5.3 *If a densely defined, linear, closed, conservative operator X generates a strongly continuous semigroup $T : \mathbf{R}_+ \to \mathcal{L}(\mathcal{E})$, then $T(t)$ is a linear isometry for all $t \in \mathbf{R}_+$.*

Proof For all $t \geq 0$, $x \in \mathcal{D}(X)$ and all $h \in \mathbf{R}$ such that $t + h \in \mathbf{R}_+$, then $T(t)x \in \mathcal{D}(X)$, and

$$T(t + h)x = T(t)x + \int_t^{t+h} XT(s)x\,ds.$$

Since the map $s \mapsto XT(s)x = T(s)Xx$ is continuous, then

$$T(t + h)x = T(t)x + hXT(t)x + o(h) \tag{3.41}$$

as $h \to 0$.

It will be shown in Section 4.3 that the set

$$\mathcal{D}^\infty = \cap\{\mathcal{D}(X^n) : n = 1, 2, \dots\}$$

is a dense linear subvariety of $\mathcal{E}$.

Hence, by Corollary 3.5.2 and by (3.41),

$$\lim_{h \to 0} \frac{\|T(t + h)x\| - \|T(t)x\|}{h} = 0$$

for all $t \geq 0$ and all $x \in \mathcal{D}^\infty$. (If $t = 0$, the limit as $h \to 0$ shall be replaced by the limit as $t \downarrow 0$). Thus the function $t \mapsto \|T(t)x\|$

is constant on $\mathbf{R}_+$. Since $T(t) = I$, this proves that $T(t)$ is a linear isometry for all $t \geq 0$. ■

It will be shown now that the converse holds.

For $x \in \mathcal{E}\backslash\{0\}$, $y \in \mathcal{E}$, $t > 0$, let $\lambda \in \Lambda(x)$, $\mu_t \in \Lambda(x + ty)$. Then, since $\Re < x, \mu_t > \leq |<x, \mu_t>| \leq \|x\|$,

$$\Re < y, \lambda > = \frac{\Re < ty, \lambda >}{t} = \frac{\Re < x + ty, \lambda > - \|x\|}{t} \leq$$

$$= \leq \frac{\|x + ty\| - \|x\|}{t}$$

$$\leq \frac{< x + ty, \mu_t > - \Re < x.\mu_t >}{t}$$

$$= \frac{\Re < x + ty, \mu_t > - \Re < x, \mu_t >}{t} = \Re < y, \mu_t > \quad (3.42)$$

Let $\{t_\nu\}$ be a sequence of positive numbers converging to 0. Since the closed unit ball of $\mathcal{E}'$ is compact for the weak star topology, the set $\{\mu_{t_\nu}\}$ has a weak star cluster point $\mu \in \mathcal{E}'$ with $\|\mu\| \leq 1$. If $< x, \mu > \neq \|x\|$, there exist some $\epsilon > 0$ and a subsequence $\{t'_\nu\}$ of the sequence $\{t_\nu\}$ such that $| < x, \mu_{t'_\nu} > - \|x\| | > \epsilon$ for all ν. Since $| < y, \mu_{t_\nu} > | \leq \|y\|$ for all ν, then

$$< x, \mu_{t'_\nu} > = < x + t'_\nu y, \mu_{t'_\nu} > - t'_\nu < y, \mu_{t'_\nu} > = \|x + t'_\nu y\| - t'_\nu < y, \mu_{t'_\nu} >,$$

which converges to $\|x\|$ as $\nu \to \infty$. This contradiction shows that $< x, \mu > = \|x\|$. Hence $\|\mu\| = 1$, and, in conclusion, $\mu \in \Lambda(x)$.

Suppose now that X generates a strongly continuous semigroup $T : \mathbf{R}_+ \to \mathcal{L}(\mathcal{E})$ of linear isometries of $\mathcal{E}$. For $x \in \mathcal{D}(X)$ and $t > 0$,

$$T(t)x = x + t\,Xx + o(t)$$

as $t \downarrow 0$.

Then

$$\|x + t\,Xx\| - \|x\| = \|T(t)x\| - \|x\| + o(t) = o(t) \quad (3.43)$$

as $t \downarrow 0$, and, setting $y = Xx$, (3.42) shows that - as is well known - $\Re < Xx, \lambda > \leq 0$ for all $\lambda \in \Lambda(x)$.

If $\Re < Xx, \mu > < 0$, arguing as before, one shows that there exist some $\epsilon > 0$ and a subsequence $\{t'_\nu\}$ of the sequence $\{t_\nu\}$ such that $\Re < Xx, \mu_{t'_\nu} > < -\epsilon$ for all ν. On the other hand, by (3.42),

$$\Re < Xx, \mu_{t'_\nu} > \geq \frac{\|x + t'_\nu Xx\| - \|x\|}{t'_\nu},$$

which, by (3.43), tends to 0 as $\nu \to \infty$. Thus $\Re < Xx, \mu > = 0$, and the following theorem holds, which extends to all Banach spaces the classical Stone theorem.

Theorem 3.5.4 *Let X be a closed, densely defined, linear operator on $\mathcal{E}$. Then X is the infinitesimal generator of a strongly continuous semigroup of linear isometries of $\mathcal{E}$ into $\mathcal{E}$, if, and only if, X is conservative and m-dissipative.*

If, and only if, also $-X$ is m-dissipative, then T is the restriction to $\mathbf{R}_+$ of a strongly continuous group $\mathbf{R} \to \mathcal{L}(\mathcal{E})$ of surjective isometries of $\mathcal{E}$.[6]

Lemma 3.2.15 yields

Corollary 3.5.5 *If X is closed, densely defined and conservative, and if the dual operator X' is dissipative, then X generates a strongly continuous semigroup of linear isometries of $\mathcal{E}$.*

If X is densely defined and m-dissipative, since $\Pi_r \subset r(X) \subset r(X')$ and, for all $\zeta \in r(X)$, $(\zeta I - X)^{-1'} = (\zeta I - X')^{-1}$, then also X' is m-dissipative. Thus, denoting by $\mathcal{F}'$ the closure of $\mathcal{D}(X')$ in $\mathcal{E}'$, the part $\tilde{X}'$ of X' in $\mathcal{F}'$, i.e., the restriction of X' to the linear space $\mathcal{D}(\tilde{X}') = \{\lambda \in \mathcal{D}(X') : X'\lambda \in \mathcal{F}'\}$, generates a strongly continuous semigroup of contractions $\tilde{T}' : \mathbf{R}_+ \to \mathcal{L}(\mathcal{F}')$.

Note that the additional hypothesis that X be conservative, whereby $-\mathbf{R}_+^* \subset r(X) \cup r\sigma(X)$, does not imply that $-X'$ is dissipative. However the stronger additional condition that also $-X$ be m-dissipative *does* imply that also $-X'$ is m-dissipative. In conclusion, the following proposition holds.

---

[6] According to a different terminology [15], in this latter case iX is called a (generalized) hermitian operator.

Proposition 3.5.6 *If X is densely defined and X and $-X$ are m-dissipative, then $\tilde{X}'$ is the infinitesimal generator of a strongly continuous group $\tilde{T}' : \mathbf{R} \to \mathcal{L}(\mathcal{F}')$ of isometries of $\mathcal{F}'$ onto $\mathcal{F}'$.*

If $\mathcal{E}$ is reflexive, then $\mathcal{D}(X')$ is dense in $\mathcal{E}'$.

Corollary 3.5.7 *Under the hypotheses of this latter Proposition, if moreover the Banach space $\mathcal{E}$ is reflexive, then X' generates a strongly continuous group T' of isometries of $\mathcal{E}'$ onto $\mathcal{E}'$.*

If the closed linear operator X is conservative, then $\Pi_r = \{\zeta \in \mathbf{C} : \Re\zeta > 0\}$ belongs entirely either to $r\sigma(X)$ or to $r(X)$, and a similar conclusion holds for Π_l. If X is also m-dissipative, then

$$r(X) \supset \Pi_r,$$

and, if moreover $-\Pi_r \subset r(X)$, also $-X$ is m-dissipative. If furthermore X is densely defined, then it generates a strongly continuous group $T : \mathbf{R} \to \mathcal{L}(\mathcal{E})$ of surjective linear isometries of $\mathcal{E}$. Therefore $\sigma(X) \neq \emptyset$ ([15], pp.212-214), and the following theorem holds.

Theorem 3.5.8 *If the closed operator X is densely defined, conservative and m-dissipative, the spectrum of X is non-empty.*

If, under the hypotheses of this latter theorem, $\sigma(X)$ is compact, then $\sigma(X) \subset i\mathbf{R}$, and X generates a strongly continuous group of surjective isometries. Hence ([15], p.214), X is a continuous operator and $T(t) = \exp tX$ for all $t \in \mathbf{R}$.[7]

3.6 Accretive operators

Except for sections 1.2 and 1.18 of the first chapter, all the operators that have been considered so far have always assumed to be linear. The final sections of this chapter will be devoted to outlining some relevant aspects of a theory of non linear semigroups acting on closed, convex subsets of a Hilbert space. Many more details and complete proofs can be found in [7]. We begin by considering a class of set-valued maps acting on a Banach space $\mathcal{E}$.

---

[7]For further informations on semigroups of linear isometries, cf. [15] and [78].

With the same notations of n. 1.18, let $A \in \Upsilon(\mathcal{E})$ be a set-valued map $A : \mathcal{D}(A) \subset \mathcal{E} \to 2^{\mathcal{E}} \backslash \{\emptyset\}$.

The map A is said to be *accretive* (or *increasing*)[8] if

$$\|x_1 - x_2\| \leq \|x_1 - x_2 + \alpha(y_1 - y_2)\| \tag{3.44}$$

for all $x_1, x_2 \in \mathcal{D}(A)$, all $y_1 \in A(x_1), y_2 \in A(x_2)$ and all $\alpha > 0$.

A first consequence of this definition is

Lemma 3.6.1 *If A is accretive, $\alpha I + A$ is injective for any $\alpha > 0$.*

Furthermore:

Lemma 3.6.2 *If A is accretive, its closure $\overline{A}$ is accretive.*

If A is linear, by Proposition3.2.2, A is accretive if, and only if, $-A$ is dissipative.

In analogy with Proposition 3.2.8 we prove now

Proposition 3.6.3 *If A is accretive, and if $r(A) \cap \mathbf{R}_-^* \neq \emptyset$, then*

$$\mathbf{R}_-^* \subset r(A) \tag{3.45}$$

and

$$p_L(\zeta I + A) \leq \frac{1}{\zeta} \quad \forall \zeta \in \mathbf{R}_+^*. \tag{3.46}$$

Proof By (3.44),

$$\|x_1 - x_2\| \leq \frac{1}{|\zeta|} \| - \zeta(x_1 - x_2) + (y_1 - y_2)\|,$$

i.e.,

$$\|x_1 - x_2\| \leq \frac{1}{|\zeta|} \|\zeta(x_1 - x_2) - (y_1 - y_2)\|$$

for all $x_1, x_2 \in \mathcal{D}(A)$, all $y_1 \in A(x_1), y_2 \in A(x_2)$ and all $\zeta < 0$. Since $r(A) \cap \mathbf{R}_-^* \neq \emptyset$, then

$$(\zeta_o I - \overline{A})(\mathcal{D}(\overline{A})) = \mathcal{E}$$

---

[8]Many deeper details on accretive operators can be found, *e.g.*, in [13] and in [14].

for some $\zeta_o < 0$. Theorem 1.18.9 implies then that (3.46) holds, and that

$$[c, +\infty) \subset -r(A) \ \forall \, c > 0.$$

Thus, also (3.45) holds, and the proof is complete. ∎

If A is accretive and $\mathbf{R}^*_- \subset r(A)$, A is called *m-accretive*.

Lemma 3.6.4 *Let A be accretive. Then A is m-accretive if, and only if,*

$$(I + A)(\mathcal{D}(A)) = \mathcal{E}. \tag{3.47}$$

Proof If (3.47) holds, the operator $(I + A$ is closed, and therefore also) A is closed, because otherwise the operator $I + \overline{A}$ could not be injective, contradicting Lemma 3.6.1. Thus $-1 \in r(A)$, and therefore A is m-accretive. The rest of the proof is obvious. ∎

Since not all linear operators are closable as linear operators (*i.e.*, such that the closures of their graphs are graphs of linear operators), Lemma 3.6.2 does not hold for any linear accretive operator $X : \mathcal{D}(X) \subset \mathcal{E} \to \mathcal{E}$. However, if X is closable as a linear operator, and therefore $\overline{X}$ is accretive, the inequality

$$\|x + \alpha \overline{X} x\| \geq \|x\|$$

holds for all $\alpha > 0$ and all $x \in \mathcal{D}(\overline{X})$.

Suppose there is some $\alpha > 0$ such that

$$\overline{(I + \alpha X)(\mathcal{D}(X))} = \mathcal{E}. \tag{3.48}$$

If X is closable, *a fortiori*,

$$\overline{(I + \alpha \overline{X})(\mathcal{D}(\overline{X}))} = \mathcal{E}.$$

Since, by Lemma 3.2.6,

$$\mathbf{R}^*_- \subset r(\overline{X}) \cup r\sigma(\overline{X}),$$

then $\mathbf{R}^*_- \subset r(\overline{X})$, *i.e.*, X is m-accretive.

In conclusion, the following lemma holds:

Lemma 3.6.5 *If the linear operator X is closable and accretive, and if (3.48) holds for some $\alpha > 0$, then $\overline{X}$ is m-accretive.*

This lemma can be extended to the non-linear case.

Proposition 3.6.6 *If the operator A is accretive and there is $\alpha > 0$ for which*

$$\overline{(I + \alpha A)(\mathcal{D}(A))} = \mathcal{E},$$

then A is m-accretive.

Proof Exercise.

An accretive operator A is said to be *maximal* if every accretive operator C such that $G_A \subset G_C$ coincides with A. By the Zorn lemma, every accretive operator is contained in a maximal accretive operator.

Theorem 3.6.7 *If A is m-accretive, A is maximal accretive.*

Proof Let C be accretive and such that $G_A \subset G_C$. If $x_0 \in \mathcal{D}(C)$ and $y_0 \in C(x_o)$, then

$$x_0 + y_0 \in (I + C)(x_o).$$

Since A is m-accretive, and therefore $-1 \in r(A)$, then

$$(I + A)(\mathcal{D}(A)) = \mathcal{E}.$$

As a consequence there is some $x_1 \in \mathcal{D}(A)$ for which

$$x_0 + y_0 = x_1 + A(x_1) \subset x_1 + C(x_1). \tag{3.49}$$

Since C is accretive, and therefore $I+C$ is injective by Lemma 3.6.1, then $x_0 = x_1$, so that, by (3.49),

$$x_0 + y_0 \in x_0 + C(x_o).$$

Thus $A = C$. ∎

Lemma 3.6.2 implies then

Corollary 3.6.8 *If A is m-accretive, A is closed.*

Theorem 3.6.9 Let $A : \mathcal{D}(A) \subset \mathcal{E} \to 2^{\mathcal{E}}\backslash\{\emptyset\}$. If there exists $\zeta_o < 0$ such that $(-\infty, \zeta_o) \subset r(A)$ and

$$p_L((\zeta I - A)^{-1}) \leq \frac{1}{|\zeta|} \ \forall\, \zeta \in (-\infty, \zeta_o)$$

then A is m-accretive.

Proof Let: $\zeta < \zeta_o$, $x_1, x_2 \in \mathcal{D}(A)$, $z_1 \in A(x_1)$, $z_2 \in A(x_2)$, $y_1 = \zeta x_1 - z_1$, $y_2 = \zeta x_2 - z_2$.

Then,

$$(\zeta I - A)(x_1) \ni \zeta x_1 - z_1 = y_1,$$
$$(\zeta I - A)(x_2) \ni \zeta x_2 - z_2 = y_2,$$

and therefore

$$x_1 = (\zeta I - A)^{-1}(y_1), \quad x_2 = (\zeta I - A)^{-1}(y_2).$$

Hence,

$$\|x_1 - x_2\| \leq p_L((\zeta I - A)^{-1})\|y_1 - y_2\| \leq \frac{1}{-\zeta}\|y_1 - y_2\|$$
$$= \frac{1}{-\zeta}\|\zeta(x_1 - x_2) - (z_1 - z_2)\|,$$

that is,

$$-\zeta\|x_1 - x_2\| \leq \|\zeta(x_1 - x_2) - (z_1 - z_2)\|$$

for all $\zeta < \zeta_o$.

Letting $\zeta_o - \zeta = \tau$, then $\tau > 0$ and the latter inequality yields

$$(\tau - \zeta_o)\|x_1 - x_2\| \leq \|(\zeta_o - \tau)(x_1 - x_2) - (z_1 - z_2)\|$$
$$= \| - \tau(x_1 - x_2) - (z_1 - z_2) + \zeta_o(x_1 - x_2)\|$$
$$\leq \|\tau(x_1 - x_2) + z_1 - z_2\| - \zeta_o\|x_1 - x_2\|,$$

whence

$$\tau\|x_1 - x_2\| \leq \|\tau(x_1 - x_2) + z_1 - z_2\|$$

for all $\tau > 0$.

That shows that A is accretive. Since $r(A) \cap \mathbf{R}^*_- \neq \emptyset$, A is m-accretive. ∎

Lemma 3.6.10 *If* $\mathcal{D}((I + \alpha A)^{-1}) = \mathcal{E}$ *and*

$$p_L((I + \alpha A)^{-1}) \leq 1$$

for all $\alpha > 0$, A *is m-accretive.*

Proof By (3.44), A is accretive, and m-accretive by Lemma 3.6.4. ∎

3.7   Monotone operators

Throughout this section $\mathcal{E}$ will be replaced by a real Hilbert space $\mathcal{H}$. Accretive operators on $\mathcal{H}$ are called also *monotone operators*.

Let $A : \mathcal{D}(A) \subset \mathcal{H} \to 2^{\mathcal{H}} \backslash \{\emptyset\}$.

Theorem 3.7.1 *The operator A is monotone if, and only if,*

$$(x_1 - x_2 | y_1 - y_2) \geq 0 \qquad (3.50)$$

for all $x_1, x_2 \in \mathcal{D}(A)$ and all $y_1 \in A(x_1)$, $y_2 \in A(x_2)$.

Proof First of all, for any $\alpha \in \mathbf{R}$

$$\|x_1 - x_2 + \alpha(y_1 - y_2)\|^2 = \|x_1 - x_2\|^2 + 2\alpha(x_1 - x_2|y_1 - y_2) + \alpha^2 \|y_1 - y_2\|^2.$$

As a consequence, if (3.50) holds,

$$\|x_1 - x_2 + \alpha(y_1 - y_2)\|^2 \geq \|x_1 - x_2\|^2$$

for all $x_1, x_2 \in \mathcal{D}(A)$ and all $y_1 \in A(x_1)$, $y_2 \in A(x_2)$, $\alpha > 0$, showing that A is accretive.

Viceversa, if A is accretive,

$$\|x_1 - x_2\|^2 \leq \|x_1 - x_2\|^2 + 2\alpha(x_1 - x_2|y_1 - y_2) + \alpha^2 \|y_1 - y_2\|^2,$$

i.e.,

$$2(x_1 - x_2|y_1 - y_2) + \alpha\|y_1 - y_2\|^2 \geq 0 \ \forall \, \alpha > 0,$$

which implies (3.50) for all $x_1, x_2 \in \mathcal{D}(A)$ and all $y_1 \in A(x_1)$, $y_2 \in A(x_2)$. ∎

If $\phi : \mathcal{H} \to (-\infty, +\infty]$ is a proper (*i.e.*, $\phi \not\equiv +\infty$) convex function, the set $D(\phi) := \{x \in \mathcal{H} : \phi(x) < \infty\}$ is non empty and convex.

By definition, the *subdifferential* of ϕ is the set-valued function $\partial\phi : \mathcal{H} \to 2^{\mathcal{H}} \setminus \{\emptyset\}$ defined by

$$\partial\phi(x) = \{y \in \mathcal{H} : \phi(z) \geq \phi(x) + (y|z - x), \ \forall z \in \mathcal{H}\}.$$

Lemma 3.7.2 *The subdifferential is monotone.*

Proof If $y_1 \in \partial(x_1)$, $y_2 \in \partial(x_2)$,

$$\phi(x_2) \geq \phi(x_1) + (y_1|x_2 - x_1)),$$

$$\phi(x_1) \geq \phi(x_2) + (y_2|x_1 - x_2)),$$

which imply (3.50). ∎

Here are other examples of monotone operators. If A is monotone, A^{-1} and tA for any $t > 0$ are monotone. If A_1 and A_2 are monotone operators, $A_1 + A_2$, with domain

$$\mathcal{D}(A_1 + A_2) = \mathcal{D}(A_1) \cap \mathcal{D}(A_2),$$

is monotone.

Lemma 3.7.3 *Let $\overline{A} : x \mapsto \overline{A(x)}$ be the (operator defined by the) closure of the graph G_A of A in $\mathcal{H} \times \mathcal{H}_w$, where $\mathcal{H}_w$ is $\mathcal{H}$ endowed with the weak topology. If A is monotone, $\overline{A}$ is monotone.*

Proof Let $(x_1, y_1), (x_2, y_2) \in \overline{A}$ and let $\{x_j^\nu\}, \{y_j^\nu\}$, with $y_j^\nu \in A(x_j^\nu)$, $j = 1, 2$, be two generalized sequences such that $x_j^\nu \to x_j$ and $y_j^\nu \rightharpoonup y_j$.

Since the generalized sequences $\{y_j^\nu\}$ are both bounded, there is $k > 0$ such that

$$\|y_2^\nu - y_1^\nu\| \leq k \ \forall \nu.$$

Since

$$(x_1 - x_2|y_1 - y_2) - (x_1^\nu - x_2^\nu|y_1^\nu - y_2^\nu) = (x_1 - x_2|y_1 - y_2)$$
$$-(x_1 - x_2|y_1^\nu - y_2^\nu) + (x_1 - x_2|y_1^\nu - y_2^\nu) - (x_1^\nu - x_2^\nu|y_1^\nu - y_2^\nu)$$

and

$$(x_1 - x_2|y_1{}^\nu - y_2{}^\nu) \to (x_1 - x_2|y_1 - y_2),$$

$$
\begin{aligned}
|(x_1 - x_2|y_1 - y_2) - (x_1{}^\nu - x_2{}^\nu|y_1{}^\nu - y_2{}^\nu)| \\
= |((x_1 - x_2) - (x_1{}^\nu - x_2{}^\nu)|y_1{}^\nu - y_2{}^\nu) \\
\leq \|((x_1 - x_2) - (x_1{}^\nu - x_2{}^\nu)\| \, \|y_1{}^\nu - y_2{}^\nu\| \\
\leq k\|((x_1 - x_2) - (x_1{}^\nu - x_2{}^\nu)\| \to 0,
\end{aligned}
$$

then

$$(x_1{}^\nu - x_2{}^\nu|y_1{}^\nu - y_2{}^\nu) \to (x_1 - x_2|y_1 - y_2),$$

and therefore

$$(x_1 - x_2|y_1 - y_2) \geq 0.$$

∎

Lemma 3.7.4 *If A is monotone, $x \mapsto \overline{\text{conv}(A(x))}$ is monotone.*

Proof Let $x_1, x_2 \in \mathcal{D}(A)$, $j = 1, 2$, $n \geq 1$, $y_{j,1}, \ldots, y_{j,n} \in A(x_j)$, $t_1, \ldots, t_n \in \mathbf{R}_+$, with $t_1 + \cdots + t_n = 1$.
    Then

$$\left(x_2 - x_1 \middle| \sum_{\alpha=1}^{n} t_\alpha(y_{2,\alpha} - y_{1,\alpha})\right) = \sum_{\alpha=1}^{n} t_\alpha(x_2 - x_1|y_{2,\alpha} - y_{1,\alpha}) \geq 0.$$

∎

Lemma 3.7.5 *Let $K \subset \mathcal{H}$ and let $f : K \to \mathcal{H}$ be a contraction. Then, $A = I - f$ is monotone.*

Proof If $x_1, x_2 \in K$, $y_j = x_j - f(x_j)$, $(j = 1, 2)$,

$$
\begin{aligned}
(x_1 - x_2|y_1 - y_2) &= (x_1 - x_2|x_1 - x_2 - (f(x_1) - f(x_2))) \\
&= \|x_1 - x_2\|^2 - (x_1 - x_2|f(x_1) - f(x_2)) \\
&\geq \|x_1 - x_2\|^2 - \|x_1 - x_2\| \, \|x_1 - x_2\| = 0.
\end{aligned}
$$

∎

Let K be now closed and convex in $\mathcal{H}$ and let Π_K be the projector from $\mathcal{H}$ to K.

Since $\mathcal{H}$ is a real Hilbert space, (1.13) and (1.15), that now read

$$(x - x'|\Pi_K(x) - \Pi_K(x')) \geq \|\Pi_K(x) - \Pi_K(x')\|^2,$$

$$(x-x'|(x-\Pi_K(x))-(x'-\Pi_K(x'))) \geq \|(x-\Pi_K(x))-(x'-\Pi_K(x'))\|^2,$$

and hold for all $x, x' \in \mathcal{H}$, yield

Lemma 3.7.6 *The maps Π_K and $I - \Pi_K$ are monotone.*

Theorem 3.7.7 *Let K be closed and convex in $\mathcal{H}$, and let A be monotone. For every $y \in \mathcal{H}$ there exists $x \in K$ such that*

$$(\eta + x|\xi - x) \geq (y|\xi - x) \ \forall \ (\xi, \eta) \in G_A.$$

Proof Replacing A by $A - y$ (i.e., $\mathcal{D}(A - y) = \mathcal{D}(A)$ and

$$\eta \in A(\xi) \Longleftrightarrow \eta - y \in A(\xi) - y),$$

we may assume $y = 0$. Hence, we have to prove that there is $x \in K$ such that

$$(\eta + x|\xi - x) \geq 0 \ \forall \ (\zeta, \eta) \in G_A.$$

For $(\xi, \eta) \in G_A$, let

$$K(\xi, \eta) = \{x \in K : (\eta + x|\xi - x) \geq 0\}.$$

a) The set $K(\xi, \eta)$ is obviously closed. If $x \in K(\xi, \eta)$,

$$\|x\|^2 \leq (\xi - \eta|x) + (\xi|\eta)$$
$$\leq \frac{1}{2}\|x\|^2 + \frac{1}{2}\|\xi - \eta\|^2 + (\xi|\eta),$$

whence

$$\|x\|^2 \leq \|\xi - \eta\|^2 + 2(\xi|\eta),$$

showing that $K(\xi, \eta)$ is bounded.

We prove now that $K(\xi, \eta)$ is convex.

For $x_1, x_2 \in K(\xi, \eta)$, $x_1 \neq x_2$, and $t \in [0,1]$, $tx_1 + (1-t)x_2 \in K$. Moreover,

$$
\begin{aligned}
q(t) &:= (\eta + tx_1 + (1-t)x_2|\xi - (tx_1 + (1-t)x_2)) \\
&= (\eta + x_2 + t(x_1 - x_2)|\xi - x_2 - t(x_1 - x_2)) \\
&= -\|x_1 - x_2\|^2 t^2 + (\xi - \eta - 2x_2|x_1 - x_2)t + (\xi - x_2|\eta + x_2).
\end{aligned}
$$

The real polynomial q, with leading coefficient < 0, is either ≤ 0 at all points of $\mathbf{R}$ or - when it has real zeros - is ≥ 0 at all points in the closed interval bounded by its roots, and ≤ 0 at all points outside that interval. Since

$$
q(0) = (\eta + x_2|\xi - x_2) \geq 0
$$

and

$$
q(1) = (\eta + x_1|\xi - x_1) \geq 0,
$$

then $q(t) \geq 0$ for all $t \in [0,1]$. That proves that $K(\xi, \eta)$ is convex.

Remark Note that we have shown that the function

$$
x \mapsto (\eta + x|\xi - x)
$$

is convex on $\mathcal{H}$.

b) We will establish the theorem proving that

$$
\cap\{K(\xi, \eta) : (\xi, \eta) \in G_A\} \neq \emptyset. \tag{3.51}
$$

The set $K(\xi, \eta)$, being bounded, closed and convex, is compact for the weak topology of $\mathcal{H}$. Hence, (3.51) is equivalent to

$$
\cap\{K(\xi_j, \eta_j) : j = 1, \dots, n\} \neq \emptyset
$$

for every positive integer n and every choice of $(\xi_1, \eta_1), \dots, (\xi_n, \eta_n)$ in G_A.

Let E be the closed, bounded, convex set of $\mathbf{R^n}$ defined by

$$
E := \{r = (r_1, \dots, r_n) \in \mathbf{R^n} : \sum_{i=1}^{n} r_i = 1\},
$$

and let $\psi : E \to \mathcal{H}$, $\phi : E \times E \to \mathbf{R}$ be defined by

$$\psi(r) = \sum_{i=1}^{n} r_i \xi_i,$$

$$\phi(r, s) = \sum_{i=1}^{n} s_i(\psi(r) + \eta_i |\psi(r) - \xi_i).$$

Since ψ is linear, the function $\phi(\bullet, s)$ is convex. Thus, ϕ is continuous, convex with respect to r and linear with respect to s.

By the minimax theorem, there is $(r^o, s^o) \in E \times E$ such that

$$\phi(r^o, s) \leq \phi(r^o, s^o) \leq \phi(r, s^o) \ \forall \ (r, s) \in E \times E.$$

As a consequence,

$$\phi(r^o, s) \leq \max_{r \in E} \phi(r, r) \ \forall \ s \in E. \tag{3.52}$$

But,

$$\phi(r, r) = \sum_{i=1}^{n} r_i(\psi(r) + \eta_i |\psi(r) - \xi_i)$$

$$= \sum_{i=1}^{n} r_i (\sum_{k=1}^{n} r_k \xi_k + \eta_i |\sum_{j=1}^{n} r_j \xi_j - \xi_i)$$

$$= \sum_{i=1}^{n} r_i (\sum_{k=1}^{n} r_k \xi_k + \eta_i |\sum_{j=1}^{n} r_j(\xi_j - \xi_i))$$

$$= \sum_{i,j=1}^{n} r_i r_j (\sum_{k=1}^{n} r_k \xi_k + \eta_i |(\xi_j - \xi_i))$$

$$= \sum_{i,j=1}^{n} r_i r_j (\eta_i |\xi_j - \xi_i) + \sum_{i,j,k=1}^{n} r_i r_j(\xi_k |\xi_j - \xi_i)$$

$$= \sum_{i,j=1}^{n} r_i r_j (\eta_i |\xi_j - \xi_i).$$

Writing

$$\eta_i = \frac{1}{2}(\eta_i - \eta_j) + \frac{1}{2}(\eta_i + \eta_j),$$

then

$$\sum_{i,j=1}^{n} r_i r_j (\eta_i | \xi_j - \xi_i) = \frac{1}{2} \sum_{i,j=1}^{n} r_i r_j (\eta_i - \eta_j) | \xi_j - \xi_i),$$

and therefore

$$\phi(r,r) = \frac{1}{2} \sum_{i,j=1}^{n} r_i r_j (\eta_i - \eta_j) | \xi_j - \xi_i) \leq 0.$$

Thus, by (3.52),

$$\phi(r^o, s) \leq 0 \ \forall \, s \in E,$$

i.e.,

$$\sum_{i=1}^{n} r_i (\psi(r^o) + \eta_i | \psi(r^o) - \xi_i) \leq 0 \ \forall \, s \in E,$$

or, equivalently,

$$\psi(r^o) \in \cap \{K(\xi_j, \eta_j) : j = 1, \dots, n\}.$$

■

Choosing $K = \mathcal{H}$, Theorem 3.7.7 can be restated in the following form:

If A is monotone, for every $y \in \mathcal{H}$ there is $x \in \mathcal{H}$ such that

$$(y - x - \eta | x - \xi) \geq 0 \ \forall (\xi, \eta) \in G_A. \tag{3.53}$$

Example Let M be a connected, C^∞ Riemannian manifold of dimension N. Let Ω^q be the vector space of all real, exterior q-forms of class C^∞ on M, and $\Omega_c^q \subset \Omega^q$ the subspace consisting of all compactly supported elements of Ω^q.

The Riemannian metric[9] defines the Hodge isomorphism

$$* : \Omega^q \xrightarrow{\sim} \Omega^{N-q}.$$

If $d\omega$ is the volume element of the Riemannian metric, for $\phi, \psi \in \Omega^q$, and for any $x \in M$, let $\langle \phi, \psi \rangle_x$ be the real scalar defined by

$$(\phi \wedge *\psi)_x = \langle \phi, \psi \rangle_x d\omega(x);$$

the map $\phi, \psi \mapsto \langle \phi, \psi \rangle_x \in \mathbf{R}$ is a positive-definite bilinear form which is a continuous function of $x \in M$. Both $\langle \phi, \psi \rangle_x$ and

$$|\phi|_x = \sqrt{\langle \phi, \phi \rangle_x}$$

(which is a norm in the space of anti-symmetric q-vectors in $\mathbf{R}^N$) do not depend on the local representations of ϕ and ψ in any neighbourhood of x.

Setting, for $\phi, \psi \in \Omega_c^q$,

$$(\phi|\psi) = \int_M \langle \phi, \psi \rangle \, d\omega \quad \text{and} \quad \|\phi\| = (\phi|\phi)^{\frac{1}{2}}$$

we define a positive-definite inner product and a norm on Ω_c^q, which thus acquires the structure of a real pre-Hilbert space. Let L^q be its completion.

If $d : \Omega^q \to \Omega^{q+1}$ is the exterior differential operator, the operator $\delta : \Omega^q \to \Omega^{q-1}$ defined by

$$\delta = (-1)^q *^{-1} \circ d \circ *$$

is the formal adjoint of d, in the sense that

$$(d\phi|\chi) = (\phi|\delta\chi)$$

for all $\phi \in \Omega_c^q$, $\chi \in \Omega_c^{q+1}$.

The Laplace-Beltrami operator

$$\Delta = d\delta + \delta d : \Omega_c^q \to \Omega_c^q$$

[9]Concerning differential forms on Riemannian manifolds, see, *e.g.*, [17] or [33]

is linear, symmetric and positive on the domain Ω_c^q:

$$(\Delta\phi|\psi) = (d\phi|d\psi) + (\delta\phi|\delta\psi) = (\phi|\Delta\psi),$$

$$(\Delta\phi|\phi) = \|d\phi\|^2 + \|\delta\phi\|^2 \geq 0$$

for all $\phi, \psi \in \Omega_c^q$.

For $p \geq 2$, the p-*Laplacian*, or p-*Laplace-Beltrami operator*,

$$\Delta_p : \Omega_c^q \to \Omega_c^q$$

is defined on any $\phi \in \Omega_c^q$ by

$$\Delta_p(\phi) = \delta\left(|d\phi|^{p-2}d\phi\right) + d\left(|\delta\phi|^{p-2}\delta\phi\right).$$

For $\phi, \psi \in \Omega_c^q$,

$$(\Delta_p(\phi)|\psi) = \left(|d\phi|^{p-2}d\phi|d\psi\right) + \left(|\delta\phi|^{p-2}\delta\phi|\delta\psi\right)$$

$$= \int_M \left[|d\phi|^{p-2}\langle d\phi|d\psi\rangle + |\delta\phi|^{p-2}\langle \delta\phi|\delta\psi\rangle\right] d\omega,$$

and, in particular,

$$(\Delta_p(\phi)|\phi) = \|d\phi\|^p + \|\delta\phi\|^p,$$

where

$$\| \bullet \| = \left(\int_M |\bullet|^p d\omega\right).$$

These identities, yield now

Theorem 3.7.8 *If* $p \geq 2$, Δ_p, *with domain* $\mathcal{D}\left(\Delta_p\right) = \Omega_c^q$, *is monotone.*

Proof If $\phi_1, \phi_2 \in \Omega_c^q$,

$$(\Delta_p(\phi_1) - \Delta_p(\phi_2)|\phi_1 - \phi_2) =$$
$$\|d\phi_1\|^p + \|\delta\phi_1\|^p + \|d\phi_2\|^p + \|\delta\phi_2\|^p -$$
$$(|d\phi_1|^{p-2}d\phi_1|d\phi_2) - (|\delta\phi_1|^{p-2}\delta\phi_1|\delta\phi_2) -$$
$$(|d\phi_2|^{p-2}d\phi_2|d\phi_1) - (|\delta\phi_2|^{p-2}\delta\phi_2|\delta\phi_1)$$
$$= \|d\phi_1\|^p + \|\delta\phi_1\|^p + \|d\phi_2\|^p + \|\delta\phi_2\|^p -$$
$$\int_M \left[\left(|d\phi_1|^{p-2} + |d\phi_2|^{p-2}\right) \langle d\phi_1|d\phi_2\rangle + \right.$$
$$\left. \left(|\delta\phi_1|^{p-2} + |\delta\phi_2|^{p-2}\right) \langle \delta\phi_1|\delta\phi_2\rangle \right] d\omega$$
$$\geq \int_M \left[|d\phi_1|^p + |d\phi_2|^p - \frac{1}{2} \left(|d\phi_1|^{p-2} + |d\phi_2|^{p-2}\right) \times \right.$$
$$\left(|d\phi_1|^2 + |d\phi_2|^2\right) + |\delta\phi_1|^p + |\delta\phi_2|^p -$$
$$\left. \frac{1}{2} \left(|\delta\phi_1|^{p-2} + |\delta\phi_2|^{p-2}\right) \times \left(|\delta\phi_1|^2 + |\delta\phi_2|^2\right) \right] d\omega$$
$$= \frac{1}{2} \int_M \left[|d\phi_1|^p + |d\phi_2|^p - |d\phi_1|^{p-2}|d\phi_2|^2 - |d\phi_2|^{p-2}|d\phi_1|^2 + \right.$$
$$\left. |\delta\phi_1|^p + |\delta\phi_2|^p - |\delta\phi_1|^{p-2}|\delta\phi_2|^2 - |\delta\phi_2|^{p-2}|\delta\phi_1|^2 \right] d\omega.$$

Since, for $p \geq 2, a \geq 0, b \geq 0$,

$$a^p + b^p - a^{p-2}b^2 - a^2b^{p-2} = \left(a^{p-2} - b^{p-2}\right)\left(a^2 - b^2\right) \geq 0,$$

then

$$(\Delta_p(\phi_1) - \Delta_p(\phi_2)|\phi_1 - \phi_2) \geq 0.$$

∎

3.8   Maximal monotone operators

A monotone operator A on $\mathcal{H}$ is said to be *maximal monotone* if G_A is maximal in the set of the graphs of all monotone operators on $\mathcal{H}$, ordered by inclusion. By the Zorn lemma, every monotone operator is contained in a maximal monotone operator.

Lemma 3.8.1 *Let A be monotone. Then A is maximal monotone if, and only if, the condition*

$$(y - \eta | x - \xi) \geq 0 \ \forall \ (\xi, \eta) \in G_A$$

implies that $x \in \mathcal{D}(A)$ and $y \in A(x)$.

Proof Exercise.

We will show now that, if K is closed and convex in $\mathcal{H}$, both Π_K and $I - \Pi_K$ are maximal monotone.

Lemma 3.8.2 *Let A be a single-valued, monotone map with $\mathcal{D}(A) = \mathcal{H}$, and such that*

$$\lim_{t \downarrow 0} A((1 - t)x + tz) = A(x).$$

Then A is maximal monotone.

Proof Let $(x, y) \in \mathcal{H} \times \mathcal{H}$ be such that

$$(A(u) - y | u - x) \geq 0 \ \forall \, u \in \mathcal{H}.$$

Proving that A is maximal monotone amounts to showing that $y = A(x)$.

For any $z \in \mathcal{H}$ and any $t \in (0, 1)$,

$$\begin{aligned}
0 &\leq (A((1 - t)x + tz) - y | (1 - t)x + tz - x) \\
&= (A((1 - t)x + tz) - y | t(z - x)) \\
&= t(A((1 - t)x + tz) - y | z - x),
\end{aligned}$$

and therefore

$$(A((1 - t)x + tz) - y | z - x) \geq 0 \ \forall \, t \in (0, 1).$$

Hence

$$(A(x) - y | z - x) = \lim_{t \downarrow 0} (A((1 - t)x + tz) - y | z - x) \geq 0$$

for all $z \in \mathcal{H}$. Thus, $A(x) = y$. ∎

Corollary 3.8.3 *Let A be a single-valued, monotone and continuous map with $\mathcal{D}(A) = \mathcal{H}$. Then A is maximal monotone.*

Since Π_K and $I - \Pi_K$ are continuous on $\mathcal{H}$, Corollary 3.8.3 yields

Theorem 3.8.4 *If K is a closed, convex subset of $\mathcal{H}$, Π_K and $I - \Pi_K$ are maximal monotone.*

Theorem 3.8.5 *The operator A is maximal monotone if, and only if, it is m-accretive.*

Proof Let A be maximal monotone. By Lemma 3.8.1 and (3.53), for every $y \in \mathcal{H}$ there is $x \in \mathcal{D}(A)$ for which

$$y = (I + A)(x),$$

showing thereby that

$$\mathcal{H} = (I + A)(\mathcal{D}(A)).$$

Since $(I + A)^{-1}$ is a contraction, then $-1 \in r(A)$. Hence, A is m-accretive.

The converse has been established with Theorem 3.6.7. ∎

Let A be maximal monotone. Since A is monotone, also its closure in $\mathcal{H} \times \mathcal{H}_w$ is monotone. Thus, A - being maximal monotone - is closed in $\mathcal{H} \times \mathcal{H}_w$. Since A^{-1} is maximal monotone, A is closed in $\mathcal{H}_w \times \mathcal{H}$. But something more can be said:

Proposition 3.8.6 *Let A be maximal monotone.*
If the sequence $\{(x_\nu, y_\nu)\} \in G_A$ is such that $x_\nu \rightharpoonup x$, $y_\nu \rightharpoonup y$ and

$$\limsup(y_\nu, x_\nu) \le (y, x), \tag{3.54}$$

then $(x, y) \in G_A$ and

$$(y, x) = \lim(y_\nu, x_\nu). \tag{3.55}$$

Proof Since $(x_\nu, y_\nu) \in G_A$ for all ν, then

$$(v - y_\nu | u - x_\nu) \geq 0 \ \forall \ (u, v) \in G_A,$$

i.e.,

$$0 \leq (v|u) - (y_\nu|u) - (v|x_\nu) + (y_\nu|x_\nu) \ \forall \ (u, v) \in G_A.$$

Passing to the limit as $\nu \to \infty$, (3.54) yields

$$0 \leq (v|u) - (y|u) - (v|x) + (y|x)$$
$$= (v - y|u - x) \ \forall \ (u, v) \in G_A,$$

showing that $(x, y) \in G_A$. Therefore

$$(y - y_\nu | x - x_\nu) \geq 0 \ \forall \ (u, v) \in G_A,$$

i.e.,

$$(y|x) - (y_\nu|x) - (y|x_\nu) + (y_\nu|x_\nu) \geq 0.$$

Passing to the limit as $\nu \to \infty$, we obtain

$$(y|x) - (y|x) - (y, x) + \liminf(y_\nu|x_\nu) \geq 0,$$

i.e,

$$(y|x) \leq \liminf(y_\nu|x_\nu),$$

which, together (3.54), yields (3.55). ∎

If A is any operator, and if $\zeta, \zeta_o \in r(A)$ with $\zeta\zeta_o \neq 0$, then

$$(\zeta I - A)^{-1} = ((\zeta - \zeta_o) I + \zeta_o I - A)^{-1}$$
$$= \left(\left(I + (\zeta - \zeta_o) (\zeta_o I - A)^{-1} \right) (\zeta_o I - A) \right)^{-1}$$
$$= (\zeta_o I - A)^{-1} \left(I + (\zeta - \zeta_o) (\zeta_o I - A)^{-1} \right)^{-1}$$

on $\mathcal{H}$.

Setting

$$\zeta = -\frac{1}{\lambda}, \quad \zeta_o = -\frac{1}{\lambda_o},$$

the above identity can be re-stated in the following form

$$(I + \lambda A)^{-1} = (I + \lambda_o A)^{-1} \left(\frac{\lambda}{\lambda_o} I + \left(1 - \frac{\lambda}{\lambda_o} \right) (I + \lambda_o A)^{-1} \right)^{-1}.$$

Let now A be monotone. For $\lambda > 0$ and $x \in \mathcal{H}$, let

$$J_\lambda = (I + \lambda A)^{-1}$$

and

$$x_\lambda = J_\lambda(x).$$

Since

$$\frac{1}{\lambda}(x - x_\lambda) = \frac{1}{\lambda}(x - J_\lambda(x))$$
$$= \frac{1}{\lambda}((I + \lambda A - I)J_\lambda(x)) \in A(x_\lambda),$$

then

$$\left(\frac{1}{\lambda}(x - x_\lambda) - v | x_\lambda - x\right) \geq 0 \ \forall \, (u, v) \in G_A,$$

i.e.,

$$\|x_\lambda\|^2 \leq (x | x_\lambda - u) + (x_\lambda | u) - \lambda \, (v | x_\lambda - u), \qquad (3.56)$$

showing that $\|x_\lambda\|$ is bounded as $\lambda \downarrow 0$.

Theorem 3.8.7 *If A is maximal monotone, $\overline{D(A)}$ is convex, and*

$$\lim_{\lambda \downarrow 0} J_\lambda(x) = \Pi_{\overline{D(A)}}(x)$$

for all $x \in \mathcal{H}$.

Proof Let $\{\lambda_\nu\}$ be a generalized sequence of positive numbers such that

$$\lambda_\nu \to 0 \ \text{ and } \ x_{\lambda_\nu} \rightharpoonup x_o.$$

By (3.56),
$$\|x_o\|^2 \leq (x | x_o - u) + (x_o | u)$$

for all $u \in D(A)$, hence for all u in the convex hull of $D(A)$ and, by continuity, in its closure, K.

Since

$$0 \geq (x - x_o | u - x_o) = -(x | x_o - u) - (x_o | u) + \|x_o\|^2$$

for all $u \in K$, and since $x_o \in K$, (1.12) yields,

$$
\begin{aligned}
0 &\geq (x - x_o | \Pi_K(x) - x_o) \\
&= (x - \Pi_K(x) + \Pi_K(x) - x_o | \Pi_K(x) - x_o) \\
&= (x - \Pi_K(x) | \Pi_K(x) - x_o) + \| \Pi_K(x) - x_o \|^2 \\
&\geq \| \Pi_K(x) - x_o \|^2,
\end{aligned}
$$

whence $x_o = \Pi_K(x)$. Thus,

$$
x_\lambda \rightharpoonup x_o \text{ as } \lambda \downarrow 0 \tag{3.57}
$$

because x_o does not depend on the choice of $\{\lambda_\nu\}$.

In view of (3.56),

$$
\limsup_{\lambda \downarrow 0} \| x_\lambda \|^2 \leq (x | x_o - u) + (x_o | u)
$$

for all $u \in \mathcal{D}(A)$ and, by continuity, for all $u \in K$.

Choosing $u = x_o$, then

$$
\limsup_{\lambda \downarrow 0} \| x_\lambda \|^2 \leq \| x_o \|^2,
$$

so that, by (3.57),

$$
\lim_{\lambda \downarrow 0} J_\lambda(x) = x_o = \Pi_K(x).
$$

Thus, if $x \in K$, then

$$
\lim_{\lambda \downarrow 0} J_\lambda(x) = x
$$

Since $J_\lambda(x) \in \mathcal{D}(A)$, then $K = \overline{\mathcal{D}(A)}$. $\blacksquare$

If A is maximal monotone, Lemma 3.7.4 implies that, for every $x \in \mathcal{D}(A)$, $A(x)$ is closed and convex in $\mathcal{H}$. Let A^o be the single-valued operator, with $\mathcal{D}(A^o) = \mathcal{D}(A)$, defined by

$$
A^o(x) = \Pi_{\overline{\mathrm{conv}(A(x))}}(0) = \Pi_{A(x)}(0).
$$

Theorem 3.8.8 *If A is maximal monotone, and if $(x, y) \in \overline{\mathcal{D}(A)} \times \mathcal{H}$ is such that*

$$(A^\circ(u) - y | u - x) \geq 0 \quad \forall u \in \mathcal{D}(A),$$

then

$$(v - y | u - x) \geq 0 \quad (u, v) \in G_A,$$

i.e., $y \in A(x)$.

Proof Let $G \subset \overline{\mathcal{D}(A)} \times \mathcal{H}$ be defined by

$$(x, y) \in G \iff (A^\circ(u) - y | u - x) \geq 0 \quad \forall u \in \mathcal{D}(A).$$

Since $G_A \subset G$, the theorem will follow from G being monotone, that is, from the implication

$$(x_1, y_1), (x_2, y_2) \in G \implies (y_1 - y_2 | x_1 - x_2) \geq 0. \tag{3.58}$$

Since $(x_j, y_j) \in G_A$, $j = 1, 2$, then, whenever $u \in \mathcal{D}(A)$,

$$(y_j - A^\circ(u) | x_j - u) \geq 0,$$

and therefore

$$(y_1 | x_1 \quad u) + (y_1 | x_1 - u) \geq (A^\circ(u) | x_1 + x_2 - 2u)$$
$$= 2 (A^\circ(u) | x - u),$$

where

$$x = \frac{1}{2} (x_1 + x_2) \in \overline{\mathcal{D}(A)},$$

because $\overline{\mathcal{D}(A)}$ is convex.

Being

$$(y_1 - y_2 | x_1 - x_2) = (y_1 - y_2 | x_1 - u - (x_2 - u))$$
$$= (y_1 | x_1 - u) - (y_1 | x_2 - u) -$$
$$(y_2 | x_1 - u) + (y_2 | x_2 - u),$$

then

$$(y_1 - y_2 | x_1 - x_2) \geq 2 (A^\circ(u) | x - u) -$$
$$(y_1 | x_2 - u) - (y_2 | x_1 - u).$$

Choosing now $u = J_\lambda(x) \in \mathcal{D}(A)$, since

$$
\begin{aligned}
x - J_\lambda(x) &= x - (I + \lambda A)^{-1}(x) \\
&= (I + \lambda A - I)(I + \lambda A)^{-1}(x) \\
&\in \lambda A (I + \lambda A)^{-1}(x) = A J_\lambda(x),
\end{aligned}
$$

then

$$
(A^\circ (J_\lambda(x)) | x - J_\lambda(x)) = \lambda (A^\circ (J_\lambda(x)) | A (J_\lambda(x))) \geq 0,
$$

and therefore

$$
(y_1 - y_2 | x_1 - x_2) \geq - [(y_1 | x_2 - J_\lambda(x)) + (y_2 | x_1 - J_\lambda(x))] . \qquad (3.59)
$$

As $\lambda \downarrow 0$,

$$
J_\lambda(x) \to \Pi_{\overline{\mathcal{D}(A)}}(x) = x.
$$

Then

$$
\begin{aligned}
\lim_{\lambda \downarrow 0} [(y_1 | x_2 - J_\lambda(x)) + (y_2 | x_1 - J_\lambda(x))] &= \\
&= (y_1 | x_2 - x) + (y_2 | x_1 - x) \\
&= \frac{1}{2} [(y_1 | x_2 - x_1) + (y_2 | x_1 - x_2)] = -\frac{1}{2} (y_1 - y_2 | x_1 - x),
\end{aligned}
$$

By (3.59), that implies (3.58). ∎

Corollary 3.8.9 *Let A_1 and A_2 be maximal monotone. If $\mathcal{D}(A_1) = \mathcal{D}(A_2)$ and $A_1^\circ = A_2^\circ$, then $A = B$.*

The following theorem characterizes projectors within the class of maximal monotone operators.

Theorem 3.8.10 *A single valued operator on $\mathcal{H}$ is a projector if, and only if, it is maximal monotone and idempotent.*

Proof By Theorem 3.8.4, we need only consider the "if" part of the statement.

Let the operator A be single valued, idempotent and maximal monotone.

Since $A^2 = A$, then $\mathcal{R}(A) \subset \mathcal{D}(A)$.

We prove now that $\mathcal{R}(A)$ is closed. Let $\{x_\nu\} \subset \mathcal{R}(A)$ be a sequence converging to some $x \in \overline{\mathcal{R}(A)}$. Let $y_\nu \in \mathcal{D}(A)$ be such that $A(y_\nu) = x_\nu$. Since

$$A(x_\nu) = A^2(y_\nu) = A(y_\nu) = x_\nu,$$

then $(A(x_\nu), x_\nu) \to (x, x)$, and therefore

$$(x, x) \in \overline{G_A} = G_A$$

(*i.e.*, $x = A(x)$) because maximal monotone operators are closed.

Since $\mathcal{R}(A) = \mathcal{D}(A^{-1})$ and A^{-1} is maximal monotone, then $\mathcal{R}(A)$ is convex by Theorem 3.8.7.

We show now that $A = \Pi_{\mathcal{R}(A)}$.

For $x, y \in \mathcal{D}(A)$,

$$\left(A(x) - \Pi_{\mathcal{R}(A)}(y) | x - y\right) = \left(A(x) - \Pi_{\mathcal{R}(A)}(y) | x - \Pi_{\mathcal{R}(A)}(y)\right) + \left(A(x) - \Pi_{\mathcal{R}(A)}(y) | \Pi_{\mathcal{R}(A)}(y) - y\right).$$

Since A is monotone,

$$\left(A(x) - \Pi_{\mathcal{R}(A)}(y) | x - \Pi_{\mathcal{R}(A)}(y)\right) = $$
$$= \left(A(x) - A\left(\Pi_{\mathcal{R}(A)}(y)\right) | x - \Pi_{\mathcal{R}(A)}(y)\right) \geq 0.$$

Furthermore,

$$\left(A(x) - A\left(\Pi_{\mathcal{R}(A)}(y)\right) | x - \Pi_{\mathcal{R}(A)}(y)\right) \geq 0$$

by Lemma 1.2.4.

Thus,

$$\left(A(x) - \Pi_{\mathcal{R}(A)}(y) | x - y\right) \geq 0,$$

and therefore $\Pi_{\mathcal{R}(A)}(y) = A(y)$, showing that the monotone operator $\Pi_{\mathcal{R}(A)}$ extends A. Since A is maximal monotone, $A = \Pi_{\mathcal{R}(A)}$. ∎

3.9 Non-linear semigroups

The similarities between maximal dissipative linear operators and maximal monotone operators in a Hilbert space, which have been outlined in the previous sections, can be pushed much further, up to establishing a link between maximal monotone operators and non-linear continuous semigroups, which may be viewed as an extension of the Lumer- Phillips theorem (Theorem 3.3.1). This extension is expressed by the following two theorems, which will be stated here, referring to [7] for further details and complete proofs.

Let $\mathcal{H}$ be a real Hilbert space and let A be a maximal monotone operator.

As before, for any $\zeta > 0$, $J_\zeta = (I + \zeta A)^{-1}$ is a contraction of $\mathcal{H}$.

Theorem 3.9.1 *For every $x_o \in \mathcal{D}(A)$ there exists a unique function $x : [0, +\infty) \to \mathcal{H}$ which satisfies the following conditions:*
i) $x(t) \in \mathcal{D}(A)$ for all $t \in (0, +\infty)$;

ii) x is lipschitz on $(0, +\infty)$ (that is to say

$$\frac{dx}{dt} \in L^\infty((0, +\infty), \mathcal{H})$$

in the sense of distributions) and

$$\left\| \frac{dx}{dt} \right\|_{L^\infty((0,+\infty),\mathcal{H})} \leq \|A^o(x_o)\| \, ;$$

iii) $-\frac{dx}{dt} \in A(x(t))$ a.e. on $(0, +\infty))$.

Furthermore:
iv) the left derivative $\frac{d^+x}{dt}$ exists at all points $t \in [0, +\infty)$, and

$$\frac{d^+x}{dt} + A^o(x(t)) = 0 \quad \forall t \in [0, +\infty);$$

v) $t \mapsto A^o(x(t))$ is continuous from the right;
vi) $t \mapsto \|A^o(x(t))\|$ is decreasing;
vii) given $y_o \in \mathcal{D}(A)$ and denoting by $y : [0, +\infty) \to \mathcal{H}$ the function defined by i), ii), iii) when x_o is replaced by y_o, then

$$\|x(t) - y(t)\| \leq \|x_o - y_o\| \quad \forall \, t \in [0, +\infty).$$

For any $t > 0$, the map $x_o \mapsto x(t)$ is a contraction $\mathcal{D}(A) \to \mathcal{D}(A)$, which can be extended by continuity, in a unique way, to a contraction

$$T(t) : K \to K,$$

where $K = \overline{\mathcal{D}(A)}$.

The map $t \mapsto T(t)$ defines a *continuous semigroup of contractions of K*, in the sense that

$$T(t_1 + t_2) = T(t_1)\,T(t_2) \ \forall\, t_1, t_2 \in \mathbf{R}_+, \tag{3.60}$$

$$T(0) = I, \tag{3.61}$$

$$\lim_{t \downarrow 0} \|T(t)(x) - x\| = 0 \ \forall\, x \in K, \tag{3.62}$$

$$\|T(t)\,(x_1) - T(t)\,(x_2)\| \leq \|x_1 - x_2\| \ \forall\, x_1, x_2 \in K \ \forall\, t \in \mathbf{R}_+. \tag{3.63}$$

Furthermore,

$$\lim_{t \downarrow 0} \frac{1}{t}(x - T(t)(x)) = A^\circ(x) \ \forall\, x \in K, \tag{3.64}$$

and, in analogy with the terminology introduced in the linear case, the semigroup T is said to be *generated by* $-A$.

Viceversa, let K be a closed, convex subset of $\mathcal{H}$, and let $\mathbf{R}_+ \ni l \mapsto T(t)$ be a continuous semigroup of contractions of K, *i.e.*, let (3.60) - (3.63) be satisfied.

Theorem 3.9.2 *There exists a unique maximal monotone operator A such that $K = \overline{\mathcal{D}(A)}$ and that (3.64) holds.*

Note Basic references for sections nn. 3.6 and 3.7-3.9 are the books [14] and [7] by G.Da Prato and H.Brézis.

The theory of non linear semigroups in general Banach spaces is still lacking an exhaustive treatment. Partial results can be found, *e.g.*, in [14], in Chapter 2 of [12] and in [13].

Chapter 4

Holomorphic semigroups

An angular sector in $\mathbf{C}$, with vertex in 0, is an additive semigroup.

Under which conditions is a strongly continuous semigroup $T :$ $\mathbf{R}_+ \to \mathcal{L}(\mathcal{E})$ the restriction to $\mathbf{R}_+$ of a semigroup defined on an angular sector containing properly $\mathbf{R}_+$?

The present chapter is aimed at answering this question, assuming that the semigroup is expressed by $\mathcal{L}(\mathcal{E})$−valued holomorphic functions defined in an angular sector.

4.1 Bounded holomorphic semigroups and their generators

Let $0 < \alpha \leq \frac{\pi}{2}$ and let Θ_α be the angular sector

$$\Theta_\alpha = \{z \in \mathbf{C} : \Re z > 0, \ |\arg z| < \alpha\}.$$

Let $T : \Theta_\alpha \cup \{0\} \to \mathcal{L}(\mathcal{E})$ be a map such that:

1) $T(0) = I$ and $T(z_1 + z_2) = T(z_1) T(z_2)$ for all $z_1, z_2 \in \Theta_\alpha$;

2) The function $z \mapsto T(z)$ is holomorphic on Θ_α;

3) For all $x \in \mathcal{E}$ and all $\epsilon \in (0, \alpha)$,

$$\lim_{\substack{z \to 0 \\ z \in \Theta_{\alpha-\epsilon}}} T(z)x = x.$$

The map T is called a *holomorphic semigroup of angle α*.

If T satisfies 1) and 3), $T_{|\mathbf{R}_+}$ is a strongly continuous semigroup. Its infinitesimal generator is a closed, densely defined, linear operator X, which - if all conditions 1) - 3) hold - is called the *infinitesimal generator of the holomorphic semigroup T*.

A similar argument to the one leading to Theorem 2.1.1 proves the following

Lemma 4.1.1 *If T satisfies conditions 1) and 3), for every $\beta \in [0, \alpha)$ there exist two real constants $a \geq 0$ and $M \geq 1$ such that*

$$\|T(z)\| \leq M e^{a \Re z} \quad \forall z \in \Theta_\beta \cup \{0\}.$$

If $T : \Theta_\alpha \cup \{0\} \to \mathcal{L}(\mathcal{E})$ satisfies, beside 1) - 3), the condition

4) *For every $\epsilon \in (0, \alpha)$ there exists a real constant $M_\epsilon \geq 1$ such that* $\|T(z)\| \leq M_\epsilon$ *for all $z \in \Theta_{\alpha - \epsilon}$*,

we say that T is a *bounded* holomorphic semigroup of angle α.

By Theorem 1.8.6, the following proposition holds.

Proposition 4.1.2 *If conditions 1) and 3) are satisfied, T is a holomorphic semigroup of angle α if, and only if, the scalar-valued function $z \mapsto\, < T(z)x, \lambda >$ is holomorphic on Θ_α for all $x \in \mathcal{E}$ and all $\lambda \in \mathcal{E}'$.*

If, moreover,

$$\sup\{| < T(z)x, \lambda > | : z \in \Theta_{\alpha - \epsilon}\} < \infty \quad x \in \mathcal{E} \, \forall \lambda \in \mathcal{E}'$$

for all $\epsilon \in (0, \alpha)$, then T is a bounded holomorphic semigroup of angle α.

The second part of this proposition can be established by similar considerations to those in the Remark following Lemma 3.3.2.

Going back to Lemma 4.1.1, the semigroup $\Theta_\beta \cup \{0\} \ni z \mapsto e^{-az} T(z)$ is uniformly bounded in norm on $\Theta_\beta \cup \{0\}$. Hence:

Corollary 4.1.3 *If X generates a holomorphic semigroup of angle $\alpha \in (0, \frac{\pi}{2}]$, for every $\beta \in (0, \alpha)$ there is some $a \in \mathbf{R}$ such that $X - aI$ is the infinitesimal generator of a bounded holomorphic semigroup of angle β.*

Theorem 4.1.4 *If T satisfies all conditions 1) - 4), then:*
  5) $\sigma(X) \subset \{\zeta \in \mathbf{C} : \alpha + \frac{\pi}{2} \leq |\arg \zeta| \leq \pi\}$;

  6) for every sufficiently small $\epsilon > 0$ there exists a real constant $N_\epsilon \geq 1$ such that

$$\|(\zeta I - X)^{-1}\| \leq \frac{N_\epsilon}{|\zeta|}$$

whenever $\zeta \neq 0$ and

$$|\arg \zeta| < \frac{\pi}{2} + \alpha - \epsilon;$$

  7) if $\zeta \in \Theta_{\alpha-\epsilon}$, one can choose $M_\epsilon = N_\epsilon$.

Remark Note that conditions 5) and 6) are equivalent to requiring, respectively, that ζ be contained in the closed angle

$$\left\{\tau \in \mathbf{C} : \Re\tau \leq 0 \text{ and } |\arg \tau - \pi| \leq \frac{\pi}{2} - \alpha\right\}$$

and be not contained in the closed angle

$$\left\{\tau \in \mathbf{C} : \Re\tau \leq 0 \text{ and } |\arg \tau - \pi| \leq \frac{\pi}{2} - \alpha + \epsilon\right\}.$$

Proof Choose $\theta \in (-\alpha, \alpha)$ and let $S : \mathbf{R}_+ \to \mathcal{L}(\mathcal{E})$ be the strongly continuous semigroup defined by

$$S(t) = T(te^{i\theta})$$

for $t \in \mathbf{R}_+$.
  Let $Z : \mathcal{D}(Z) \subset \mathcal{E} \to \mathcal{E}$ be the infinitesimal generator of S. Setting, for $x \in \mathcal{D}(Z)$, $f(z) = T(z)x$, $f : \Theta_\alpha \to \mathcal{E}$ is a holomorphic function.

Hence, for any $s \in \mathbf{R}_+$,

$$
\begin{aligned}
S(s)Zx &= \lim_{t \downarrow 0} \frac{1}{t} \left(S(s)S(t)x - S(s)x \right) \\
&= \lim_{t \downarrow 0} \frac{1}{t} \left(S(s+t)x - S(s)x \right) \\
&= \lim_{t \downarrow 0} \frac{1}{t} \left(T((s+t)e^{i\theta})x - T(se^{i\theta})x \right) \\
&= \lim_{t \downarrow 0} \frac{1}{t} \left(f(se^{i\theta} + te^{i\theta}) - f(se^{i\theta}) \right) \\
&= e^{i\theta} f'(se^{i\theta}) = e^{i\theta} \lim_{t \downarrow 0} \frac{1}{t} \left(f(se^{i\theta} \dotplus t) - f(se^{i\theta}) \right) \\
&= e^{i\theta} \lim_{t \downarrow 0} \frac{1}{t} \left(T(t)(S(s)x) - S(s)x \right) \\
&= e^{i\theta} X S(s)x.
\end{aligned}
$$

As a consequence, $S(s)x \in \mathcal{D}(X)$, and the above equalities hold for all $x \in \mathcal{D}(Z)$, all $s > 0$ and all $\theta \in (-\alpha, \alpha)$. Since

$$
\lim_{s \downarrow 0} S(s)x = x
$$

and

$$
Zx = \lim_{s \downarrow 0} S(s)Zx = \lim_{s \downarrow 0} e^{i\theta} X S(s)x,
$$

the fact that X is closed implies that $x \in \mathcal{D}(X)$ and $Zx = e^{i\theta} Xx$ for all $x \in \mathcal{D}(Z)$, that is, $Z \subset e^{i\theta} X$. Since Z is closed and generates a strongly continuous semigroup, if $Z \neq e^{i\theta} X$, there exists $k \in \mathbf{R}_+^*$ such that, whenever $\Re\zeta > k$, then $e^{-i\theta}\zeta \in p\sigma(X)$, contradicting the hypothesis whereby X generates a strongly continuous semigroup. Thus,

$$
Z = e^{i\theta} X. \tag{4.1}
$$

Since $\|S(t)\| \leq M_\epsilon$ for all $t \geq 0$, then

$$
\sigma(Z) \subset \{\zeta \in \mathbf{C} : \Re\zeta \leq 0\}
$$

and

$$
\|(\zeta I - Z)^{-1}\| \leq \frac{M_\epsilon}{\Re\zeta} \quad \text{whenever } \Re\zeta > 0. \tag{4.2}
$$

By (4.1), $\zeta \in \sigma(X)$ if, and only if, $e^{i\theta}\zeta \in \sigma(Z)$. If $\zeta \in \sigma(X)$, $\Re(e^{i\theta}\zeta) \leq 0$ - i.e. $|\arg\zeta + \theta| \geq \frac{\pi}{2}$ - for all $\theta \in (-(\alpha - \epsilon), \alpha - \epsilon)$. Since $\epsilon > 0$ is arbitrary, then, if $\zeta \in \sigma(X)$,

$$|\arg\zeta - \alpha| \geq \frac{\pi}{2} \quad \text{and} \quad |\arg\zeta - \alpha| \geq \frac{\pi}{2}.$$

Hence, if $\zeta \in \sigma(X)$, then

$$|\arg\zeta| \geq \frac{\alpha}{2} + \frac{\pi}{2},$$

showing that condition 5) is fulfilled.

As for condition 6), let

$$|\arg\zeta| < \frac{\pi}{2} + \alpha - \epsilon. \tag{4.3}$$

Let $\theta \in (-\alpha, \alpha)$ be such that

$$|\arg\zeta + \theta| < \frac{\pi}{2} - \frac{\epsilon}{2}$$

(Such a θ exists. Proof: exercise). Then,

$$\cos(\arg\zeta + \theta) > \cos\left(\frac{\pi}{2} - \frac{\epsilon}{2}\right) = \sin\frac{\epsilon}{2},$$

i.e.,

$$\Re(e^{i\theta}\zeta) \geq |\zeta| \sin\frac{\epsilon}{2}.$$

Thus,

$$\|(\zeta I - X)^{-1}\| = \left\|(e^{i\theta}\zeta I - Z)^{-1}\right\|$$

$$\leq \frac{M_{\epsilon/2}}{\Re(e^{i\theta}\zeta)} \leq \frac{N_\epsilon}{|\zeta|},$$

where $N_\epsilon = \frac{M_{\epsilon/2}}{\sin\frac{\epsilon}{2}}$.

Finally, if $|\arg\zeta| < \alpha - \epsilon$, choosing $\theta = -\arg\zeta$, then $\Re(e^{i\theta}\zeta) = |\zeta|$, and therefore

$$\|(\zeta I - X)^{-1}\| = \left\|(e^{i\theta}\zeta I - Z)^{-1}\right\| = \frac{M_\epsilon}{|\zeta|}. \qquad \blacksquare$$

272

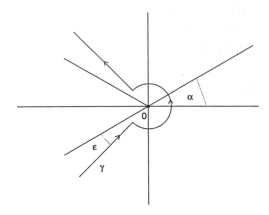

Figure 4.1:

We shall now prove the converse.

Let $0 < \alpha \leq \frac{\pi}{2}$, and let $X : \mathcal{D}(X) \subset \mathcal{E} \to \mathcal{E}$ be a closed linear operator satisfying conditions 5) and 6) of Theorem 4.1.4.

Choose $\epsilon \in (0, \frac{\alpha}{2})$, and let $z \in \Theta_{\alpha-2\epsilon}$.

Let γ be the path defined by

$$\gamma = \begin{cases} re^{-i(\alpha-\epsilon+\frac{\pi}{2})} & \text{if } 1 \leq r < +\infty, \\ e^{is} & \text{if } s \in [-(\alpha-\epsilon+\frac{\pi}{2}), \alpha-\epsilon+\frac{\pi}{2}], \\ re^{i(\alpha-\epsilon+\frac{\pi}{2})} & \text{if } 1 \leq r < +\infty. \end{cases}$$

Lemma 4.1.5 *Let $0 < \epsilon < \frac{\alpha}{2}$. There is a constant $k > 0$ such that, if $z \in \Theta_{\alpha-2\epsilon}$, and $\zeta \neq 0$ with $\frac{\pi}{2} + \alpha - \epsilon \leq |\arg \zeta| \leq \pi$, then*

$$\Re(z\,\zeta) \leq -k|z|\,|\zeta|. \tag{4.4}$$

Proof Setting $z = |z|e^{i\theta}$ and $\zeta = |\zeta|e^{i\phi}$, θ and ϕ can be chosen in such a way that

$$-\alpha + 2\epsilon < \theta < \alpha - 2\epsilon$$

and

$$\alpha + \frac{\pi}{2} - \epsilon \leq \phi \leq \frac{3\pi}{2} - \alpha + \epsilon,$$

Then,
$$\frac{\pi}{2} + \epsilon < \phi + \theta < \frac{3\pi}{2} - \epsilon,$$

and therefore
$$\cos(\phi + \theta) < \cos\left(\frac{\pi}{2} + \epsilon\right) = -\sin\epsilon.$$

We can choose
$$k = \sin\epsilon.$$

∎

The integral of $e^{z\zeta}(\zeta I - X)^{-1}$, extended to any one of the two rectilinear branches of γ, is such that
$$\left\| \int e^{z\zeta}(\zeta I - X)^{-1} d\zeta \right\| \leq \int_1^{+\infty} e^{-k|z|r} \frac{N_\epsilon}{r} dr < \infty.$$

Thus, setting, for $z \in \Theta_{\alpha - 2\epsilon}$,
$$T(z) = \frac{1}{2\pi i} \int_\gamma e^{z\zeta}(\zeta I - X)^{-1} d\zeta, \tag{4.5}$$

the integral is norm convergent, and $T(z)$ is a linear operator for which $T(z) \in \mathcal{L}(\mathcal{E})$ whenever $z \in \Theta_{\alpha - 2\epsilon}$.

To establish a uniform estimate for $\|T(z)\|$, we will begin by showing that, for all $R > 0$,
$$T(z) = \frac{1}{2\pi i} \int_\gamma \frac{1}{R} e^{\frac{z\zeta}{R}} \left(\frac{\zeta}{R} I - X\right)^{-1} d\zeta. \tag{4.6}$$

Consider the bottom rectilinear branch of γ. Since
$$\frac{1}{R} \int_1^{+\infty} e^{\frac{z\zeta}{R}} \left(\frac{\zeta}{R} I - X\right)^{-1} d\zeta = \int_{\frac{1}{R}}^{+\infty} e^{z\zeta}(\zeta I - X)^{-1} d\zeta,$$

then, assuming $R > 1$,
$$\frac{1}{R} \int_1^{+\infty} e^{\frac{z\zeta}{R}} \left(\frac{\zeta}{R} I - X\right)^{-1} d\zeta - \int_1^{+\infty} e^{z\zeta}(\zeta I - X)^{-1} d\zeta$$
$$= \left(\int_{\frac{1}{R}}^{+\infty} - \int_1^{+\infty} \right) e^{z\zeta}(\zeta I - X)^{-1} d\zeta$$
$$= \int_{\frac{1}{R}}^1 e^{z\zeta}(\zeta I - X)^{-1} d\zeta.$$

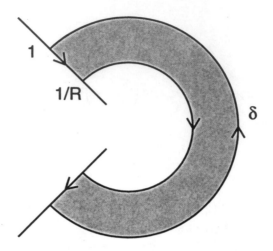

Figure 4.2:

A similar computation for the top rectilinear branch of γ yields then

$$\int_\gamma \frac{\mathrm{e}^{\frac{z\zeta}{R}}}{R}\left(\frac{\zeta}{R}I - X\right)^{-1}d\zeta - \int_\gamma \mathrm{e}^{z\zeta}(\zeta I - X)^{-1}d\zeta$$

$$= \int_\delta \mathrm{e}^{z\zeta}(\zeta I - X)^{-1}d\zeta,$$

where δ is the oriented boundary of the shadowed domain in the figure.

Since the function $\zeta \mapsto \mathrm{e}^{z\zeta}(\zeta I - X)^{-1}$ is holomorphic in an open neighbourhood of the closure of the domain, the Cauchy integral theorem yields

$$\int_\delta \mathrm{e}^{z\zeta}(\zeta I - X)^{-1}d\zeta = 0,$$

and therefore (4.6) holds when $R > 1$. Similar manipulations lead to the same conclusion when $0 < R \leq 1$. For $R = |z|$,

$$T(z) = \frac{1}{2\pi i}\int_\gamma \frac{\mathrm{e}^{\zeta\frac{z}{|z|}}}{|z|}\left(\frac{\zeta}{|z|}I - X\right)^{-1}d\zeta,$$

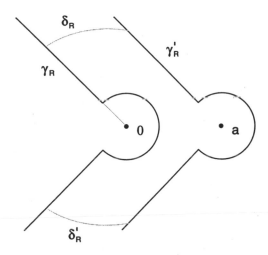

Figure 4.3:

and therefore, by condition 6) and (4.4),

$$\|T(z)\| \leq \frac{1}{2\pi} \int_\gamma \left| \frac{e^{\zeta \frac{z}{|z|}}}{z} \right| \frac{N_\epsilon}{|\frac{\zeta}{z}|} |d\zeta|$$

$$\leq \frac{1}{2\pi} N_\epsilon \int_\gamma \frac{e^{\Re(\zeta \frac{z}{|z|})}}{|\zeta|} |d\zeta| \leq M_\epsilon$$

for some finite $M_\epsilon > 0$ and for all $z \in \Theta_{\alpha - 2\epsilon}$.

Let γ' be the image of γ by the horizontal translation $\zeta \mapsto \zeta + a$ of amplitude $a \in \mathbf{R}_+$ so large that γ and γ' are disjoint. As in Figure 4.3, γ' lies on the right-hand side of γ. Let $R \gg 0$ and let δ_R be the arc, contained in the upper half plane, cut on the circle with center 0 and radius R by the strip defined by γ and γ'. If ω_R is the angle with vertex in 0 defined by the arc δ_R, then $\lim_{R \to +\infty} \omega_R = 0$. Hence, if $0 < \beta < \epsilon$ and if $\zeta \in \gamma'$, there is $R_o > 0$ such that

$$\frac{\pi}{2} + \alpha - (\epsilon - \beta) < |\arg \zeta| \leq \pi$$

whenever $|\zeta| > R_o$.

Since, by Lemma 4.1.5, if $R \gg 0$,

$$\left\| \int_{\delta_R} e^{z\zeta}(\zeta I - X)^{-1} d\zeta \right\| \leq N_\epsilon \int_{\delta_R} \frac{e^{-k'R|z|}}{R} |d\zeta| = N_\epsilon e^{-k'R|z|} \omega_R,$$

where $k' = \sin(\epsilon - \beta)$, then

$$\lim_{R \to +\infty} \int_{\delta_R} e^{z\zeta}(\zeta\, I - X)^{-1} d\zeta = 0.$$

Similarly,

$$\lim_{R \to +\infty} \int_{\delta'_R} e^{z\zeta}(\zeta\, I - X)^{-1} d\zeta = 0,$$

where δ'_R is the image of δ_R by the symmetry around the real axis.

Since the function $\zeta \mapsto e^{z\zeta}(\zeta\, I - X)^{-1}$ is holomorphic in an open neighbourhood of the closure of the intersection of the disc with center 0 and radius $R \gg 0$ with the strip defined by γ and γ', by the Cauchy integral theorem the integral of the function $\zeta \mapsto e^{z\zeta}(\zeta\, I - X)^{-1}$ along the boundary of the domain vanishes. Hence

$$\int_\gamma e^{z\zeta}(\zeta\, I - X)^{-1} d\zeta = \int_{\gamma'} e^{z\zeta'}(\zeta'\, I - X)^{-1} d\zeta'.$$

On the basis of this latter equality we show now that, if z and w are any two points in $\Theta_{\alpha-2\epsilon}$, then

$$T(z + w) = T(z)\, T(w).$$

With the same notations as before,

$$T(z)\, T(w) = \frac{1}{(2\pi i)^2} \int_\gamma \int_{\gamma'} e^{z\zeta + w\zeta'}(\zeta\, I - X)^{-1}(\zeta'\, I - X)^{-1} d\zeta\, d\zeta'$$

$$= \frac{1}{(2\pi i)^2} \int_\gamma \int_{\gamma'} \frac{e^{z\zeta + w\zeta'}}{\zeta' - \zeta} \times$$

$$\left[(\zeta\, I - X)^{-1} - (\zeta'\, I - X)^{-1}\right] d\zeta\, d\zeta'. \tag{4.7}$$

By Lemma 4.1.5, there is a constant $h > 0$ such that, if $\zeta \in \gamma$ and $\zeta' \in \gamma'$ with $|\zeta| \gg 0$ and $|\zeta'| \gg 0$, and if $z, w \in \Theta_{\alpha-2\epsilon}$, then

$$\left|e^{z\zeta}\right| \le e^{-h|z||\zeta|}, \quad \left|e^{w\zeta'}\right| \le e^{-h|w||\zeta'|}.$$

Since both the right-hand sides tend to zero when $|\zeta| \to \infty$ and $|\zeta'| \to \infty$, the Cauchy integral theorem and the Cauchy integral formula yield

$$\int_\gamma \frac{e^{z\zeta}}{\zeta' - \zeta} d\zeta = 0 \ \forall\ z \in \Theta_{\alpha-2\epsilon}, \ \forall\ \zeta' \in \gamma'$$

and

$$\frac{1}{2\pi i}\int_{\gamma'}\frac{e^{w\zeta'}}{\zeta'-\zeta}d\zeta' = e^{w\zeta} \ \forall \ w \in \Theta_{\alpha-2\epsilon}, \ \forall \zeta \in \gamma.$$

Then, as a consequence of (4.7),

$$T(z)\,T(w) = \frac{1}{(2\pi i)^2}\int_{\gamma}e^{z\zeta}\left(\int_{\gamma'}\frac{e^{w\zeta'}}{\zeta'-\zeta}d\zeta'\right)(\zeta\,I-X)^{-1}d\zeta -$$
$$\frac{1}{(2\pi i)^2}\int_{\gamma'}e^{w\zeta'}\left(\int_{\gamma}\frac{e^{z\zeta}}{\zeta'-\zeta}d\zeta\right)(\zeta'\,I-X)^{-1}d\zeta'$$
$$= \frac{1}{2\pi i}\int_{\gamma}e^{(z+w)\zeta}(\zeta\,I-X)^{-1}d\zeta = T(z+w).$$

The function $z \mapsto T(z)$ is a holomorphic map of $\Theta_{\alpha-2\epsilon}$ into $\mathcal{L}(\mathcal{E})$, and

$$\frac{d}{dz}T(z) = \frac{1}{2\pi i}\int_{\gamma}\zeta\,e^{z\zeta}(\zeta\,I-X)^{-1}d\zeta. \qquad (4.8)$$

If $R > 1$, the function $\zeta \mapsto \frac{1}{\zeta}(\zeta\,I-X)^{-1} \in \mathcal{L}(\mathcal{E})$ is holomorphic in a neighbourhood of the closure of the shadowed region in the figure.

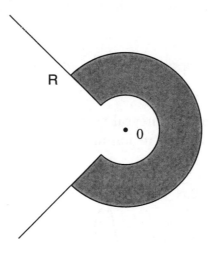

Figure 4.4:

278

As a consequence, the integral of $\frac{1}{\zeta}(\zeta I - X)^{-1}$ along the boundary vanishes. Since, by condition 6),

$$\left\| \int_{-(\alpha-\epsilon+\frac{\pi}{2})}^{\alpha-\epsilon+\frac{\pi}{2}} \frac{1}{Re^{it}}(Re^{it}I - X)^{-1}d(Re^{it}) \right\|$$

$$= \left\| \int_{-(\alpha-\epsilon+\frac{\pi}{2})}^{\alpha-\epsilon+\frac{\pi}{2}} (Re^{it}I - X)^{-1}dt \right\|$$

$$\leq \int_{-(\alpha-\epsilon+\frac{\pi}{2})}^{\alpha-\epsilon+\frac{\pi}{2}} \|(Re^{it}I - X)^{-1}\|dt$$

$$\leq \frac{N_\epsilon}{R} 2(\alpha - \epsilon + \frac{\pi}{2}) \to 0$$

as $R \to +\infty$, then

$$\int_\gamma \frac{1}{\zeta}(\zeta I - X)^{-1}d\zeta = 0. \tag{4.9}$$

If $z = |z|e^{i\theta} \in \Theta_{\alpha-2\epsilon}$,

$$\int_{\delta_2} \frac{e^{z\zeta}}{\zeta}d\zeta = \int \frac{e^{|z|Re^{i(\theta+t)}}}{Re^{it}} iRe^{it}dt,$$

where the integral is taken on the arc $(\alpha - \epsilon + \frac{\pi}{2}, -\alpha + \epsilon + \frac{3\pi}{2})$. Thus, by Lemma 4.1.5,

$$\lim_{R\to+\infty} \int_{\delta_2} \frac{e^{z\zeta}}{\zeta}d\zeta = 0. \tag{4.10}$$

Since, by the Cauchy integral theorem, the integral of $\frac{e^{z\zeta}}{\zeta}$ over the boundary of the shadowed region in the figure is zero, then

$$1 = \frac{1}{2\pi i} \int_{\partial\Delta} \frac{e^{z\zeta}}{\zeta}d\zeta =$$

$$= \frac{1}{2\pi i}\left(\int_{\delta_0} - \int_{\delta_1}\right)\frac{e^{z\zeta}}{\zeta}d\zeta$$

$$= \frac{1}{2\pi i}\left(\int_{\delta_0} + \int_{\delta_4} + \int_{\delta_3} + \int_{\delta_2}\right)\frac{e^{z\zeta}}{\zeta}d\zeta.$$

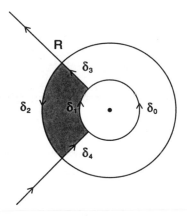

Figure 4.5:

Letting $R \to +\infty$, (4.10) yields then

$$\frac{1}{2\pi i} \int_\gamma \frac{e^{z\zeta}}{\zeta} d\zeta = 1 \qquad (4.11)$$

for all $z \in \Theta_{\alpha-2\epsilon}$.

Using this result, we show now that

$$\lim_{\substack{z \to 0 \\ z \in \Theta_{\alpha-2\epsilon}}} T(z)x = x \quad \forall x \in \mathcal{E}. \qquad (4.12)$$

Let $y \in \mathcal{D}(X)$. By (4.9) and (4.11),

$$T(z)y - y = \frac{1}{2\pi i} \int_\gamma e^{z\zeta} \left((\zeta I - X)^{-1} - \frac{1}{\zeta} \right) y d\zeta$$

$$= \frac{1}{2\pi i} \int_\gamma e^{z\zeta} (\zeta I - X)^{-1} \left(I - \frac{1}{\zeta}(\zeta I - X) \right) y d\zeta$$

$$= \frac{1}{2\pi i} \int_\gamma \frac{e^{z\zeta}}{\zeta} (\zeta I - X)^{-1} X y d\zeta$$

$$= \frac{1}{2\pi i} \int_\gamma \frac{e^{z\zeta} - 1}{\zeta} (\zeta I - X)^{-1} X y d\zeta.$$

Hence, condition 6) and (4.4) yield

$$\|T(z)y - y\| \le \frac{N_\epsilon}{2\pi} \int_\gamma \frac{|e^{z\zeta} - 1|}{|\zeta|^2} |d\zeta| \, \|Xy\|$$

$$\le \frac{N_\epsilon}{2\pi} \int_\gamma \frac{1 + e^{-k|z||\zeta|}}{|\zeta|^2} |d\zeta| \, \|Xy\|,$$

and therefore

$$\lim_{\substack{z \to 0 \\ z \in \Theta_{\alpha - 2\epsilon}}} \|T(z)y - y\| = 0$$

by the dominated convergence theorem.

If $x \in \mathcal{E}$, for any $\eta > 0$ there is some $y \in \mathcal{D}(X)$ for which $\|x - y\| < \eta$. Once such an y has been chosen there is $\rho > 0$ such that, if $z \in \Theta_{\alpha - 2\epsilon}$ and $|z| < \rho$, then

$$\|T(z)y - y\| < \eta.$$

Hence

$$\|T(z)x - x\| \le \|T(z)(x - y)\| + \|x - y\|$$
$$+ \|T(z)y - y\| \le (M_\epsilon + 2)\,\eta;$$

that is to say, (4.12) holds, showing thereby that $T : \{0\} \cup \Theta_\alpha \to \mathcal{L}(\mathcal{E})$ is a bounded holomorphic semigroup of angle α.

As a consequence, the semigroup $T_{|\mathbf{R}_+} : t \mapsto T(t)$ is strongly continuous.

We are left to prove that, if $Y : \mathcal{D}(Y) \subset \mathcal{E} \to \mathcal{E}$ is the infinitesimal generator of $T_{|\mathbf{R}_+} : \mathbf{R}_+ \to \mathcal{L}(\mathcal{E})$, then $Y = X$. We begin by showing that X commutes with $T(t)$ on $\mathcal{D}(X)$. That is to say: if $x \in \mathcal{D}(X)$, $T(t)x \in \mathcal{D}(X)$, and

$$XT(t)x = T(t)Xx. \tag{4.13}$$

Since, for $x \in \mathcal{D}(X)$ and $\zeta \in r(X)$,

$$X(\zeta I - X)^{-1}x = -x + \zeta(\zeta I - X)^{-1}x = (\zeta I - X)^{-1}Xx,$$

the function $\zeta \mapsto X(\zeta I - X)^{-1}x$ is continuous on $r(X)$.

A standard approximation procedure yields a sequence of points $\zeta_\nu \in \gamma$ such that $T(t)$ is the limit in $\mathcal{L}(\mathcal{E})$ of a sequence of finite linear combinations $\sum a_\nu Z_\nu$ of the linear operators

$$Z_\nu = \frac{1}{2\pi i} e^{t\zeta_\nu} (\zeta_\nu I - X)^{-1}.$$

Since $X(\zeta I - X)^{-1} = (\zeta I - X)^{-1} X$ on $\mathcal{D}(X)$ for all $\zeta \in r(X)$, then

$$X \sum a_\nu Z_\nu x = \sum a_\nu X Z_\nu x = \sum a_\nu Z_\nu X x$$

for all $x \in \mathcal{D}(X)$. Because X is closed, then $T(t)x \in \mathcal{D}(X)$ and $XT(t)x = T(t)Xx$, , i.e., (4.13) holds.

Hence, for $x \in \mathcal{D}(X)$ and $t > 0$,

$$XT(t)x = T(t)Xx = \frac{1}{2\pi i} \int_\gamma e^{t\zeta} (\zeta I - X)^{-1} X x \, d\zeta$$

$$= \frac{1}{2\pi i} \int_\gamma e^{t\zeta} (-I + \zeta(\zeta I - X)^{-1}) x \, d\zeta. \tag{4.14}$$

By the Cauchy integral theorem,

$$\int_\gamma e^{t\zeta} d\zeta = \left(\int_{\delta_0} + \int_{\gamma_1} + \int_{\gamma_2} \right) e^{t\zeta} d\zeta$$

$$= \left(\int_{\delta_1} + \int_{\gamma_1} + \int_{\gamma_2} \right) e^{t\zeta} d\zeta$$

$$= \int_{\delta_1} e^{t\zeta} d\zeta + \int_R^{+\infty} \exp\left(t\, r\, e^{i\left(\alpha - \epsilon + \frac{\pi}{2}\right)} \right) e^{i\left(\alpha - \epsilon + \frac{\pi}{2}\right)} dr +$$

$$+ \int_{+\infty}^R \exp\left(t\, r\, e^{-i\left(\alpha - \epsilon + \frac{\pi}{2}\right)} \right) e^{-i\left(\alpha - \epsilon + \frac{\pi}{2}\right)} dr$$

$$= \int_{\delta_1} e^{t\zeta} d\zeta + e^{i\left(\alpha - \epsilon + \frac{\pi}{2}\right)} \int_R^{+\infty} \exp\left(t\, r\, e^{i\left(\alpha - \epsilon + \frac{\pi}{2}\right)} \right) dr +$$

$$- e^{-i\left(\alpha - \epsilon + \frac{\pi}{2}\right)} \int_R^{+\infty} \exp\left(t\, r\, e^{-i\left(\alpha - \epsilon + \frac{\pi}{2}\right)} \right) dr.$$

If $t > 0$ and $\zeta \in \delta_2$, Lemma 4.1.5 yields

$$|e^{t\zeta}| \leq e^{-kt|\zeta|} = e^{-ktR}.$$

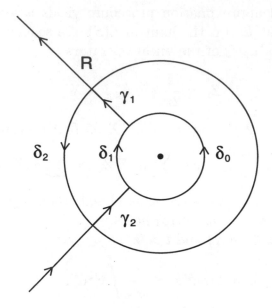

Figure 4.6:

Therefore

$$\left| \int_{\delta_2} e^{t\zeta} d\zeta \right| \le (\pi - 2\alpha + 2\epsilon) R e^{-ktR} \to 0$$

as $R \to +\infty$. Furthermore,

$$\left| \int_R^{+\infty} \exp\left(t\, r\, e^{i\left(\alpha - \epsilon + \frac{\pi}{2}\right)} \right) dr \right| \le \int_R^{+\infty} e^{t r \cos\left(\alpha - \epsilon + \frac{\pi}{2}\right)} dr =$$

$$\int_R^{+\infty} e^{-t r \sin(\alpha - \epsilon)} dr = \frac{e^{-t R \sin(\alpha - \epsilon)}}{t \sin(\alpha - \epsilon)} \to 0$$

as $R \to +\infty$ because $\sin(\alpha - \epsilon) > 0$. Similarly,

$$\lim_{R \to +\infty} \int_R^{+\infty} \exp\left(tr e^{-i\left(\alpha - \epsilon + \frac{\pi}{2}\right)} \right) dr = 0.$$

In conclusion,

$$\int_\gamma e^{t\zeta} d\zeta = 0,$$

and, (4.14) and (4.8) yield,

$$XT(t)x = \frac{1}{2\pi i} \int_\gamma \zeta e^{t\zeta}(\zeta I - X)^{-1} x d\zeta = \frac{d}{dt}T(t)x.$$

Consequently, for $x \in \mathcal{D}(X)$,

$$\left\| \frac{1}{t}(T(t)x - x) - Xx \right\| \le$$

$$\le \left\| \frac{1}{t}(T(t)x - x) - T(t)Xx \right\| + \|T(t)Xx - Xx\|$$

$$= \left\| \frac{1}{t} \int_0^t \left(\frac{d}{ds}T(s)x - T(t)Xx \right) ds \right\| + \|T(t)Xx - Xx\|$$

$$= \left\| \frac{1}{t} \int_0^t (T(s)Xx - T(t)Xx)ds \right\| + \|T(t)Xx - Xx\|$$

$$\le \frac{1}{t} \int_0^t \|T(s)Xx - T(t)Xx\|ds + \|T(t)Xx - Xx\| \to 0$$

as $t \downarrow 0$ because of (4.12) and because the function $s \mapsto \|T(s)Xx - T(t)Xx\|$ is continuous. Therefore, $x \in \mathcal{D}(X)$ and

$$Yx = \lim_{t \downarrow 0} \frac{1}{t}(T(t) - I)x = Xx$$

for all $x \in \mathcal{D}(X)$, i. e. $X \subset Y$. Since X is closed, if $X \ne Y$, then $r(X) \subset p\sigma(Y)$, and therefore $\{\zeta \in \mathbf{C} : \Re\zeta \ge 0\} \subset p\sigma(Y)$, contradicting the fact that Y is the infinitesimal generator of a strongly continuous semigroup.

In conclusion, the following theorem holds.

Theorem 4.1.6 *If X is a closed, densely defined, linear operator that satisfies both conditions 5) and 6) of Theorem 4.1.4 for some $\alpha \in (0, \frac{\pi}{2}]$, X is the infinitesimal generator of bounded holomorphic semigroup of angle α.*

Since the function $\Theta_{\alpha-2\epsilon} \to \mathcal{L}(\mathcal{E})$ defined by (4.5) is holomorphic, and therefore norm-continuous, the following proposition holds.

Proposition 4.1.7 *If the strongly continuous semigroup $T : \mathbf{R}_+ \to \mathcal{L}(\mathcal{E})$ can be extended to a bounded holomorphic semigroup of angle α, for some $\alpha \in (0, \frac{\pi}{2}]$, the function $t \mapsto \|T(t)\|$ is continuous on $\mathbf{R}_+^*$.*

Theorem 2.8.3 implies then

Corollary 4.1.8 *Under the hypothesis of Proposition 4.1.7,*

$$\sigma(T(t))\backslash\{0\} = e^{t\sigma(X)}$$

for all $t \in \mathbf{R}_+$.

Example The linear operator X defined by (3.20) on the domain $\mathcal{D}(X) \subset l^1$ given by (3.21) generates a strongly continuous semigroup $T : \mathbf{R}_+ \to \mathcal{L}(l^1)$ of linear contractions. The operator $T(t)$ is represented by the matrix

$$T(t) = \begin{pmatrix} 1 & 0 & 0 & \cdots \\ 0 & e^{-t} & 0 & \cdots \\ 0 & 0 & e^{-2t} & \cdots \\ . & . & . & \cdots \\ . & . & . & \cdots \\ . & . & . & \cdots \end{pmatrix}$$

which acts on $x = (x_0, x_1, x_2, \dots) \in l^1$ by

$$T(t)x = x_0 + e^{-t}x_1 + e^{-2t}x_2 + \cdots.$$

Lemma 4.1.9 *The operator X is the infinitesimal generator of a bounded holomorphic semigroup of angle $\frac{\pi}{2}$.*

Proof Since $\sigma(X) = \{0, -1, -2, \dots\}$, condition 5) of Theorem 4.1.4 is satisfied. For $0 < \epsilon < \frac{\pi}{2}$ le r be the hypotenuse of a right triangle having a cathetus and its opposite angle equal respectively to 1 and to ϵ: $r \sin \epsilon = 1$. If $[s]$ stands for the integral part of the real number s, for all $\zeta \in \Theta_{\pi-\epsilon}$, $|\zeta - [-r]| \geq 1$. Thus, (3.24) implies that

$$\|(\zeta I - X)^{-1}\| \leq \frac{1}{\min\{|\zeta|, |\zeta + 1|, \dots, |\zeta - [-r] - 1|\}}$$

for all $\zeta \in \Theta_{\pi-\epsilon}$, showing thereby that condition 6) of Theorem 4.1.4 holds. ■

Denoting by the same symbol T the holomorphic semigroup, for $z \in \Theta_{\frac{\pi}{2}} \cup \{0\}$, then

$$T(z) = \begin{pmatrix} 1 & 0 & 0 & \cdots \\ 0 & e^{-z} & 0 & \cdots \\ 0 & 0 & e^{-2z} & \cdots \\ \cdot & \cdot & \cdot & \cdots \\ \cdot & \cdot & \cdot & \cdots \\ \cdot & \cdot & \cdot & \cdots \end{pmatrix},$$

i. e.

$$T(z)x = x_0 + c^{-z}x_1 + e^{-2z}x_2 + \cdots .$$

An important class of bounded holomorphic semigroups is described by the following proposition.

Proposition 4.1.10 *If the linear, closed, densely defined operator X is the infinitesimal generator of a uniformly bounded, strongly continuous group $G : \mathbf{R} \to \mathcal{L}(\mathcal{E})$, then $X^2 : \mathcal{D}(X^2) \subset \mathcal{E} \to \mathcal{E}$ is the infinitesimal generator of a bounded holomorphic semigroup of angle $\frac{\pi}{2}$.*

Proof Let $\beta \in (0, \frac{\pi}{2})$ and let $\zeta = re^{i\tau}$ with $r > 0$ and $-(\beta + \frac{\pi}{2}) < \tau < \beta + \frac{\pi}{2}$, and therefore

$$-\frac{1}{2}(\beta + \frac{\pi}{2}) < \frac{\tau}{2} < \frac{1}{2}(\beta + \frac{\pi}{2}).$$

Thus, ζ can be written $\zeta = \mu^2$, with

$$|\arg \mu| < \frac{1}{2}\left(\beta + \frac{\pi}{2}\right).$$

Since $\sigma(X) \subset i\mathbf{R}$, then $\{-\mu, \mu\} \subset r(X)$, and therefore $\zeta = \mu^2 \in r(X^2)$ (because $\mu I - X$ and $\mu I + X$ are both invertible in $\mathcal{L}(\mathcal{E})$, and $\zeta I - X^2 = \mu^2 I - X^2 = (\mu I + X)(\mu I - X) = (\mu I - X)(\mu I + X)$).
Hence,

$$\sigma(X^2) \subset \left\{\zeta \in \mathbf{C} : \beta + \frac{\pi}{2} \leq |\arg \zeta| \leq \pi\right\}$$

for all $\beta \in (0, \frac{\pi}{2})$, and therefore

$$\sigma(X^2) \subset -\mathbf{R}_+.$$

Since G is bounded, there is a finite constant $M \geq 1$, for which

$$\|(\kappa I - X)^{-1}\| \leq \frac{M}{\Re\kappa}, \quad \|(\kappa I + X)^{-1}\| \leq \frac{M}{\Re\kappa}$$

for all $\kappa \in \mathbf{C}$ with $\Re\kappa > 0$. As a consequence,

$$\|(\zeta I - X^2)^{-1}\| \leq \|(\mu I + X)^{-1}\| \, \|(\mu I - X)^{-1}\|$$
$$\leq \frac{M^2}{(\Re\mu)^2} = \frac{M^2}{\left(|\mu| \cos \frac{\tau}{2}\right)^2}$$
$$\leq \frac{M^2}{\left(|\mu| \cos \left(\frac{1}{2}(\beta + \frac{\pi}{2})\right)\right)^2}$$
$$= \frac{N}{|\mu|^2} = \frac{N}{|\zeta|},$$

where

$$N = \frac{M^2}{\left(\cos \left(\frac{1}{2}(\beta + \frac{\pi}{2})\right)\right)^2}.$$

Since $(\zeta I - X^2)^{-1} \in \mathcal{L}(\mathcal{E})$, and therefore is closed, then $(\zeta I - X^2$ is closed, and) X^2 is closed.

In conclusion, the closed, densely defined operator X^2 generates a bounded holomorphic semigroup of angle $\frac{\pi}{2}$. ∎

If T is a bounded holomorphic semigroup of angle $\alpha \in (0, \frac{\pi}{2}]$, and if $z \in \Theta_\alpha$, then (4.5) shows that, for all $n = 0, 1, 2, \ldots$,

$$z \mapsto \frac{d^n}{dz^n} T(z)$$

is a holomorphic map of the domain Θ_α into $\mathcal{L}(\mathcal{E})$. Since

$$(T(t) - I)T(z)x = (T(z + t) - T(z))x,$$

then

$$\lim_{t \downarrow 0} \frac{1}{t}(T(t) - I)T(z)x = \left(\frac{d}{dz}T(z)\right)(x) \ \forall \, x \in \mathcal{E}.$$

As a consequence,

$$x \in \mathcal{E} \Rightarrow T(z)x \in \mathcal{D}(X) \ \forall \, z \in \Theta_\alpha.$$

In particular, $T(t)x \in \mathcal{D}(X)$ for all $t > 0$ and all $x \in \mathcal{E}$, and therefore

$$X\,T(t)x = T(t)\,Xx \in \mathcal{D}(X).$$

Iteration of this argument and Corollary 4.1.3 yield

Theorem 4.1.11 *If T is a bounded holomorphic semigroup of angle $\alpha \in (0, \frac{\pi}{2}]$, then $T(t)x \in \mathcal{D}(X^n)$ for all $x \in \mathcal{E}$ and all $n = 1, 2, \ldots$.*

Theorem 4.1.12 *If $-X$ is a positive self-adjoint operator in a complex Hilbert space $\mathcal{H}$, X is the infinitesimal generator of a bounded holomorphic semigroup of angle $\frac{\pi}{2}$, whose restriction to $\mathbf{R}_+$ is a contraction semigroup.*

Proof For $\zeta \in \mathbf{C}\backslash\{0\}$, write $\zeta = |\zeta|e^{i\theta}$. For any fixed $\epsilon \in (0, \frac{\pi}{2})$, the union of the three sets

$$D_1 = \left\{\zeta \in \mathbf{C}\backslash\{0\} : \frac{\pi}{2} \leq |\theta| < \pi - \epsilon\right\}$$

$$D_2 = \left\{\zeta \in \mathbf{C}\backslash\{0\} : \frac{\pi}{4} \leq |\theta| \leq \frac{\pi}{2}\right\}$$

$$D_3 = \left\{\zeta \in \mathbf{C}\backslash\{0\} : |\theta| \leq \frac{\pi}{4}\right\}$$

cover the domain

$$\Xi_\epsilon = \{\zeta \in \mathbf{C}\backslash\{0\} : |\arg\zeta| < \pi - \epsilon\}.$$

Since X is self-adjoint, then

$$\|(\zeta I - X)^{-1}\| \leq \frac{1}{|\Im\zeta|} \quad \text{whenever } \Im\zeta \neq 0. \tag{4.15}$$

If $\zeta \in D_1$, $|\sin\theta| > \sin\epsilon$, and therefore

$$|\Im\zeta| = |\zeta|\,|\sin\theta| > |\zeta|\sin\epsilon.$$

Hence, (4.15) yields

$$\|(\zeta I - X)^{-1}\| < \frac{1}{|\zeta| \sin \epsilon} \quad \forall \zeta \in D_1. \tag{4.16}$$

If $\zeta \in D_2$, then $|\sin \theta| \geq \frac{1}{\sqrt{2}}$,

$$|\Im \zeta| = |\zeta| |\sin \theta| \geq \frac{|\zeta|}{\sqrt{2}},$$

and therefore (4.15) yields

$$\|(\zeta I - X)^{-1}\| \leq \frac{\sqrt{2}}{|\zeta|} \quad \forall \zeta \in D_2. \tag{4.17}$$

Since the self-adjoint operator $-X$ is positive, and therefore X generates a strongly continuous semigroup T of linear contractions of $\mathcal{H}$, by Theorem 2.4.1,

$$\|(\zeta I - X)^{-1} x\| \leq \int_0^{+\infty} |e^{-t\zeta}| \, dt \, \|x\|$$

$$= \int_0^{+\infty} e^{-t\Re \zeta} \, dt \, \|x\| = \frac{1}{\Re \zeta} \|x\|$$

for all $x \in \mathcal{H}$ and all $\zeta \in \mathbf{C}$ with $\Re \zeta > 0$, i.e.,

$$\|(\zeta I - X)^{-1}\| \leq \frac{1}{\Re \zeta} \quad \text{whenever } \Re \zeta > 0.$$

Since, for $\zeta \in D_3$,

$$|\Re \zeta| = |\zeta| |\cos \theta| \geq \frac{|\zeta|}{\sqrt{2}},$$

then

$$\|(\zeta I - X)^{-1}\| \leq \frac{\sqrt{2}}{|\zeta|} \quad \forall \zeta \in D_3. \tag{4.18}$$

In view of Theorem 4.1.6, (4.16), (4.17) and (4.18) yield the conclusion. ∎

Remark With the same notations of Theorem 4.1.12, the operator $Y = -iX$ is such that

$$\sigma(Y) = -i\sigma(X) \subset i\mathbf{R}_+,$$

and

$$Y^* = iX^* = iX = -Y.$$

This situation motivates the following theorem.

Theorem 4.1.13 *Let $\mathcal{E}$ be a complex Banach space. If the linear operator $X : \mathcal{D}(X) \subset \mathcal{E} \to \mathcal{E}$ is densely defined and m-dissipative, and both iX and $-iX$ are m-dissipative, then X generates a strongly continuous semigroup of linear contractions of $\mathcal{E}$ which is the restriction to $\mathbf{R}_+$ of a bounded holomorphic semigroup of angle $\frac{\pi}{2}$.*

Sketch of proof. The fact that both iX and $-iX$ are m-dissipative, and therefore iX generates a strongly continuous group $\mathbf{R} \to \mathcal{L}(\mathcal{E})$ of linear isometries of $\mathcal{E}$, implies, by Theorem 2.4.1, that

$$\|(\zeta I - iX)^{-1}\| \leq \frac{1}{|\Re\zeta|} \quad \text{whenever } \Re\zeta \neq 0.$$

Replacing ζ by $i\zeta$, this latter inequality becomes

$$\|(\zeta I - X)^{-1}\| \leq \frac{1}{|\Im\zeta|} \quad \text{whenever } \Im\zeta \neq 0.$$

On the other hand, since X generates a strongly continuous semigroup of linear contractions of $\mathcal{E}$, then

$$\|(\zeta I - X)^{-1}\| \leq \frac{1}{\Re\zeta} \quad \text{whenever } \Re\zeta > 0.$$

From this point on, the argument follows the lines of the proof of Theorem 4.1.12.

4.2 Holomorphic semigroups

We shall investigate now holomorphic semigroups that are not necessarily bounded.

We begin by establishing the following elementary lemma.

Lemma 4.2.1 *Let $\beta > 0$ and let $b > 0$. There is $k > 0$ such that, if $|\tau| > k$ and $|\arg \tau| < \beta$, then $|\arg(\tau + b)| < \beta$.*

Proof Let $\tau \in \mathbf{C} \backslash \mathbf{R}$, and let γ be the angle in τ of the triangle Ξ with vertices $0, \tau, b$. Then

$$1 \geq \cos \gamma = \frac{|\tau|^2 + |\tau - b|^2 - b^2}{2|\tau| \, |\tau - b|}$$

$$\geq \frac{|\tau|^2 + |\tau - b|^2 - b^2}{|\tau|^2 + |\tau - b|^2} = 1 - \frac{b^2}{|\tau|^2 + |\tau - b|^2}$$

$$\geq 1 - \frac{b^2}{|\tau|^2} \to 1$$

as $|\tau| \to \infty$. Hence, for any $\delta > 0$ there is $k > 0$ such that, if $|\tau| > k$, then $\gamma < \delta$; that is to say, the exterior angle in b to the triangle Ξ - *i.e.* $\arg(\tau - b)$ - is larger than the angle in 0 *i.e.* $\arg \tau$ - but less than this latter angle increased by δ. Thus, for any $\beta > 0$, there is $k > 0$ such that

$$|\arg \tau| < \beta \iff |\arg(\tau + b)| < \beta$$

whenever $|\tau| > k$. ∎

Now, let T be a holomorphic semigroup of angle $\alpha \in (0, \frac{\pi}{2}]$, and let X be its infinitesimal generator. By Corollary 4.1.3, for every $\alpha_1 \in (0, \alpha)$, there is some $a \in \mathbf{R}_+$ such that $Y = X - a\,I$ generates a bounded holomorphic semigroup of angle α_1. By Theorem 4.1.4,

$$\sigma(X) = \sigma(Y + a\,I) = \left\{ \zeta + a : \zeta \in \mathbf{C}, \alpha_1 + \frac{\pi}{2} \leq |\arg \zeta| \leq \pi \right\}$$

$$= \left\{ \zeta : \zeta \in \mathbf{C}, \alpha_1 + \frac{\pi}{2} \leq |\arg \zeta| \leq \pi \right\} + a,$$

and, for every $\epsilon \in (0, \alpha_1)$, there exists a finite constant $N_\epsilon \geq 1$ such that

$$\|(\zeta I - Y)^{-1}\| \leq \frac{N_\epsilon}{|\zeta|},$$

that is to say,

$$\|((\zeta + a) I - X)^{-1}\| \leq \frac{N_\epsilon}{|\zeta|} \tag{4.19}$$

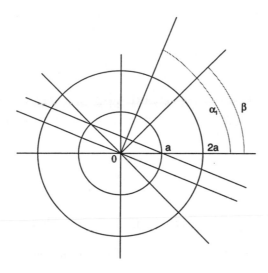

Figure 4.7:

whenever $\zeta \neq 0$ and $|\arg \zeta| \leq \frac{\pi}{2} + \alpha_1 - \epsilon$.

Consider the circle with center in 0 and radius a, and the half lines joining 0 with the intersections of the circle with the parallel lines, passing through $(a, 0)$, to the half lines

$$\left\{ \Re\zeta < 0 : \arg \zeta = \alpha_1 + \frac{\pi}{2} \right\},$$

and

$$\left\{ \Re\zeta < 0 : \arg \zeta = -\left(\alpha_1 + \frac{\pi}{2}\right) \right\}.$$

If $\beta + \frac{\pi}{2}$ and $-(\beta + \frac{\pi}{2})$, with $0 < \beta \leq \frac{\pi}{2}$, are the angles between each of the two half lines and the positive real axis, then $0 < \beta < \alpha_1$, and

$$\sigma(X) \subset \left\{ \zeta \in \mathbf{C} : \beta + \frac{\pi}{2} \leq |\arg \zeta| \leq \pi \right\} \cup \overline{\Delta_a}, \qquad (4.20)$$

where $\Delta_a = \{\zeta \in \mathbf{C} : |\zeta| < a\}$ and $\overline{\Delta_a}$. A fortiori,

$$\sigma(X) \subset \left\{ \zeta \in \mathbf{C} : \beta + \frac{\pi}{2} \leq |\arg \zeta| \leq \pi \right\} \cup \overline{\Delta_{2a}}.$$

If $|\zeta| > 2a$, then

$$\left| \frac{\zeta + a}{\zeta} \right| = \left| 1 + \frac{a}{\zeta} \right| \leq 1 + \frac{a}{|\zeta|} < 1 + \frac{a}{2a} = \frac{3}{2},$$

and therefore, by (4.19),

$$\|((\zeta + a)\,I - X)^{-1}\| \leq \frac{3}{2}\frac{N_\epsilon}{|\zeta + a|} \qquad (4.21)$$

whenever $|\arg\zeta| < \alpha_1 - \epsilon + \frac{\pi}{2}$ and $|\zeta| > 2a$. By Lemma 4.2.1, where now $\tau = \zeta + a$ and $b = a$, there is $k > 2a$ such that, if $|\zeta| > k$ and $|\arg\zeta| < \alpha_1 - \epsilon + \frac{\pi}{2}$, then

$$\|(\zeta\,I - X)^{-1}\| \leq \frac{3}{2}\frac{N_\epsilon}{|\zeta|}.$$

That proves the first part of the following theorem

Theorem 4.2.2 *If X is the infinitesimal generator of holomorphic semigroup T, of angle $\alpha \in (0, \frac{\pi}{2}]$, then:*
  there is a constant $a > 0$ such that (4.20) holds;
  for any $\epsilon \in (0, \alpha)$ there exist two constants $N > 0$ and $c \geq a$ such that, if $|\zeta| > c$ and if $|\arg\zeta| < \alpha - \epsilon + \frac{\pi}{2}$, then

$$\|(\zeta\,I - X)^{-1}\| \leq \frac{N}{|\zeta|}. \qquad (4.22)$$

Viceversa, any closed, densely defined, linear operator $X : \mathcal{D}(X) \subset \mathcal{E} \to \mathcal{E}$ satisfying the two conditions stated above for $\alpha \in (0, \frac{\pi}{2}]$, is the infinitesimal generator of a holomorphic semigroup of angle α.

To establish the second part of the theorem, note that, by Lemma 4.2.1, there is a constant $k \geq c$ such that, if $|\zeta| > k$ and $|\arg\zeta| < \alpha - \epsilon + \frac{\pi}{2}$, then $|\zeta + c| > k$ and $|\arg(\zeta + c)| < \alpha - \epsilon + \frac{\pi}{2}$.

Let $Y = X - cI$. If $|\zeta| > 2k \geq 2c$ and $|\arg\zeta| < \alpha - \epsilon + \frac{\pi}{2}$, then, by (4.22),

$$\|(\zeta I - Y)^{-1}\| = \|((\zeta + c)I - X)^{-1}\|$$
$$\leq \frac{N}{|\zeta + c|} = \frac{N}{|\zeta|}\frac{|\zeta|}{|\zeta + c|} < \frac{2N}{|\zeta|}$$

because, being $|\zeta| > 2c$,

$$1 - \frac{c}{|\zeta|} > \frac{1}{2},$$

and therefore

$$\frac{|\zeta|}{|\zeta + c|} \le \frac{|\zeta|}{|\zeta| - c} < 2.$$

By Theorem 4.1.6, the proof of Theorem 4.2.2 is complete.

We will extend now Proposition 4.1.10 to the case of strongly continuous groups wich are not necessarily uniformly bounded.

Lemma 4.2.3 *Let X be a closed linear operator for which there exist two real constants $M \ge 1$ and $a \ge 0$ such that*

$$\{\zeta \in \mathbf{C} : \Re\zeta > 0\} \subset r(\pm X - aI) \tag{4.23}$$

and

$$\|(\zeta I - (\pm X - aI))^{-1}\| \le \frac{M}{\Re\zeta} \tag{4.24}$$

for all $\zeta \in \mathbf{C}$ for which $\Re\zeta > 0$. Then, X^2 is the infinitesimal generator of a holomorphic semigroup of angle $\frac{\pi}{2}$.

Proof 1) We begin by showing that, for every $\alpha \in (0, \frac{\pi}{2})$, there exist $r_o \ge 0$ and $\beta \in (0, \frac{\pi}{2})$ such that

$$\Theta_{\alpha + \frac{\pi}{2}} \backslash \overline{\Delta_{r_o}} \subset \{z^2 : z - a \in \Theta_\beta\}. \tag{4.25}$$

For $\theta \in [0, \alpha)$, consider the half-line

$$\{z = a + e^{i\theta} = a + r\cos\theta + i\sin\theta : r \ge 0\},$$

for which

$$z^2 = (a + r\cos\theta)^2 - (r\sin\theta)^2 + 2ir(a + r\cos\theta)\sin\theta,$$

and therefore

$$\begin{aligned}
\frac{\Im z^2}{\Re z^2} &= \frac{2r(a + r\cos\theta)\sin\theta}{(a + r\cos\theta)^2 - (r\sin\theta)^2} \\
&= \frac{2\left(\frac{a}{r} + \cos\theta\right)\sin\theta}{\left(\frac{a}{r} + \cos\theta\right)^2 - (\sin\theta)^2} \longrightarrow \\
&\longrightarrow \frac{2\cos\theta\sin\theta}{(\cos\theta)^2 - (\sin\theta)^2} = \tan 2\theta
\end{aligned}$$

as $r \to +\infty$. If $r_o > a$ and if $\beta \in (0, \frac{\pi}{2})$ is such that $2\beta > \alpha + \frac{\pi}{2}$, i. e.

$$\frac{\alpha}{2} + \frac{\pi}{4} < \beta < \frac{\pi}{2},$$

(4.25) is satisfied whenever $|\theta| \leq \beta$.

2) Let $\zeta \in \Theta_{\alpha + \frac{\pi}{2}} \backslash \overline{\Delta_{r_o}}$. If $\theta \in (-\beta, \beta)$ and $r > 0$ are such that

$$\zeta = (re^{i\theta} + a)^2,$$

then

$$\zeta I - X^2 = ((re^{i\theta} + a) I - X)((re^{i\theta} + a) I + X).$$

If $r > 0$ is such that

$$r \cos\left(\frac{\alpha}{2} + \frac{\pi}{4}\right) + a > 0,$$

then,

$$r \cos \theta + a > r \cos\left(\frac{\alpha}{2} + \frac{\pi}{4}\right) + a > 0,$$

and, by (4.23), $\zeta \in r(X^2)$.

Furthermore,

$$(\zeta I - X^2)^{-1} = ((re^{i\theta} + a) I - X)^{-1}((re^{i\theta} + a) I + X)^{-1}$$
$$= (re^{i\theta} I - (X - a I))^{-1}(re^{i\theta} - (X - a I))^{-1},$$

and, by (4.24),

$$|\zeta| \, \|(\zeta I - X^2)^{-1}\| \leq \frac{M^2 |\zeta|}{(r \cos \theta)^2} \leq \frac{M^2 |\zeta|}{(r \cos \beta)^2}.$$

Thus, $|\zeta| \, \|(\zeta I - X^2)^{-1}\|$ is uniformly bounded on $\Theta_{\alpha + \frac{\pi}{2}} \backslash \overline{\Delta_{r_o}}$, and therefore X^2 generates a holomorphic semigroup of angle α for every $\alpha \in (0, \frac{\pi}{2})$, that is, a holomorphic semigroup of angle $\frac{\pi}{2}$. ■

Theorem 4.2.4 *If X is the infinitesimal generator of a strongly continuous group $T : \mathbf{R} \to \mathcal{L}(\mathcal{E})$, X^2 generates a holomorphic semigroup of angle $\frac{\pi}{2}$.*

Proof Let $M \geq 1$ and $a \geq 0$ be such that

$$\|T(t)\| \leq M e^{a|t|},$$

that is to say,

$$\|e^{-a|t|} T(t)\| \leq M$$

for all $t \in \mathbf{R}$. The operators $X - aI$ and $-X - aI$ are the infinitesimal generators of two strongly continuous semigroups uniformly bounded on $\mathbf{R}_+$. Hence, (4.23) and (4.24) are satisfied, and Lemma 4.2.3 yields the proof of the theorem. ∎

Note The proofs of Proposition 4.1.10 and of Theorem 4.2.4 follow those exposed by R. Nagel in [1] (A-II).

4.3 Differentiable and analytic vectors

Let $T : \mathbf{R}_+ \to \mathcal{L}(\mathcal{E})$ be a strongly continuous semigroup, and let $X : \mathcal{D}(X) \subset \mathcal{E} \to \mathcal{E}$ be its infinitesimal generator.

A vector $x \in \mathcal{E}$ such that the function $t \mapsto T(t)x$ be C^∞ on $\mathbf{R}_+$ is called a *differentiable vector* of T. The set $\mathcal{D}^\infty$ of all differentiable vectors of T is a linear subspace of $\mathcal{E}$.

If $x \in \mathcal{D}^\infty$, $\lim_{h \downarrow 0} \frac{1}{t}(T(h) - I)x$ exists. Therefore $x \in \mathcal{D}(X)$. Furthermore $T(t)x \in \mathcal{D}(X)$ and

$$\frac{d}{dt}T(t)x = T(t)Xx \tag{4.26}$$

for all $t \in \mathbf{R}_+$[1]. As a consequence, $Xx \in \mathcal{D}^\infty$ and

$$\frac{d^n}{dt^n}T(t)x = T(t)X^n x \tag{4.27}$$

for all $x \in \mathcal{D}^\infty$, $t \in \mathbf{R}_+$, $n = 0, 1, \ldots$. Thus, $x \in \mathcal{D}(X^n)$ for $n = 0, 1, \ldots$.

---

[1]As before, in this and in the following equalities, if $t = 0$, the derivative $\frac{d}{dt}$ is a right derivative.

Viceversa, let $x \in \mathcal{E}$ be such that $x \in \mathcal{D}(X^n)$ for $n = 0, 1, \ldots$. Then, $T(t)x \in \mathcal{D}(X)$, and (4.26) implies that the function $t \mapsto T(t)x$ is of class C^1 on $\mathbf{R}_+$. Since $x \in \mathcal{D}(X^2)$, a further differentiation yields (4.27) for $n = 2$, and proves that the function $t \mapsto T(t)x$ is of class C^2 on $\mathbf{R}_+$. Iteration of this argument shows that the function $t \mapsto T(t)x$ is of class C^∞ on $\mathbf{R}_+$, that is, $x \in \mathcal{D}^\infty$. In conclusion,

$$\mathcal{D}^\infty = \cap\{\mathcal{D}(X^n) : n = 1, 2, \ldots\}. \tag{4.28}$$

Theorem 4.3.1 *The space $\mathcal{D}^\infty$ is a core of X.*

Proof 1) We show first that $\mathcal{D}^\infty$ is dense in $\mathcal{E}$.

For $n = 1, 2, \ldots$, let $\phi_n : \mathbf{R}_+ \to [0, 1]$ be a C^∞ function whose support supp $\phi_n \subset (0, \frac{1}{n})$ and is such that

$$\int_0^{+\infty} \phi_n(t)dt = 1.$$

For $x \in \mathcal{E}$ and $n \geq 1$, let

$$x_n = \int_0^{+\infty} \phi_n(t)T(t)xdt.$$

Then

$$\|x_n - x\| = \left\|\int_0^{+\infty} \phi_n(t)(T(t) - I)xdt\right\|$$
$$\leq \int_0^{+\infty} \phi_n(t)\|T(t)x - x\|dt,$$

whence

$$\lim_{n \to +\infty} \|x_n - x\| = 0.$$

That proves that the vectors x_n span a dense linear space in $\mathcal{E}$. We show now that $x_n \in \mathcal{D}(X)$ for all n. First of all,

$$\lim_{h \downarrow 0} \frac{1}{h}(T(h)x_n - x_n) =$$

$$= \lim_{h \downarrow 0} \frac{1}{h} \left\{ \int_0^{+\infty} \phi_n(t)T(t+h)x\,dt - \int_0^{+\infty} \phi_n(t)T(t)x\,dt \right\}$$

$$= \lim_{h \downarrow 0} \frac{1}{h} \left\{ \int_h^{+\infty} \phi_n(t-h)T(t)x\,dt - \int_0^{+\infty} \phi_n(t)T(t)x\,dt \right\}$$

$$= \lim_{h \downarrow 0} \frac{1}{h} \left\{ \int_0^{+\infty} (\phi_n(t-h) - \phi_n(t))T(t)x\,dt - \int_0^h \phi_n(t-h)T(t)x\,dt \right\}$$

$$= \lim_{h \downarrow 0} \frac{1}{h} \int_0^{+\infty} (\phi_n(t-h) - \phi_n(t))T(t)x\,dt,$$

because

$$\operatorname{supp} \phi_n(\bullet - h) \subset \left(h, h + \frac{1}{n} \right),$$

and therefore

$$(0, h] \cap \operatorname{supp} \phi_n(\bullet - h) = \emptyset, \tag{4.29}$$

whence

$$\int_0^h \phi_n(t-h)T(t)x\,dt = 0.$$

Hence, $x_n \in \mathcal{D}(X)$ and

$$X x_n = \lim_{h \downarrow 0} \frac{1}{h}(T(h)x_n - x_n) = -\int_0^{+\infty} \dot{\phi}_n(t)T(t)x\,dt.$$

As a consequence,

$$\lim_{h \downarrow 0} \frac{1}{h}(T(h)X x_n - X x_n) =$$

$$= -\lim_{h \downarrow 0} \frac{1}{h} \left\{ \int_0^{+\infty} \dot{\phi}_n(t)T(t+h)x\,dt - \int_0^{+\infty} \dot{\phi}_n(t)T(t)x\,dt \right\}$$

$$= -\lim_{h \downarrow 0} \frac{1}{h} \left\{ \int_0^{+\infty} (\dot{\phi}_n(t-h) - \dot{\phi}_n(t))T(t)x\,dt - \int_0^h \dot{\phi}_n(t-h)T(t)x\,dt \right\}$$

$$= -\lim_{h \downarrow 0} \frac{1}{h} \int_0^{+\infty} (\dot{\phi}_n(t-h) - \dot{\phi}_n(t))T(t)x\,dt$$

$$= \int_0^{+\infty} \ddot{\phi}_n(t)T(t)x\,dt,$$

because, (4.29) implies

$$\int_0^h \dot{\phi}_n(t-h)T(t)xdt = 0.$$

Hence $x_n \in \mathcal{D}(X^2)$. Iterating this argument we see that $x_n \in \mathcal{D}^\infty$, and therefore that $\mathcal{D}^\infty$ is dense in $\mathcal{E}$.

2) If $x \in \mathcal{D}^\infty$ and $t \in \mathbf{R}_+$, the function $s \mapsto T(s)T(t)x = T(s+t)x$ is of class C^∞ on $\mathbf{R}_+$. Hence $T(t)x \in \mathcal{D}^\infty$, and therefore $T(t)\mathcal{D}^\infty \subset \mathcal{D}^\infty$ for all $t \in \mathbf{R}_+$. That completes the proof of the theorem. ∎

If, for $x \in \mathcal{E}$, the function $t \mapsto T(t)x$ is the restriction to $\mathbf{R}_+^*$ of a holomorphic function on a neighbourhood of $\mathbf{R}_+^*$ in $\mathbf{C}$, x is called an *analytic vector* of the semigroup T. If T is the retriction to $\mathbf{R}_+$ of a holomorphic semigroup, every $x \in \mathcal{E}$ is an analytic vector of T.

If x is an analytic vector and the function $t \mapsto T(t)x$ is the restriction to $\mathbf{R}_+$ of a holomorphic function $\mathbf{C} \to \mathcal{E}$, x is called an *entire vector* of the semigroup T.

Similar definitions introduce differentiable, analytic and entire vectors of strongly continuous groups.

The following theorem of I.M.Gelfand, [30], [15], exhibits a class of entire vectors.

Theorem 4.3.2 *If $T : \mathbf{R} \to \mathcal{L}(\mathcal{E})$ is a strongly continuous group, the set of all entire vectors of T is a core of T.*

Proof 1) Let $M \geq 1$ and $a \in \mathbf{R}$ be such that

$$\|T(t)\| \leq Me^{a|t|}$$

for all $t \in \mathbf{R}$. If $\xi, \eta, t \in \mathbf{R}$, $z = \xi + i\eta$, $n \in \mathbf{N}^*$, then

$$\|e^{-\frac{n}{2}(t-z)^2}T(t)\| \leq Me^{-\frac{n}{2}\Re(t-z)^2+a|t|},$$

and

$$-\frac{n}{2}\Re(t-z)^2 + a|t| = -\frac{n}{2}\left[(t-\xi)^2 - \eta^2 - \frac{2a}{n}|t|\right]$$

$$= -\frac{n}{2}\left[t^2 - 2\left(\xi \pm \frac{a}{n}\right)t + \xi^2 - \eta^2\right]$$

$$= -\frac{n}{2}\left[\left(t - \left(\xi \pm \frac{a}{n}\right)\right)^2 - \left(\eta^2 + \frac{a^2}{n^2} \pm \frac{2a\xi}{n}\right)\right],$$

with $+$ if $t > 0$ and $-$ if $t < 0$. Since

$$\int_{-\infty}^{+\infty} e^{-t^2} dt = \sqrt{\pi},$$

and therefore

$$\int_{-\infty}^{+\infty} e^{-\frac{nt^2}{2}} dt = \sqrt{\frac{2\pi}{n}}$$

for $n = 1, 2, \ldots$, then, for any $x \in \mathcal{E}$,

$$\left\| \int_{-\infty}^{+\infty} e^{-\frac{n(t-z)^2}{2}} T(t)x \, dt \right\| \leq$$

$$= M e^{\frac{n}{2}(\eta^2 + \frac{a^2}{n^2} \pm \frac{2a\xi}{n})} \int_{-\infty}^{+\infty} e^{-\frac{n}{2}(t - (\xi \pm \frac{a}{n}))^2} dt \| x \|$$

$$\leq M e^{\frac{n}{2}(\eta^2 + \frac{a^2}{n^2} + \frac{|2a\xi|}{n})} \sqrt{\frac{2\pi}{n}} \| x \|.$$

Thus, setting

$$y_n(z) = \sqrt{\frac{n}{2\pi}} \int_{-\infty}^{+\infty} e^{-\frac{n(t-z)^2}{2}} T(t)x \, dt, \quad (n = 1, 2, \ldots)$$

the integral converges and defines a vector $y_n(z) \in \mathcal{E}$. Since,

$$\frac{\partial}{\partial \bar{z}} y_n(z) = 0,$$

$y_n : \mathbf{C} \to \mathcal{E}$ is a holomorphic function.

2) Setting

$$x_n = y_n(0) = \sqrt{\frac{n}{2\pi}} \int_{-\infty}^{+\infty} e^{-\frac{nt^2}{2}} T(t)x \, dt,$$

then

$$\| x_n - x \| \leq \sqrt{\frac{n}{2\pi}} \int_{-\infty}^{+\infty} e^{-\frac{nt^2}{2}} \|(T(t) - I)x\| \, dt,$$

whence

$$\lim_{n \to +\infty} x_n = x. \tag{4.30}$$

For $s \in \mathbf{R}$,

$$T(s)x_n = \sqrt{\frac{n}{2\pi}} \int_{-\infty}^{+\infty} e^{-\frac{nt^2}{2}} T(t+s)x \, dt$$

$$= \sqrt{\frac{n}{2\pi}} \int_{-\infty}^{+\infty} e^{-\frac{n(t-s)^2}{2}} T(t)x \, dt = y_n(s).$$

Hence, x_n is an entire vector, and (4.30) shows that the set of all entire vectors is dense in $\mathcal{E}$.

3) If, for $x \in \mathcal{E}$, $t \mapsto T(t)x$ is the restriction to $\mathbf{R}$ of a holomorphic function $\mathbf{C} \to \mathcal{E}$, for any given $s \in \mathbf{R}$ also the function $t \mapsto T(t)T(s)x = T(t+s)x$ is the restriction to $\mathbf{R}$ of a holomorphic function $\mathbf{C} \to \mathcal{E}$. Thus, if x is an entire vector, also $T(s)x$ is an entire vector for all $s \in \mathbf{R}$.

That shows that the space of all entire vectors of T is invariant under the action of $T(s)$, for any $s \in \mathbf{R}$, and completes the proof of the theorem. ∎

If $A : \mathcal{D}(A) \subset \mathcal{E} \to \mathcal{E}$ is any linear operator, the elements of $\cap\{\mathcal{D}(A^n) : n = 1, 2, \ldots\}$ are called *differentiable vectors* of A. By (4.28), the differentiable vectors of the semigroup T are the differentiable vectors of its infinitesimal generator X.

An *analytic vector* of A is a differentiable vector x of A for which there is some $r > 0$ such that

$$\sum_{n=0}^{+\infty} \frac{r^n}{n!} \|A^n x\| < \infty,$$

that is, such that the power series $\sum_{n=0}^{+\infty} \frac{s^n}{n!} \|A^n\|$ has a positive radius of convergence. If the series converges for every $s > 0$, that is to say, if the series $\sum_{n=0}^{+\infty} \frac{z^n}{n!} \|A^n\|$ converges on $\mathbf{C}$, the vector x is called an *entire vector* of A.

The families of the differentiable, analytic or entire vectors of A are linear subspaces of $\mathcal{E}$.

The analytic or entire vectors of X are analytic or, respectively, entire vectors of the semigroup T generated by X.

Theorem 4.3.3 *Let X be a linear, densely defined, dissipative operator in a complex Hilbert space $\mathcal{H}$. If there is a dense set of analytic vectors of X, the closure $\overline{X}$ of X is maximal dissipative.*

Proof We prove first that $\overline{\mathcal{R}(I - X)} = \mathcal{H}$, showing that, if $y \perp \mathcal{R}(I-X)$, then $y = 0$.

If $x \in \mathcal{D}$, where $\mathcal{D}$ is the space of all analytic vectors of X, then also $X^n x \in \mathcal{D}$ and

$$(X^n x|y) = (X\, X^{n-1} x|y) = (X^{n-1} x|y) = \cdots = (Xx|y) = (x|y)$$

for $n = 1, 2, \ldots$.

Let $Y : \mathcal{D}(Y) \subset \mathcal{H} \to \mathcal{H}$ be a maximal dissipative extension of X. Then Y is the infinitesimal generator of a strongly continuous semigroup $S : \mathbf{R}_+ \to \mathcal{L}(\mathcal{H})$. Let the function $\phi : \mathbf{R}_+ \to \mathbf{C}$ be defined by $\phi(t) = (S(t)x|y)$. Since

$$\left(\frac{d^n \phi}{dt^n}\right)_{t=0} = \left(\left(\frac{d^n}{dt^n} S(t)x\right)_{t=0} \Big| y\right)$$
$$= (X^n x|y) = (x|y) = \phi(0),$$

for $n = 0, 1, \ldots$, then

$$\phi(t) = \phi(0)e^t.$$

At this point, the chain of inequalities

$$|\phi(t)| = |(S(t)x|y)| \leq \|S(t)x\|\, \|x\| \leq \|x\|\, \|y\|$$

implies that $\phi(0) = 0$, that is, $y \perp x$ for all $x \in \mathcal{D}$. Because $\overline{\mathcal{D}} = \mathcal{H}$, then $y = 0$, showing that $\mathcal{R}(I - X)$ is dense in $\mathcal{H}$.

Hence, $1 \notin r\sigma(X)$ and, by Theorem 1.9.6, $1 \notin r\sigma(\overline{X})$. Since $\overline{X}$ is dissipative, then $1 \in r(\overline{X})$, i.e., $\overline{X}$ is m-dissipative, and therefore, by Proposition 3.2.10, maximal dissipative. ∎

Let the closed, densely defined linear operator $X : \mathcal{D}(X) \subset \mathcal{H} \to \mathcal{H}$ be such that iX is a symmetric operator, i. e., $\mathcal{D}(X) \subset \mathcal{D}(X^*)$ and $X + X^* = 0$ on $\mathcal{D}(X)$. Thus, both X and $-X$ are dissipative. Furthermore, they have the same analytic vectors. By Theorem 4.3.3 X and $-X$ are maximal dissipative. Thus $\sigma(X) \subset i\mathbf{R}$, that is, iX is self-adjoint, in view of Theorem 4.3.2, and the following theorem [59] holds.

Theorem 4.3.4 *If iX is a closed, symmetric operator, iX is self-adjoint if, and only if, X has a dense set of analytic vectors.*

The second part of the theorem follows from the Stone Theorem in Section 3.4 and from the Gelfand Theorem in this section.

4.4 The Gårding space

A class of differentiable vectors arises in the theory of linear representations of Lie groups, as was shown by L.Gårding in [28]. For all notions concerning Lie groups and linear representations of locally compact groups required here, see, *e.g.*, [47] and [57].

Let G be a Lie group, $\check{G}$ its Lie algebra, dg a left invariant Haar measure on G (defined up to an arbitrary positive constant factor), C_c the Banach sspace of all compactly supported, complex valued, continuous functions on G.

For $v \in \check{G}$, let $\mathbf{R} \ni t \mapsto \exp tv$ be the one-parameter group generated by v in G. Let $U : G \to \mathcal{L}(\mathcal{E})$ be a continuous linear representation of G in a complex Banach space $\mathcal{E}$. For $t \in \mathbf{R}$, let

$$T_v(t) = U \circ \exp tv.$$

Then $T_v : \mathbf{R} \to \mathcal{L}(\mathcal{E})$ is a strongly continuous group because, for all $x \in \mathcal{E}$,

$$\lim_{t \downarrow 0} T_v(t)x = \lim_{t \downarrow 0} U \circ \exp tv \, x = U(1)x = x,$$

where 1 is the identity element of G. The infinitesimal generator of the group T_v will be denoted by $dU(v)$, and the map $v \mapsto dU(v)$ will be called the *differential* of U. For $f \in C_c(G)$, let $U_f \in \mathcal{L}(\mathcal{E})$ be the operator defined by the integral

$$U_f = \int_G f(g)U(g)dg.$$

Let $C_c^\infty(G)$ be the the linear dense submanifold of $C_c(G)$ consisting of all compactly supported functions $G \to \mathbf{C}$ of class C^∞ on G. For $f \in C_c^\infty(G)$. let $D_v f \in C_c^\infty(G)$ be the C^∞ function defined by

$$D_v f(g) = \frac{d}{dt} f((\exp tv)g)_{|t=0}.$$

We will prove that, for all $f \in C_c^\infty(G)$, $\mathcal{R}(U_f) \subset \mathcal{D}(X_v)$. For $x \in \mathcal{E}$ and $t \neq 0$,

$$\frac{1}{t}(T_v(t)U_f x - U_f x) =$$

$$\frac{1}{t}(U(\exp tv) - I) \int_G f(g)U(g)x dg$$

$$= \frac{1}{t}\left(\int_G f(g)U((\exp tv)g)x dg - \int_G f(g)U(g)x dg\right)$$

$$= \frac{1}{t}\left(\int_G f((\exp -tv)g)U(g)x dg - \int_G f(g)U(g)x dg\right)$$

$$= \frac{1}{t}\int_G (f((\exp -tv)g) - f(g))U(g)x dg$$

$$= \frac{1}{t}\int_G (\varphi_g(t) - \varphi_g(0))U(g)x dg, \tag{4.31}$$

where

$$\varphi_g(t) := f((\exp -tv)g)$$

defines, for every $g \in G$, a C^∞ function $\varphi : G \to \mathbf{C}$. For this function,

$$\varphi_g(t) - \varphi_g(0) = \int_0^t \dot{\varphi}_g(s)ds = -\int_0^t \dot{\varphi}_g(s)d(t - s)$$

$$= -(t - s)\dot{\varphi}_g(s)|_0^t + \int_0^t (t - s)\ddot{\varphi}_g(s)ds$$

$$= t\dot{\varphi}_g(0) + \int_0^t (t - s)\ddot{\varphi}_g(s)ds, \tag{4.32}$$

$$\dot{\varphi}_g(0) = \frac{d}{ds}f((\exp -sv)g)|_{s=0} = D_{-v}f(g) \tag{4.33}$$

and

$$\ddot{\varphi}_g(s) = \frac{d^2}{ds^2}f((\exp -sv)g). \tag{4.34}$$

Let W be an open, relatively compact neighbourhood of the compact set $K = \mathrm{supp} f$. There is some $\delta > 0$ such that, if $|t| < \delta$ and

$g \notin W$, then $(\exp -tv)g \notin K$. It follows that, if $|s| < \delta$, the function $g \mapsto f((\exp -sv)g)$, and therefore also the function

$$g \mapsto \frac{d^2}{ds^2} f((\exp -sv)g),$$

have compact support. If $|t| < \delta$, (4.31), (4.32), (4.33) and (4.34) imply

$$\left\| \frac{1}{t} (T_v(t)U_f x - U_f x) - U_{D_{-v}f} x \right\|$$

$$= \left\| \frac{1}{t} (T_v(t)U_f x - U_f x) - \int_G D_{-v}f(g)U(g)x \, dg \right\|$$

$$= \left\| \frac{1}{t} \int_G \left(\int_0^t (t-s) \frac{d^2}{ds^2} f((\exp -sv)g) ds \right) U(g)x \, dg \right\|$$

$$\leq \frac{1}{|t|} \int_G \left(\int_0^{|t|} |t-s| \left| \frac{d^2}{ds^2} f((\exp -sv)g) \right| |ds| \right) \|U(g)x\| dg$$

$$= \frac{1}{|t|} \int_0^{|t|} |t-s| \left(\int_G \left| \frac{d^2}{ds^2} f((\exp -sv)g) \right| \|U(g)x\| dg \right) |ds|$$

$$\leq \frac{|t|}{|t|} \int_0^{|t|} \left(\int_G \left| \frac{d^2}{ds^2} f((\exp -sv)g) \right| \|U(g)x\| dg \right) |ds|$$

$$= \int_0^{|t|} \left(\int_G \left| \frac{d^2}{ds^2} f((\exp -sv)g) \right| \|U(g)x\| dg \right) |ds|,$$

and therefore

$$\lim_{t \to 0} \left\| \frac{1}{t} (T_v(t)U_f x - U_f x) - U_{D_{-v}f} x \right\| = 0.$$

That proves

Proposition 4.4.1 *For all $f \in C_c^\infty(G)$ and all $v \in \check{G}$,*

$$\mathcal{R}(U_f) \subset \mathcal{D}(dU(v))$$

and

$$dU(v)U_f = U_{D_{-v}f}.$$

As a consequence, the linear span $\mathcal{F}$ of $\mathcal{R}(U_f)$, when f varies in $C_c^\infty(G)$, is invariant under the action of $dU(v)$ for all $v \in \check{G}$.

For $g_o \in G$, the left translation $g \mapsto g_o^{-1}g$ of G defines a linear map $\theta(g_o)$ of $C_c(G)$ into itself, $\theta(g_o)f(g) = f(g_o^{-1}g)$, for which

$$\theta(g_o)\, C_c^\infty(G) \subset C_c^\infty(G).$$

Since, for $f \in C_c(G)$ and $x \in \mathcal{E}$,

$$U(g_o)U_f x = \int_G f(g)U(g_o g)x\,dg$$

$$= \int_G f(g_o^{-1}g)U(g)x\,dg = U_{\theta(g_o)f},$$

then

$$U(g_o)\mathcal{F} \subset \mathcal{F}.$$

If λ is a continuous linear form on $\mathcal{E}$, for which $< \mathcal{F}, \lambda >= \{0\}$ (that is,

$$\int f(g) < U(g)x, \lambda > dg = 0$$

for all $f \in C_c^\infty(G)$), then

$$< U(g)\mathcal{E}, \lambda >= \{0\}\ \forall g \in G.$$

If there exists a cyclic vector x for U, i.e., if the linear span of $U(g)x$, when g varies in G, is dense in $\mathcal{E}$, then $\lambda = 0$.

That proves

Theorem 4.4.2 *If the linear representation U has a cyclic vector, $\mathcal{F}$ is a core of $dU(v)$ for all $v \in \check{G}$.*

The elements of the space $\mathcal{F}$ - which is called the *Gårding space* of U - are all differentiable vectors of T_v for all $v \in \check{G}$.

Corollary 4.4.3 *If the linear representation U is irreducible, the Gårding space of U is dense in $\mathcal{E}$.*

Chapter 5

Perturbations and approximations

In this final chapter we will introduce two basic questions in the theory of strongly continuous semigroups of linear operators.

The first one may be considered, in some sense, a stability problem, and, as such, has fundamental applications to concrete problems in analysis. It consists in the search of sufficient conditions under which an additive perturbation of the infinitesimal generator of a strongly continuous semigroup still generates a semigroup of the same sort.

The second question concerns the approximation of semigroups by semigroups, and is expected to lead to important implications in what might be called the theory of "Lie semigroups".

The two topics will be only touched upon in these notes. Further developments and extended bibliographical references (especially to the first question) can be found in the books listed at the end.

5.1 Bounded perturbations

Let $T : \mathbf{R}_+ \to \mathcal{L}(\mathcal{E})$ be a strongly continuous semigroup satisfying (2.6) for some real constants a and $M \geq 1$, and all $t \in \mathbf{R}_+$. Let $X : \mathcal{D}(X) \subset \mathcal{E} \to \mathcal{E}$ be the infinitesimal generator of T.

Theorem 5.1.1 *If $A \in \mathcal{L}(\mathcal{E})$, $X + A : \mathcal{D}(X) \to \mathcal{E}$ is the infinitesimal*

308

generator of a strongly continuous semigroup $S : \mathbf{R}_+ \to \mathcal{L}(\mathcal{E})$ such that

$$\|S(t)\| \leq Me^{(a+M\|A\|)t} \ \forall t \in \mathbf{R}_+. \tag{5.1}$$

Proof For $x \in \mathcal{E}$ consider the formal series

$$T(t)x + \int_0^t T(t-t_1)AT(t_1)xdt_1 +$$

$$\int_0^t \int_0^{t_1} T(t-t_1)AT(t_1-t_2)AT(t_2)xdt_1\,dt_2 +$$

$$\int_0^t \int_0^{t_1} \int_0^{t_2} T(t-t_1)AT(t_1-t_2)AT(t_2-t_3)AT(t_3)xdt_1\,dt_2\,dt_3 + \cdots +$$

$$\int_0^t \int_0^{t_1} \cdots \int_0^{t_{n-1}} T(t-t_1)A \cdots \times$$

$$\times AT(t_{n-1}-t_n)AT(t_n)xdt_1 \cdots dt_n + \cdots . \tag{5.2}$$

Since, by (2.6),

$$\left\| \int_0^t \int_0^{t_1} \cdots \int_0^{t_{n-1}} T(t-t_1)A \cdots AT(t_{n-1}-t_n)AT(t_n)xdt_1 \cdots dt_n \right\| \leq$$

$$\int_0^t \int_0^{t_1} \cdots \int_0^{t_{n-1}} \|T(t-t_1)A \cdots AT(t_{n-1}-t_n)AT(t_n)x\|dt_1 \cdots dt_n$$

$$\leq M^{n+1}\|A\|^n \int_0^t \int_0^{t_1} \cdots \int_0^{t_{n-1}} e^{a[(t-t_1)+\cdots+(t_{n-1}-t_n)+t_n]}dt_1 \cdots dt_n \, \|x\|$$

$$= M^{n+1}\|A\|^n e^{at} \int_0^t \int_0^{t_1} \cdots \int_0^{t_{n-2}} t_{n-1}dt_1 \cdots dt_{n-1} \, \|x\|$$

$$= M^{n+1}\|A\|^n e^{at} \frac{1}{2} \int_0^t \int_0^{t_1} \cdots \int_0^{t_{n-3}} t_{n-2}{}^2 dt_1 \cdots dt_{n-2} \, \|x\|$$

$$= M^{n+1}\|A\|^n e^{at} \frac{1}{2 \cdot 3} \int_0^t \int_0^{t_1} \cdots \int_0^{t_{n-4}} t_{n-3}{}^3 dt_1 \cdots dt_{n-3} \, \|x\| = \cdots$$

$$= M^{n+1}\|A\|^n e^{at} \frac{t^n}{n!} \, \|x\|, \tag{5.3}$$

the series (5.2) is dominated in norm by

$$Me^{at} \left(1 + M\|A\|t + \cdots + \frac{(M\|A\|t)^n}{n!} + \cdots \right) \|x\| =$$

$$Me^{at} e^{M\|A\|t}\|x\| = Me^{(a+M\|A\|)t}\|x\|. \tag{5.4}$$

The sum, $S(t)x$ of the series (5.2) defines an operator $S(t) \in \mathcal{L}(\mathcal{E})$ for which (5.1) holds. A direct computation (exercise) shows that

$$S(t_1 + t_2) = S(t_1)\,S(t_2)\ \forall t_1 \geq 0,\ t_2 \geq 0.$$

Since $S(0) = I$, then $S : \mathbf{R}_+ \to \mathcal{L}(\mathcal{E})$ is a semigroup, which is strongly continuous because, in view of (5.2), (5.3) and (5.4),

$$\|S(t)x - x\| \leq \|T(t)x - x\| + Me^{at}\sum_{1}^{\infty} \frac{(M\|A\|t)^n}{n!}\|x\|$$

$$= \|T(t)x - x\| + Me^{at}(e^{M\|A\|t} - 1)\|x\|,$$

and therefore

$$\lim_{t\downarrow 0} \|S(t)x - x\| = 0$$

for all $x \in \mathcal{E}$. For $h > 0$, again by (5.2), (5.3) and (5.4),

$$\left\| \frac{1}{h}(S(h) - I)x - \frac{1}{h}(T(h) - I)x - Ax \right\| = \tag{5.5}$$

$$= \left\| \frac{1}{h}(S(h) - T(h))x - Ax \right\|$$

$$\leq \left\| \frac{1}{h}\int_0^h T(h - t)AT(t)x\,dt - Ax \right\| +$$

$$\left\| \frac{1}{h}\sum_{2}^{\infty}\int_0^h\int_0^{t_1}\cdots\int_0^{t_n} T(h - t_1)AT(t_1 - t_2)A \times \cdots \times \right.$$
$$\left. AT(t_{n-1} - t_n)AT(t_n)\,x\,dt\,dt_1\cdots dt_n \right\|$$

$$\leq \left\| \frac{1}{h}\int_0^h T(h - t)AT(t)x\,dt - Ax \right\| +$$

$$\frac{1}{h}Me^{ah}\left(e^{M\|A\|h} - 1 - M\|A\|h\right)\|x\|. \tag{5.6}$$

Since, for all $x \in \mathcal{E}$,

$$\lim_{h\downarrow 0} \frac{1}{h}\int_0^h T(h - t)AT(t)x\,dt = Ax$$

and

$$\lim_{h\downarrow 0} \frac{1}{h}Me^{ah}\left(e^{M\|A\|h} - 1 - M\|A\|h\right) = 0,$$

it follows from (5.5) that, if Y is the infinitesimal generator of the semigroup S, then $x \in \mathcal{D}(Y)$ if, and only if, $x \in \mathcal{D}(X)$, and that, if $x \in \mathcal{D}(X) = \mathcal{D}(Y)$, then $Yx = Xx + Ax$. ∎

The series (5.2) is called a *perturbation series* or a *Dyson series*, [20], or a *Dyson-Phillips series*, [65], of T.

By (5.2), S satisfies the integral equation

$$S(t)x = T(t)x + \int_0^t T(t-s)AS(s)x ds$$

for all $x \in \mathcal{E}$ and all $t \geq 0$. By (5.1),

$$\|(S(t) - T(t))x\| \leq \int_0^t \|T(t-s)\| \|A\| \|S(s)\| ds \, \|x\|$$

$$\leq M^2 \|A\| \int_0^t e^{a(t-s)} e^{(a+M\|A\|)s} ds \, \|x\|$$

$$= Me^{at} \left(e^{M\|A\|t} - 1\right) \|x\|.$$

Thus, the following corollary holds.

Corollary 5.1.2 *Under the hypotheses of Theorem 5.1.1,*

$$\|S(t) - T(t)\| \leq Me^{at} \left(e^{M\|A\|t} - 1\right)$$

for all $t \in \mathbf{R}_+$.

5.2 Dissipative perturbations

Let $A : \mathcal{D}(A) \subset \mathcal{E} \to \mathcal{E}$ be a linear closed operator such that

$$\mathcal{D}(X) \subset \mathcal{D}(A) \tag{5.7}$$

. Since X is closed, $\mathcal{D}(X)$ is a Banach space for the graph norm $x \mapsto \|x\| = \|x\| + \|Xx\|$. Le $\{x_\nu\}$ be a sequence in $\mathcal{D}(X)$, converging to some point $x \in \mathcal{E}$, for the graph norm, i.e.,

$$\lim_{\nu \to \infty} \|x_\nu - x\| = 0,$$

and such that $\{Ax_\nu\}$ converges to some $y \in \mathcal{E}$. Since $\mathcal{D}(X)$ is complete for the graph norm, then $x \in \mathcal{D}(X)$. Because this latter norm dominates the norm of $\mathcal{E}$, then $\lim_{\nu \to \infty} \|x - x_\nu\| = 0$. The fact that A is closed implies then that $Ax = y$. Hence the linear operator

$$A_{|\mathcal{D}(X)} : \mathcal{D}(X)_{\| \|} \to \mathcal{E}$$

is closed. The closed graph theorem implies then that

$$A_{|\mathcal{D}(X)} \in \mathcal{L}(\mathcal{D}(X)_{\| \|}, \mathcal{E})$$

(where $\mathcal{D}(X)_{\| \|}$ indicates the space $\mathcal{D}(X)$ endowed with the norm $\| \|$), and the following lemma holds.

Lemma 5.2.1 *If A and X are closed operators for which (5.7) holds, there is a positive constant k such that*

$$\|Ax\| \le k(\|Xx\| + \|x\|)$$

for all $x \in \mathcal{D}(X)$.

Theorem 5.2.2 *Let $T(t)$ be a linear contraction of $\mathcal{E}$ for all $t \ge 0$, and let A be a linear, closed, dissipative operator satisfying (5.7) and for which there exist two constants $a \in [0, 1)$ and $b \in \mathbf{R}_+$ such that*

$$\|Ax\| \le a\|Xx\| + b\|x\| \ \forall x \in \mathcal{D}(X). \tag{5.8}$$

Then $X + A : \mathcal{D}(X) \to \mathcal{E}$ generates a strongly continuous semigroup of linear contractions of $\mathcal{E}$.

Proof 1. Let $a \in [0, \frac{1}{2})$.

For every $x \in \mathcal{D}(X)$ there is $\lambda \in \mathcal{E}'$ for which $< x, \lambda >= \|x\|$, $\|\lambda\| = 1$ and $\mathfrak{R} < Ax, \lambda >\le 0$. Since, by Theorem 3.3.1, $\mathfrak{R} < Xx, \lambda >\le 0$, then $X + A$ is dissipative.

To show that $X + A$ is m-dissipative (and therefore, being densely defined, generates a strongly continuous semigroup of linear contractions of $\mathcal{E}$), it suffices to prove that $r(X + A) \cap \mathbf{R}_+^* \ne \emptyset$, that is to say, $\mathcal{R}(\zeta I - (X + A)) = \mathcal{E}$ for some $\zeta \in \mathbf{R}_+^*$.

Since $\mathcal{D}(X) \subset \mathcal{D}(A)$, if $\zeta > 0$

$$\mathcal{R}(\zeta I - (X + A)) \supset \mathcal{R}[(\zeta I - (X + A))(\zeta I - X)^{-1}]$$
$$\supset \mathcal{R}[I - A(\zeta I - X)^{-1}].$$

By (5.8), $\|(\zeta I - X)^{-1}x\| \leq \frac{\|x\|}{\zeta}$ whenever $x \in \mathcal{E}$ and $\zeta > 0$. Therefore

$$
\begin{aligned}
\|A(\zeta I - X)^{-1}\| &\leq a\|X(\zeta I - X)^{-1}x\| + b\|(\zeta I - X)^{-1}x\| \\
&= a\| -x + \zeta(\zeta I - X)^{-1}x\| + b\|(\zeta I - X)^{-1}x\| \\
&\leq a\|x\| + a\zeta\|(\zeta I - X)^{-1}x\| + b\|(\zeta I - X)^{-1}x\| \\
&\leq a\|x\| + a\zeta\frac{1}{\zeta}\|x\| + \frac{b}{\zeta}\|x\| \\
&= \left(2a + \frac{b}{\zeta}\right)\|x\|.
\end{aligned}
$$

Since $2a < 1$, if $\zeta \gg 0$, then $2a + \frac{b}{\zeta} < 1$. Therefore

$$
\|A(\zeta I - X)^{-1}\| < 1,
$$

and, as a consequence, $I - A(\zeta I - X)^{-1}$ is invertible in $\mathcal{L}(\mathcal{E})$. Thus $\mathcal{R}(I - A(\zeta I - X)^{-1}) = \mathcal{E}$, and, a fortiori, $\mathcal{R}(\zeta I - (X + A)) = \mathcal{E}$. Since, for $\zeta \gg 0$, $\zeta I - (X + A) = (I - A(\zeta I - X)^{-1})(\zeta I - X)$, then $(\zeta I - (X+A))^{-1} \in \mathcal{L}(\mathcal{E})$; therefore $\zeta I - (X+A)$ is closed when $\zeta \gg 0$, and thus also $X + A$ is closed.

2. Let $a \in [0, 1)$. For $0 \leq r \leq 1$ and $x \in \mathcal{D}(X)$, (5.8) yields

$$
\begin{aligned}
\|(X + rA)x\| &\geq \|Xx\| - r\|Ax\| \\
&\geq \|Xx\| - \|Ax\| \\
&\geq (1 - a)\|Xx\| - b\|x\|. \quad\quad\quad (5.9)
\end{aligned}
$$

If the integer $n > 2$ is such that $\frac{a}{n} < \frac{1-a}{4}$, (5.8) and (5.9) yield

$$
\begin{aligned}
\left\|\frac{1}{n}Ax\right\| &\leq \frac{a}{n}\|Xx\| + \frac{b}{n}\|x\| \\
&\leq \frac{1-a}{4}\|Xx\| + \frac{b}{n}\|x\| \\
&\leq \frac{1}{4}\|(X + rA)x\| + \left(\frac{b}{4} + \frac{b}{n}\right)\|x\|.
\end{aligned}
$$

In view of what has been seen in 1., if $X + rA$ generates a strongly continuous semigroup of linear contractions, also $X + rA + \frac{1}{n}A$ generates a semigroup of the same kind. Starting from $r = 0$, one proves then that $X+\frac{1}{n}A$, $X+\frac{2}{n}A$, ... , $X+\frac{n}{n}A = X+A$ are infinitesimal generators of strongly continuous semigroups of linear contractions. ∎

Example Theorem 5.2.2 fails when $a = 1$, as the present example shows [35].

Let $X : \mathcal{D}(X) \subset \mathcal{H} \to \mathcal{H}$ be a linear, self adjoint, unbounded operator on a complex Hilbert space $\mathcal{H}$, and let $A = -X$. Then, $X + A = 0$, but $X + A$ is not closed because otherwise $\mathcal{D}(X)$ would be closed for the graph norm of $X + A$, which coincides with the norm of $\mathcal{E}$; thus $\mathcal{D}(X)$ would be closed, and therefore $\mathcal{D}(X) = \mathcal{E}$. But then, by the closed graph theorem the closed operator X would be bounded on $\mathcal{E}$.

On the other hand, $X + A = 0$ is closable. Since it is dissipative, its closure $\overline{X + A}$ is dissipative. Because

$$\sigma(X + A) = \sigma(\overline{X + A}) = \{0\},$$

$\overline{X + A}$ is m-dissipative, and thus generates a strongly continuous semigroup of linear contractions of $\mathcal{H}$.

Theorem 5.2.2 implies the first part of Theorem 5.1.1. In fact, with the same notations of Theorem 5.2.2, by (5.8) $X - aI$ generates a strongly continuous, uniformly bounded semigroup. Since

$$X + A = X - aI + A + aI,$$

and $A + aI \in \mathcal{L}(\mathcal{E})$, we suppose that the semigroup T is uniformly bounded. In view of what we have seen at the beginning of section 3.3 there is a norm $\|\| \ \|\|$ in $\mathcal{E}$ which is equivalent to the norm $\| \ \|$ and such that $T(t)$ is a contraction for $\|\| \ \|\|$ for all $t \geq 0$. For $x \in \mathcal{E}$, let $\lambda \in \mathcal{E}'$ be such that $< x, \lambda >= \|x\|$, $\|\lambda\| = 1$. Since

$$\Re < (A - \|A\| \, I)x, \lambda >= \Re < Ax, \lambda > - \|A\| < x, \lambda >$$
$$\leq \|A\| \, \|x\| \, \|\lambda\| - \|A\| \, \|x\| = 0,$$

$A - \|A\| \, I$ is dissipative. Since

$$\|(A - \|A\| \, I)\| \leq \|Ax\| + \|A\| \, \|x\|$$
$$\leq 2\|A\| \, \|x\|$$

for all $x \in \mathcal{E}$, (5.8) holds with $a = 0$ and $b = 2\|A\|$. By Theorem 5.2.2, $X + A - \|A\| I$ generates a strongly continuous semigroup $R : \mathbf{R}_+ \rightarrow \mathcal{L}(\mathcal{E})$. As a consequence. $X + A$ generates the strongly continuous semigroup $t \mapsto e^{\|A\| t} R(t)$.

The following theorem has been established in [79].

Theorem 5.2.3 *Let X be the infinitesimal generator of a strongly continuous group of linear isometries of $\mathcal{E}$. Let A be a closed, conservative operator which has a chore contained in $\mathcal{D}(X)$ and has relative bound $a \in [0, 1)$ with respect to X on a chore of X contained in $\mathcal{D}(A)$.*

Then, A is the infinitesimal generator of a strongly continuos group of linear isometries of $\mathcal{E}$, and $\mathcal{D}(X) \subset \mathcal{D}(A)$.

Let X be the infinitesimal generator of a holomorphic semigroup, and let A be a linear closed operator with $\mathcal{D}(X) \subset \mathcal{D}(A)$, for which there exist two real constants a and b such that (5.8) holds. By Theorem 4.2.2, there are two constants $c \geq 0$ and $N > 0$ such that, if $\Re\zeta > 0$ and $|\zeta| > c$, then $\zeta \in r(X)$ and

$$\|(\zeta I - X)^{-1}\| \leq \frac{N}{|\zeta|}. \tag{5.10}$$

Since $\mathcal{R}((\zeta I - X)^{-1}) \subset \mathcal{D}(X)$, (5.8) yields

$$\|A(\zeta I - X)^{-1} x\| \leq a\|X(\zeta I - X)^{-1} x\| + b\|(\zeta I - X)^{-1} x\|$$

for all $x \in \mathcal{E}$. Being $X(\zeta I - X)^{-1} = -I + \zeta((\zeta I - X)^{-1}$, then, by (5.10),

$$\|A(\zeta I - X)^{-1} x\| \leq \left[a(1 + N) + \frac{bN}{|\zeta|} \right] \|x\|$$

for all $x \in \mathcal{E}$. If $a(1 + N) < \frac{1}{2}$, there exists $k \in (0, 1)$ such that, if $\Re\zeta > 0$ and $|\zeta| > \max\{c, 2bN\}$, then

$$\|A(\zeta I - X)^{-1} x\| \leq k\|x\|$$

for all $x \in \mathcal{E}$, i. e.,

$$\|A(\zeta I - X)^{-1}\| \leq k$$

whenever $\Re\zeta > 0$ and $|\zeta| > \max\{c, 2bN\}$. Writing

$$(\zeta I - (X + A))^{-1} = (\zeta I - X)^{-1}I - A(\zeta I - X)^{-1})^{-1},$$

for $\Re\zeta > 0$ and $|\zeta| > \max\{c, 2bN\}$, then

$$\|(\zeta I - (X + A))^{-1}\| \leq \|(\zeta I - X)^{-1}\| \sum_{n=0}^{+\infty} \|A(\zeta I - X)^{-1}\|^n \leq \frac{N'}{|\zeta|},$$

where $N' = N/(1 - k)$. By Theorem 4.2.2, $X + A$ is the infinitesimal generator of a holomorphic semigroup. In conclusion, the following theorem holds.

Theorem 5.2.4 *Let X be the infinitesimal generator of a holomorphic semigroup, and let A be a closed linear operator such that $\mathcal{D}(X) \subset \mathcal{D}(A)$ and for which there exist two real constants a and b satisfying (5.8). There is a positive number α such that, if $0 \leq a \leq \alpha$, then $X + A$ generates a holomorphic semigroup.*

Exercise Prove that, if the semigroup generated by X is a bounded holomorphic semigroup, $X + A$ generates a bounded holomorphic semigroup.

5.3 Approximation of semigroups by semigroups

The central topic of this and the following section will be that of describing how a strongly continuous semigroup can be defined as the limit of a sequence, or of a continuous family, of strongly continuous semigroups.

For $\varsigma > 0$, let $T^\varsigma : \mathbf{R}_+ \to \mathcal{L}(\mathcal{E})$ be a strongly continuous semigroup, generated by $X^\varsigma : \mathcal{D}(X^\varsigma) \subset \mathcal{E} \to \mathcal{E}$. Let $\mathcal{E}_o$ be a closed, linear subspace of $\mathcal{E}$ and let $T : \mathbf{R}_+ \to \mathcal{L}(\mathcal{E}_o)$ be a strongly continuous semigroup generated by $X : \mathcal{D}(X) \subset \mathcal{E}_o \to \mathcal{E}_o$; let $\mathcal{D}$ be dense in $\mathcal{D}(X)$. Suppose there are real constants a and $M \geq 1$ such that

$$\|T(t)\| \leq M\,e^{at} \ \forall t \in \mathbf{R}_+, \tag{5.11}$$

$$\|T^\varsigma(t)\| \le M\, e^{at} \quad \forall t \in \mathbf{R}_+, \ \forall \varsigma > 0, \tag{5.12}$$

and consider the following statements:

i) There exists $t_o \in \mathbf{R}_+^*$ such that

$$\limsup_{\varsigma \downarrow 0}\{\|T^\varsigma(t)x - T(t)x\| : 0 \le t \le t_o\} = 0 \ \forall x \in \mathcal{E}_o.$$

ii) This latter equation holds for all $t_o \in \mathbf{R}_+^*$.
iii) There exists $\xi \in (a, +\infty)$ such that

$$\lim_{\varsigma \downarrow 0}(\xi\, I - X^\varsigma)^{-1}x = (\xi\, I - X)^{-1}x \ \forall x \in \mathcal{E}_o.$$

iv) This latter equation holds for all $\xi \in (a, +\infty)$.
v) For every $x \in \mathcal{D}$ and every $\varsigma > 0$ there exists $x_\varsigma \in \mathcal{D}(X^\varsigma)$ such that

$$\lim_{\varsigma \downarrow 0} x_\varsigma = x, \ \ \lim_{\varsigma \downarrow 0} X^\varsigma x_\varsigma = Xx.$$

Clearly $ii) \Rightarrow i)$ and $iv) \Rightarrow iii)$.

Theorem 5.3.1 *(Trotter-Kato-Neveu theorem) The five conditions i), ... , v) are equivalent.*

The logical pattern of the proof, which apart from minor modifications is that exposed in [15], is the following

$$
\begin{array}{ccc}
ii) & & \\
\Updownarrow & \Longleftrightarrow \ iv) \Leftarrow v) \\
i) & \Downarrow \\
& iii) \Rightarrow v).
\end{array}
$$

We prove that $i) \Rightarrow ii)$ by showing that, in $i)$, t_o can be replaced by $2t_o$.

If $0 \le t \le t_o$ and $x \in \mathcal{E}_o$, then

$$\|T^\varsigma(t + t_o)x - T(t + t_o)x\| =$$
$$\|T^\varsigma(t + t_o)x - T^\varsigma(t)T(t_o)x + T^\varsigma(t)T(t_o)x - T(t + t_o)x\|$$
$$\le \|T^\varsigma(t)(T^\varsigma(t_o) - T(t_o))x\| + \|(T^\varsigma(t) - T(t))T(t_o)x\|$$
$$\le \|T^\varsigma(t)\|\, \|(T^\varsigma(t_o) - T(t_o))x\| + \|(T^\varsigma(t) - T(t))T(t_o)x\|$$
$$\le M e^{at}\, \|(T^\varsigma(t_o) - T(t_o))x\| + \|(T^\varsigma(t) - T(t))T(t_o)x\|.$$

Hence, by i) applied to x and to $T(t_o)x \in \mathcal{E}_o$,

$$\limsup_{\varsigma \downarrow 0} \{\|T^\varsigma(t+t_o)x - T(t+t_o)x\| : 0 \leq t \leq t_o\} = 0.$$

Therefore i) holds for $0 \leq t \leq 2t_o$.

For all $x \in \mathcal{E}_o$ and all $\xi > a$,

$$[(\xi I - X^\varsigma)^{-1} - (\xi I - X)^{-1}]x = \int_0^{+\infty} e^{-\xi t}(T^\varsigma(t) - T(t))x\,dt.$$

Hence, for any $t_o > 0$,

$$\|[(\xi I - X^\varsigma)^{-1} - (\xi I - X)^{-1}]x\| \leq$$

$$\int_0^{t_o} e^{-\xi t}\|(T^\varsigma(t) - T(t))x\|dt +$$

$$\int_{t_o}^{+\infty} e^{-\xi t}[\|T^\varsigma(t)x\| + \|T(t))x\|]dt$$

$$\leq \int_0^{t_o} \|(T^\varsigma(t) - T(t))x\|dt + 2M\int_{t_o}^{+\infty} e^{at}e^{-\xi t}\|x\|dt$$

$$\leq \int_0^{t_o} \|(T^\varsigma(t) - T(t))x\|dt + \frac{2M}{\xi - a}e^{(a-\xi)t_o}\|x\|.$$

Given $\epsilon > 0$, let $t_o > 0$ be such that

$$\frac{2M}{\xi - a}e^{(a-\xi)t_o}\|x\| < \frac{\epsilon}{2}.$$

If ii) holds there exists $\varsigma_o > 0$ such that, whenever $0 < \varsigma < \varsigma_o$,

$$\sup\{\|(T^\varsigma(t) - T(t))x\| : 0 \leq t \leq t_o\} < \frac{\epsilon}{2t_o}.$$

Thus, if ii) holds, for all $x \in \mathcal{E}_o$ and all $\xi > a$ there exists some $\varsigma_o > 0$ such that, whenever $0 < \varsigma < \varsigma_o$,

$$\|[(\xi I - X^\varsigma)^{-1} - (\xi I - X)^{-1}]x\| < \epsilon$$

whenever $0 < \varsigma < \varsigma_o$. That proves that ii) $\Rightarrow iv$).

Lemma 5.3.2 *If iii) holds, then iii) holds uniformly with respect to x varying on compact subsets of $\mathcal{E}_o$.*

Proof For $x \in \mathcal{E}$ and $\epsilon > 0$, let

$$B(x, \epsilon) = \{y \in \mathcal{E}_o : \|x - y\| < \epsilon\}$$

be the open ball in $\mathcal{E}_o$ with center x and radius ϵ. Let

$$0 < \rho < \frac{\epsilon}{2} \frac{\xi - a}{2M}.$$

By *iii)* there exists $\delta > 0$ such that, whenever $0 < \varsigma < \delta$,

$$\|(\xi I - X^\varsigma)^{-1}x - (\xi I - X)^{-1}x\| < \frac{\epsilon}{2}.$$

For $y \in B(x, \rho)$ and $0 < \varsigma\delta$,

$$
\begin{aligned}
\|(\xi I &- X^\varsigma)^{-1}y - (\xi I - X)^{-1}y\| = \\
&\|(\xi I - X^\varsigma)^{-1}x - (\xi I - X)^{-1}x + \\
&(\xi I - X^\varsigma)^{-1}(y - x) - (\xi I - X)^{-1}(y - x)\| \\
\leq &\|(\xi I - X^\varsigma)^{-1}x - (\xi I - X)^{-1}x\| + \\
&(\|(\xi I - X^\varsigma)^{-1}\| + \|(\xi I - X)^{-1}\|)\|y - x\| \\
\leq &\|(\xi I - X^\varsigma)^{-1}x - (\xi I - X)^{-1}x\| + 2\frac{M}{\xi - a}\frac{\epsilon}{2}\frac{\xi - a}{2M}
\end{aligned}
$$

because, for $\xi > a$,

$$\|(\xi I - X^\varsigma)^{-1}\| \leq \frac{M}{\xi - a}, \quad \|(\xi I - X)^{-1}\| \leq \frac{M}{\xi - a}.$$

Hence

$$\|(\xi I - X^\varsigma)^{-1}y - (\xi I - X)^{-1}y\| < \frac{\epsilon}{2} + \frac{\epsilon}{2} = \epsilon$$

whenever $0 < \varsigma < \delta$ and for all $y \in B(x, \rho)$. A standard compactness argument completes the proof. ∎

We prove now that $iii) \Rightarrow v)$.

Let $\xi > a$ satisfy iii). For $x \in \mathcal{D}$, let $y = (\xi I - X)x \in \mathcal{E}_o \subset \mathcal{E}$, $x_\varsigma = (\xi I - X^\varsigma)^{-1}y$. Then, by iii),

$$\lim_{\varsigma \downarrow 0} x_\varsigma = \lim_{\varsigma \downarrow 0}(\xi I - X^\varsigma)^{-1}y = (\xi I - X)^{-1}y = x.$$

Since $(\xi I - X^\varsigma)x_\varsigma = y$, then

$$\lim_{\varsigma \downarrow 0} X^\varsigma x_\varsigma = \xi \lim_{\varsigma \downarrow 0} x_\varsigma - y = \xi x - y = Xx.$$

That proves that $v)$ holds.

Let $\xi > a$. Then $(\xi I - X)(\mathcal{D}(X)) = \mathcal{E}_o$, $(\xi I - X^\varsigma)(\mathcal{D}(X^\varsigma)) = \mathcal{E}$, $\xi I - X$, $\xi I - X^\varsigma$ are injective, and $(\xi I - X)^{-1} \in \mathcal{L}(\mathcal{E}_o)$, $(\xi I - X^\varsigma)^{-1} \in \mathcal{L}(\mathcal{E})$.

For $x \in \mathcal{D}$, let $y = (\xi I - X)x$. Similarly, for $x_\varsigma \in \mathcal{D}(X^\varsigma)$, let $y_\varsigma = (\xi I - X^\varsigma)x_\varsigma$.

The following lemma does not depend on any of the conditions $i), \dots, v)$.

Lemma 5.3.3 *The condition*

$$\lim_{\varsigma \downarrow 0} x_\varsigma = x, \quad \lim_{\varsigma \downarrow 0} X^\varsigma x_\varsigma = Xx \tag{5.13}$$

is equivalent to

$$\lim_{\varsigma \downarrow 0} y_\varsigma = y, \quad \lim_{\varsigma \downarrow 0}(\xi I - X^\varsigma)^{-1}y_\varsigma = (\xi I - X)^{-1}y. \tag{5.14}$$

Proof (5.13) implies

$$\begin{cases} \lim_{\varsigma \downarrow 0} y_\varsigma & = \xi x - Xx = y \\ \lim_{\varsigma \downarrow 0}(\xi I - X^\varsigma)^{-1}y_\varsigma = x & = (\xi I - X)^{-1}y. \end{cases} \tag{5.15}$$

Viceversa, by (5.15),

$$\lim_{\varsigma \downarrow 0} x_\varsigma = \lim_{\varsigma \downarrow 0}(\xi I - X^\varsigma)^{-1}y_\varsigma = (\xi I - X)^{-1}y = x,$$

and

$$\lim_{\varsigma \downarrow 0} X^\varsigma x_\varsigma = \lim_{\varsigma \downarrow 0}(\xi x_\varsigma - y_\varsigma)$$
$$= \lim_{\varsigma \downarrow 0}\left(\xi(\xi I - X^\varsigma)^{-1}y_\varsigma - y_\varsigma\right)$$
$$= \xi x - y = Xx.$$

Hence (5.13) and (5.15) are equivalent.
Now, since for $\xi > a$,

$$\|(\xi I - X)^{-1}\| \leq \frac{M}{\xi - a},$$

and therefore

$$\|(\xi I - X^{\varsigma})^{-1}y - (\xi I - X^{\varsigma})^{-1}y_{\varsigma}\| \leq \|(\xi I - X^{\varsigma})^{-1}\| \, \|y - y_{\varsigma}\|$$
$$\leq \frac{M}{\xi - a}\|y - y_{\varsigma}\|,$$

then the top line of (5.15) yields

$$\lim_{\varsigma \downarrow 0}(\xi I - X^{\varsigma})^{-1}(y - y_{\varsigma}) = 0.$$

Hence, by the bottom line of (5.15),

$$\lim_{\varsigma \downarrow 0}(\xi I - X^{\varsigma})^{-1}y = \lim_{\varsigma \downarrow 0}(\xi I - X^{\varsigma})^{-1}(y - y_{\varsigma}) + \lim_{\varsigma \downarrow 0}(\xi I - X^{\varsigma})^{-1}y_{\varsigma}$$
$$= (\xi I - X)^{-1}y.$$

That proves that (5.15) is equivalent to (5.14). ∎

Let $\xi > a$. In view of the second equation in (5.14), if v) holds, then iv) holds, $i.e.$,
$$\lim_{\varsigma \downarrow 0}(\xi I - X^{\varsigma})^{-1}y = (\xi I - X)^{-1}y$$
for all $y \in (\xi I - X)(\mathcal{D})$.

Since $\xi \in r(X)$, $\xi I - X$ is injective, $(\xi I - X)(\mathcal{D}(X)) = \mathcal{E}_o$ and $(\xi I - X)^{-1} \in \mathcal{L}(\mathcal{E}_o)$. If $(\xi I - X)(\mathcal{D})$ is not dense in $\mathcal{E}_o$, there exists a non-empty open set $A \subset \mathcal{E}_o$ such that $(\xi I - X)(\mathcal{D}) \cap A = \emptyset$. But then

$$\emptyset = (\xi I - X)^{-1}((\xi I - X)(\mathcal{D}) \cap A)$$
$$= \mathcal{D} \cap (\xi I - X)^{-1}(A).$$

Since $(\xi I - X)^{-1}(A)$ is open, $\mathcal{D}$ is not dense. This contradiction shows that $(\xi I - X)(\mathcal{D})$ is dense in $\mathcal{E}_o$. Thus, the second equation in (5.14) holds for all $y \in \mathcal{E}_o$, proving thereby that v) $\Rightarrow iv$).

We show now that $iv) \Rightarrow ii)$.

For $x \in \mathcal{E}_o$, $\xi > a$ and $0 \le s \le t$,

$$\frac{d}{ds}T^{\varsigma}(t-s)[(\xi I - X^{\varsigma})^{-1}T(s)(\xi I - X)^{-1}x] =$$
$$-T^{\varsigma}(t-s)X^{\varsigma}(\xi I - X^{\varsigma})^{-1}T(s)(\xi I - X)^{-1}x +$$
$$T^{\varsigma}(t-s)(\xi I - X^{\varsigma})^{-1}T(s)X(\xi I - X)^{-1}x$$
$$= -T^{\varsigma}(t-s)(-I + \xi(\xi I - X^{\varsigma})^{-1})T(s)(\xi I - X)^{-1}x +$$
$$T^{\varsigma}(t-s)(\xi I - X^{\varsigma})^{-1}T(s)(-I + \xi(\xi I - X)^{-1})x$$
$$= T^{\varsigma}(t-s)T(s)(\xi I - X)^{-1}x - T^{\varsigma}(t-s)(\xi I - X^{\varsigma})^{-1}T(s)x$$
$$= T^{\varsigma}(t-s)(\xi I - X)^{-1}T(s)x - T^{\varsigma}(t-s)(\xi I - X^{\varsigma})^{-1}T(s)x$$
$$= T^{\varsigma}(t-s)((\xi I - X)^{-1} - (\xi I - X^{\varsigma})^{-1})T(s)x.$$

Integration from 0 to t yields

$$\left((\xi I - X^{\varsigma})^{-1}T(t)(\xi I - X)^{-1} - T^{\varsigma}(t)(\xi I - X^{\varsigma})^{-1}(\xi I - X)^{-1}\right)x =$$
$$\int_0^t T^{\varsigma}(t-s)((\xi I - X)^{-1} - (\xi I - X^{\varsigma})^{-1})T(s)x\,dt.$$

Since

$$\|T^{\varsigma}(t-s)\| \le Me^{a(t-s)}$$

and

$$T^{\varsigma}(t)(\xi I - X^{\varsigma})^{-1} = (\xi I - X^{\varsigma})^{-1}T^{\varsigma}(t) \quad \text{on } \mathcal{D}(X^{\varsigma}),$$

then

$$\|(\xi I - X^{\varsigma})^{-1}(T(t) - T^{\varsigma}(t))(\xi I - X)^{-1}x\| \le$$
$$\int_0^t Me^{a(t-s)}\|((\xi I - X)^{-1} - (\xi I - X^{\varsigma})^{-1})T(s)x\|\,ds.$$

Setting $y = (\xi I - X)^{-1}x$, this inequality becomes

$$\|(\xi I - X^{\varsigma})^{-1}(T(t) - T^{\varsigma}(t))y\| \le$$
$$\int_0^t Me^{a(t-s)}\|((\xi I - X)^{-1} - (\xi I - X^{\varsigma})^{-1})T(s)x\|\,ds.$$

Since

$$\|((\xi I - X)^{-1} - (\xi I - X^{\varsigma})^{-1})T(s)x\| \le \frac{2M}{\xi - a}Me^{as}\|x\|,$$

then

$$e^{a(t-s)}\|((\xi I - X)^{-1} - (\xi I - X^\varsigma)^{-1})T(s)x\| \leq \frac{2M^2}{\xi - a}e^{at}\|x\|.$$

Since $iv) \Rightarrow iii)$, and, by Lemma 5.3.2, $iii)$ holds uniformly on all compact subsets of $\mathcal{E}_o$, then

$$\lim_{\varsigma \downarrow 0} \|(\xi I - X^\varsigma)^{-1}(T(t) - T^\varsigma(t))y\| = 0 \quad \forall y \in \mathcal{D}(X) \qquad (5.16)$$

uniformly with respect to $t \in [0, t_o]$ for any $t_o \in \mathbf{R}_+^*$.

It will be shown now that (5.16) holds for all $y \in \mathcal{E}_o$. First of all, since $\mathcal{D}(X)$ is dense in $\mathcal{E}_o$, for any $y \in \mathcal{E}_o$ and any $\delta > 0$ there is some $z \in \mathcal{D}(X)$ such that $\|z - x\| < \delta$. Being

$$\|(T(t) - T^\varsigma(t))(z - y)\| \leq 2Me^{at}\|z - x\| < 2Me^{at}\delta$$

and thus

$$\|(\zeta I - X^\varsigma)^{-1}(T(t) - T^\varsigma(t))(z - y)\| < \frac{2M^2e^{at}}{\xi - a}\delta,$$

then

$$\|(\zeta I - X^\varsigma)^{-1}(T(t) - T^\varsigma(t))y\| <$$
$$\|(\zeta I - X^\varsigma)^{-1}(T(t) - T^\varsigma(t))z\| + \frac{2M^2e^{at}}{\xi - a}\delta.$$

In view of (5.16), for any $\epsilon > 0$ there exists some $\sigma > 0$ such that, whenever $0 < \varsigma < \sigma$,

$$\|(\zeta I - X^\varsigma)^{-1}(T(t) - T^\varsigma(t))z\| < \frac{\epsilon}{2}.$$

If $\delta > 0$ is such that

$$\frac{2M^2e^{at}}{\xi - a}\delta < \frac{\epsilon}{2},$$

then

$$\|(\zeta I - X^\varsigma)^{-1}(T(t) - T^\varsigma(t))y\| < \frac{\epsilon}{2} + \frac{\epsilon}{2} = \epsilon,$$

proving thereby that (5.16) holds for all $y \in \mathcal{E}_o$ uniformly with respect to $t \in [0, t_o]$.

Since

$$\|(\xi I - X^\varsigma)^{-1} T^\varsigma(t) y - T^\varsigma(t)(\xi I - X)^{-1} y\| =$$
$$\|T^\varsigma(t)((\xi I - X^\varsigma)^{-1} - (\xi I - X)^{-1}) y\|$$
$$\leq M e^{at} \|((\xi I - X^\varsigma)^{-1} - (\xi I - X)^{-1}) y\|$$

and since $T(t)$ and $(\xi I - X)^{-1}$ commute, then, by iv),

$$\lim_{\varsigma \downarrow 0} \|(\xi I - X^\varsigma)^{-1} T^\varsigma(t) y - T^\varsigma(t)(\xi I - X)^{-1} y\| = 0 \qquad (5.17)$$

and

$$\lim_{\varsigma \downarrow 0} \|((\xi I - X^\varsigma)^{-1} T(t) - T(t)(\xi I - X)^{-1}) y\| = 0. \qquad (5.18)$$

Being

$$\|(T^\varsigma(t) - T(t))(\xi I - X)^{-1} y\| =$$
$$\|T^\varsigma(t)(\xi I - X)^{-1} y - (\xi I - X^\varsigma)^{-1} T^\varsigma(t) y + (\xi I - X^\varsigma)^{-1} T^\varsigma(t) y -$$
$$(\xi I - X^\varsigma)^{-1} T(t) y + (\xi I - X^\varsigma)^{-1} T(t) y - T(t)(\xi I - X)^{-1} y\|$$
$$\leq \|(T^\varsigma(t)(\xi I - X)^{-1} - (\xi I - X^\varsigma)^{-1} T^\varsigma(t)) y\| +$$
$$+ \|(\xi I - X^\varsigma)^{-1} (T^\varsigma(t) - T(t)) y\|$$
$$+ \|((\xi I - X^\varsigma)^{-1} T(t) - T(t)(\xi I - X)^{-1}) y\|,$$

(5.16), (5.17) and (5.18) yield

$$\lim_{\varsigma \downarrow 0} \|(T^\varsigma(t) - T(t)) y\| = 0 \qquad (5.19)$$

for all $y \in \mathcal{R}((\xi I - X)^{-1}) = \mathcal{D}(X)$. If $y_o \in \mathcal{E}_o$, for any $\epsilon > 0$ there is $y \in \mathcal{D}(X)$ with $\|y - y_o\| < \epsilon$. Since

$$\|(T^\varsigma(t) - T(t)) y_o\| \leq \|(T^\varsigma(t) - T(t)) y\| +$$
$$\|T^\varsigma(t)(y - y_o)\| + \|T(t)(y - y_o)\|$$
$$\leq \|(T^\varsigma(t) - T(t)) y\| +$$
$$2 M e^{at} \|y - y_o\|$$
$$\leq \|(T^\varsigma(t) - T(t)) y\| +$$
$$+ 2 M e^{at} \epsilon,$$

then (5.19) holds for all $y_o \in \mathcal{E}_o$ uniformly on all compact subsets of $\mathcal{E}_o$ and uniformly with respect to $t \in [0, t_o]$.

That shows that $iv) \Rightarrow ii)$ and completes the proof of Theorem 5.3.1.

The following proposition is a direct consequence of (the implication $v) \Rightarrow ii)$ in) Theorem 5.3.1.

Proposition 5.3.4 *Let $\mathcal{E}_o = \mathcal{E}$. If, for every $x \in \mathcal{D}$ there exists $\delta > 0$ such that $x \in \mathcal{D}(X^\varsigma)$ whenever $0 < \varsigma < \delta$, and if*

$$\lim_{\varsigma \downarrow 0} X^\varsigma x = X x, \qquad (5.20)$$

then

$$\lim_{\varsigma \downarrow 0} \sup\{\|T^\varsigma(t)x - T(t)x\| : 0 \le t \le t_o\} = 0$$

for all $x \in \mathcal{E}$ and all $t_o \in \mathbf{R}_+$.

Theorem 5.3.5 *Let $\varsigma > 0$ and let $T^\varsigma : \mathbf{R}_+ \to \mathcal{L}(\mathcal{E})$ be a strongly continuous semigroup satisfying (5.12) for some real constants a and $M \ge 1$. Suppose that, for some $\xi > a$ the limit*

$$\lim_{\varsigma \downarrow 0} (\xi I - X^\varsigma)^{-1} = J_\xi(x)$$

exists for all $x \in \mathcal{E}$, and that the linear operator $J_\xi \in \mathcal{L}(\mathcal{E})$ has dense range.

There exists a strongly continuous semigroup $T : \mathbf{R}_+ \to \mathcal{L}(\mathcal{E})$ such that

$$\lim_{\varsigma \downarrow 0} \sup\{\|T^\varsigma(t)x - T(t)x\| : 0 \le t \le t_o\} = 0$$

for all $x \in \mathcal{E}$ and all $t_o \in \mathbf{R}_+$.

Proof [15] The inequality

$$\|(\xi I - X^\varsigma)^{-n}\| \le \frac{M}{(\xi - a)^n} \qquad (5.21)$$

holds for $n = 1, 2, \ldots$. For any $\zeta \in \mathbf{C}$ with $\Re\zeta > a$,

$$(\zeta I - X^\varsigma)^{-1} = ((\zeta - \xi) I - (\xi I - X^\varsigma))^{-1} =$$
$$(I - (\xi - \zeta)(\xi I - X^\varsigma)^{-1})^{-1}(\xi I - X^\varsigma)^{-1},$$

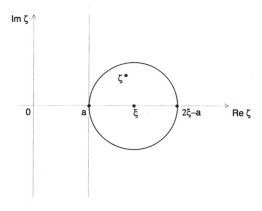

Figure 5.1:

and, at least formally,

$$(I - (\xi - \zeta)(\xi I - X^{\varsigma})^{-1})^{-1}(\xi I - X^{\varsigma})^{-1} =$$
$$\left(\sum_{n=0}^{+\infty}(\xi - \zeta)^n(\xi I - X^{\varsigma})^{-n}\right)(\xi I - X^{\varsigma})^{-1}. \qquad (5.22)$$

Since, by (5.21),

$$\|((\xi - \zeta)(\xi I - X^{\varsigma})^{-1})^n\| \leq \frac{M|\xi - \zeta|^n}{(\xi - a)^n},$$

the power series on the right hand side of (5.22) converges to the left hand side whenever

$$|\zeta - \xi| < \xi - a. \qquad (5.23)$$

The convergence is uniform on all compact subsets of the open disc in $\mathbf{C}$ with center ξ and radius $\xi - a$.

Thus

$$(\zeta I - X^{\varsigma})^{-1} = \left(\sum_{n=0}^{+\infty}(\xi - \zeta)^n(\xi I - X^{\varsigma})^{-n}\right)(\xi I - X^{\varsigma})^{-1}$$

for the norm topology, whenever ζ satisfies (5.23).

Since

$$\|(\xi I - X^\varsigma)^{-2}x - J_\xi^2 x\| \le \|(\xi I - X^\varsigma)^{-1}((\xi I - X^\varsigma)^{-1} - J_\xi)x\| +$$
$$\|((\xi I - X^\varsigma)^{-1} - J_\xi)J_\xi x\|$$
$$\le \frac{M}{\xi - a}\|((\xi I - X^\varsigma)^{-1} - J_\xi)x\| +$$
$$+\|((\xi I - X^\varsigma)^{-1} - J_\xi)J_\xi x\|,$$

then

$$\lim_{\varsigma \downarrow 0}(\xi I - X^\varsigma)^{-2}x = J_\xi^2 x,$$

and an inductive argument yields

$$\lim_{\varsigma \downarrow 0}(\xi I - X^\varsigma)^{-n}x = J_\xi^n x \quad \forall x \in \mathcal{E}, \quad n = 1, 2, \ldots . \tag{5.24}$$

For any ζ satisfying (5.23) and any integer $N \ge 1$,

$$\|\sum_{n=N}^{+\infty}(\xi - \zeta)^n(\xi I - X^\varsigma)^{-n}\| \le M \sum_{n=N}^{+\infty}\left(\frac{|\xi - \zeta|}{\xi - a}\right)^n$$
$$= M\left(\frac{|\xi - \zeta|}{\xi - a}\right)^N \frac{1}{1 - \frac{|\xi - \zeta|}{\xi - a}}$$
$$= M\left(\frac{|\xi - \zeta|}{\xi - a}\right)^N \frac{\xi - a}{\xi - a - |\xi - \zeta|}.$$

Given $\epsilon > 0$, let $N \ge 1$ be such that

$$M\left(\frac{|\xi - \zeta|}{\xi - a}\right)^N \frac{\xi - a}{\xi - a - |\xi - \zeta|} < \epsilon.$$

Then, for $\varsigma_1, \varsigma_2 > 0$,

$$\|((\zeta I - X^{\varsigma_1})^{-1} - (\zeta I - X^{\varsigma_2})^{-1})x\| \le$$
$$\left\|\sum_{n=0}^{+\infty}(\xi - \zeta)^n(\xi I - X^{\varsigma_1})^{-(n+1)} - \sum_{n=0}^{+\infty}(\xi - \zeta)^n(\xi I - X^{\varsigma_2})^{-(n+1)}\right\|$$
$$\le \|(\xi I - X^{\varsigma_1})^{-1} - (\xi I - X^{\varsigma_2})^{-1}\| +$$
$$(\xi - a)\|(\xi I - X^{\varsigma_1})^{-2} - (\xi I - X^{\varsigma_2})^{-2}\| + \cdots +$$
$$(\xi - a)^{N-2}\|(\xi I - X^{\varsigma_1})^{-(N-1)} - (\xi I - X^{\varsigma_2})^{-(N-1)}\| + 2\epsilon.$$

This proves that $\lim_{\varsigma \downarrow 0}(\zeta I - X^\varsigma)^{-1}x$ exists for all $x \in \mathcal{E}$, whenever $\Re\zeta > a$. The limit - which is a bounded linear operator on $\mathcal{E}$ - will be denoted by J_ζ.

As for J_ξ, one shows that, if ζ satisfies (5.23),

$$\lim_{\varsigma \downarrow 0}(\zeta I - X^\varsigma)^{-n}x = J_\zeta^n x \quad \forall x \in \mathcal{E}, \ n = 1, 2, \ldots . \tag{5.25}$$

In view of (5.12),

$$\|(\zeta I - X^\varsigma)^{-n}\| \le \frac{M}{|\zeta - a|^n}$$

for all $\varsigma > 0$, $n = 1, 2, \ldots$ and all $\zeta \in \mathbf{C}$ with $\Re\zeta > a$. Hence, by (5.25),

$$\|J_\zeta^n\| \le \frac{M}{|\zeta - a|^n}$$

for $n = 1, 2, \ldots$ and ζ satisfying (5.23). Choose now $\zeta \in \mathbf{R}$, in which case (5.23) yields

$$a < \zeta < 2\xi - a.$$

Replacing ξ by a point in the interval $(\xi, 2\xi - a)$ and iterating the procedure, we define, in conclusion, an operator $J_\zeta \in \mathcal{L}(\mathcal{E})$ satisfying (5.25), and therefore also the inequality

$$\|J_\zeta^n\| \le \frac{M}{(\zeta - a)^n} \quad \forall \zeta \in (a, +\infty), \ n = 1, 2, \ldots . \tag{5.26}$$

For $\zeta_1, \zeta_2 \in (a, +\infty)$ and $x \in \mathcal{E}$,

$$\begin{aligned}
\|(\zeta_1 I - X^\varsigma)^{-1}(\zeta_2 I - X^\varsigma)^{-1}x - J_{\zeta_1}J_{\zeta_2}x\| &= \\
= \|(\zeta_1 I - X^\varsigma)^{-1}((\zeta_2 I - X^\varsigma)^{-1}x - J_{\zeta_2}x) + \\
((\zeta_1 I - X^\varsigma)^{-1} - J_{\zeta_1})J_{\zeta_2}x\| \\
\le \|(\zeta_1 I - X^\varsigma)^{-1}\| \, \|((\zeta_2 I - X^\varsigma)^{-1} - J_{\zeta_2})x\| + \\
\|((\zeta_1 I - X^\varsigma)^{-1} - J_{\zeta_1})J_{\zeta_2}x\| \\
\le \frac{M}{\zeta_1 - a}\|((\zeta_2 I - X^\varsigma)^{-1} - J_{\zeta_2})x\| + \\
\|((\zeta_1 I - X^\varsigma)^{-1} - J_{\zeta_1})J_{\zeta_2}x\|,
\end{aligned}$$

and therefore

$$\lim_{\varsigma \downarrow 0}(\zeta_1 I - X^\varsigma)^{-1}(\zeta_2 I - X^\varsigma)^{-1}x = J_{\zeta_1}J_{\zeta_2}x$$

for all $x \in \mathcal{E}$. Thus,

$$
\begin{aligned}
(J_{\varsigma_1} - J_{\varsigma_2})x &= \lim_{\varsigma \downarrow 0}[((\varsigma_1 I - X^\varsigma)^{-1} - (\varsigma_2 I - X^\varsigma)^{-1})x] \\
&= \lim_{\varsigma \downarrow 0}(\varsigma_2 - \varsigma_1)(\varsigma_1 I - X^\varsigma)^{-1}(\varsigma_2 I - X^\varsigma)^{-1}x \\
&= (\varsigma_2 - \varsigma_1)J_{\varsigma_1}J_{\varsigma_2}x,
\end{aligned}
$$

proving that $\varsigma \mapsto J_\varsigma$ is a pseudo-resolvent. In view of (5.26) and of the fact $\overline{\mathcal{R}(J_\varsigma)} = \mathcal{E}$, Theorem 1.13.3 shows that there is a linear, closed, densely defined operator $X : \mathcal{D}(X) \subset \mathcal{E} \to \mathcal{E}$ such that

$$
J_\varsigma = (\varsigma I - X)^{-1}
$$

for all $\varsigma \in (a, +\infty)$. By (5.26), X is the infinitesimal generator of a strongly continuous semigroup $T : \mathbf{R}_+ \to \mathcal{L}(\mathcal{E})$. Since

$$
\lim_{\varsigma \downarrow 0}(\xi I - X^\varsigma)^{-1}x = (\xi I - X)^{-1}x \quad \forall x \in \mathcal{E},
$$

the implication $iii) \Rightarrow ii)$ in Theorem 5.3.1 completes the proof. ∎

5.4 Additioning infinitesimal generators

Let G be a Lie group and let $\breve{G}$ be its Lie algebra. If X and Y are two elements of $\breve{G}$, the one-parameter subgroups generated by X, Y and $X + Y$ are related by the *perturbation formula* (see *e.g.* [39], [80]):

$$
\exp(t(X + Y)) = \lim_{n \to +\infty}\left[\exp\left(\frac{t}{n}X\right) \cdot \exp\left(\frac{t}{n}Y\right)\right]^n.
$$

The extension of this perturbation formula to one-parameter semigroups of linear operators is the basic motivation of the *Trotter product formula*. This section will be devoted to this formula, to some of its consequences and to further developments.

We begin by establishing a lemma, owing to P.R Chernoff. Let $F : \mathbf{R}_+ \to \mathcal{L}(\mathcal{E})$ such that $F(0) = 0$ and that $F(t)$ is a contraction for all $t \geq 0$. Let X be the infinitesimal generator of a strongly continuous semigroup $T : \mathbf{R}_+ \to \mathcal{L}(\mathcal{E})$ and let $\mathcal{D}$ be a core of X.

Lemma 5.4.1 *If*

$$\lim_{t\downarrow 0} \frac{1}{t}(F(t)x - x) = Xx$$

for all $x \in \mathcal{D}$, then

$$\lim_{n\to+\infty} \left(F\left(\frac{t}{n}\right)\right)^n x = T(t)x \tag{5.27}$$

for all $x \in \mathcal{E}$ and all $t \in \mathbf{R}_+$.

Proof [15] For $n \geq 1$ and $t > 0$, the operator $X_n(t) \in \mathcal{L}(\mathcal{E})$ defined by

$$X_n(t) = \frac{n}{t}\left[F\left(\frac{t}{n}\right) - I\right]$$

is dissipative. Indeed, for any $x \in \mathcal{E}$ and any $\lambda \in \mathcal{E}'$ with $\|\lambda\| = 1$, $\|x\| = < x, \lambda >$,

$$\begin{aligned}
\Re < X_n(t)x, \lambda > &= \frac{n}{t}\left[\Re < F\left(\frac{t}{n}\right)x, \lambda > - < x, \lambda >\right] \\
&= \frac{n}{t}\left[\Re < F\left(\frac{t}{n}\right)x, \lambda > - \|x\|\right] \\
&\leq \frac{n}{t}\left(\|\lambda\|\left\|F\left(\frac{t}{n}\right)x\right\| - \|x\|\right) \\
&\leq \frac{n}{t}\left(\left\|\frac{t}{n}x\right\| - \|x\|\right) \leq \frac{n}{t}(\|x\| - \|x\|) = 0.
\end{aligned}$$

For any $x \in \mathcal{D}$,

$$\begin{aligned}
\lim_{n\to+\infty} X_n(t)x &= \lim_{n\to+\infty} \frac{n}{t}\left(F\left(\frac{t}{n}\right) - I\right)x \\
&= \lim_{t\downarrow 0} \frac{1}{t}(F(t)x - x) = Xx. \tag{5.28}
\end{aligned}$$

Hence, by Proposition 5.3.4 (applied to the uniformly continuous semi-group $s \to \exp sX_n(t)$ generated by $X_n(t)$) we obtain

$$\lim_{n\to+\infty} \exp tX_n(t)x = T(t)x$$

for all $t \in \mathbf{R}_+$ and all $x \in \mathcal{E}$, the convergence being uniform on all compact subsets of $\mathbf{R}_+$. By Lemma 3.3.5

$$\left\| \exp tX_n(t)x - \left(F\left(\frac{t}{n}\right) \right)^n x \right\| =$$
$$\left\| \exp\left(n\left(F\left(\frac{t}{n}\right) - I \right) \right) x - \left(F\left(\frac{t}{n}\right) \right)^n x \right\|$$
$$\leq \sqrt{n} \left\| \left(F\left(\frac{t}{n}\right) - I \right) x \right\|$$
$$\leq \frac{t}{\sqrt{n}} \frac{n}{t} \left\| \left(F\left(\frac{t}{n}\right) - I \right) x \right\| = \frac{t}{\sqrt{n}} \| X_n(t)x \|.$$

By (5.28), if $x \in \mathcal{D}$,

$$\lim_{n \to +\infty} \frac{t}{\sqrt{n}} \| X_n(t)x \| = 0.$$

Thus,

$$\lim_{n \to +\infty} \left\| \exp tX_n(t)x - \left(F\left(\frac{t}{n}\right) \right)^n x \right\| = 0 \qquad (5.29)$$

for all $x \in \mathcal{D}$. If $y \in \mathcal{E}$ and $\epsilon > 0$, there is $x \in \mathcal{D}$ such that $\| y - x \| < \epsilon$. Since

$$\left\| \exp tX_n(t)y - \left(F\left(\frac{t}{n}\right) \right)^n y \right\| \leq \left\| \exp tX_n(t)x - \left(F\left(\frac{t}{n}\right) \right)^n x \right\| +$$
$$\| \exp tX_n(t)(y - x) \| +$$
$$\left\| \left(F\left(\frac{t}{n}\right) \right)^n (y - x) \right\|$$
$$\leq \left\| \exp tX_n(t)x - \left(F\left(\frac{t}{n}\right) \right)^n x \right\| + 2\epsilon,$$

then (5.29) holds for all $x \in \mathcal{E}$. ∎

The convergence in (5.27) is uniform on all compact subsets of $\mathbf{R}_+$. (Proof: exercise).

Corollary 5.4.2 *If the derivative $F'(0)x$ exists for all x in a dense space $\mathcal{D}$ and if the closure $\overline{F'(0)}$ generates a strongly continuous semigroup $T : \mathbf{R}_+ \to \mathcal{L}(\mathcal{E})$, (5.27) holds for all $t \in \mathbf{R}_+$ and all $x \in \mathcal{E}$.*

Theorem 5.4.3 *(The Trotter product formula) Let X, Y and Z be infinitesimal generators of strongly continuous contraction semigroups $T : \mathbf{R}_+ \to \mathcal{E}$, $S : \mathbf{R}_+ \to \mathcal{E}$ and $R : \mathbf{R}_+ \to \mathcal{E}$. Let $\mathcal{D}$ be a core of Z. If $\mathcal{D} \subset \mathcal{D}(X) \cap \mathcal{D}(Y)$ and if*

$$Zx = Xx + Yx \;\; \forall x \in \mathcal{D}, \tag{5.30}$$

then

$$\lim_{n \to +\infty} \left(T\left(\frac{t}{n}\right) S\left(\frac{t}{n}\right) \right)^n x = R(t)x \; \forall x \in \mathcal{E}, \; \forall t \in \mathbf{R}_+. \tag{5.31}$$

Proof [15] For $t \geq 0$, let

$$F(t) = T(t)\, S(t).$$

Then $F(t) \in \mathcal{L}(\mathcal{E})$ is a contraction, and $F(0) = I$. Moreover, if $x \in \mathcal{D}$,

$$\lim_{t \downarrow 0} \frac{1}{t}(F(t)x - x) = \lim_{t \downarrow 0} \frac{1}{t}(T(t)S(t)x - x)$$

$$= \lim_{t \downarrow 0}(T(t)\frac{1}{t}(S(t) - I)x + \lim_{t \downarrow 0} \frac{1}{t}(T(t) - I)x$$

$$= Yx + Xx = Zx.$$

Corollary 5.4.2 yields the conclusion. ∎

As before, the convergence in (5.31) is uniform on all compact subsets of $\mathbf{R}_+$.

Corollary 5.4.4 *Under the same hypotheses of Theorem 5.4.3 on X, Y, T and S, if the closure $\overline{X+Y}$ of $X+Y$ generates a contractive semigroup R, then (5.31) holds for all $x \in \mathcal{E}$ and all $t \in \mathbf{R}_+$.*

Corollary 5.4.5 *Let X, Y and Z be the infinitesimal generators of three strongly continuous semigroups $T : \mathbf{R}_+ \to \mathcal{L}(\mathcal{E})$, $S : \mathbf{R}_+ \to \mathcal{L}(\mathcal{E})$ and $R : \mathbf{R}_+ \to \mathcal{L}(\mathcal{E})$. Let $\mathcal{D}$ be a core of Z such that $\mathcal{D} \subset \mathcal{D}(X) \cap \mathcal{D}(Y)$ and that (5.30) holds. If*

$$\|T(t)\| \le e^{at}, \ \|S(t)\| \le e^{bt} \ \forall t \in \mathbf{R}_+ \tag{5.32}$$

and for some a, b in $\mathbf{R}$, then (5.31) holds.

Proof The strongly continuous semigroups of contractions $\tilde{T} : t \mapsto e^{-at}T(t)$ and $\tilde{S} : t \mapsto e^{-bt}S(t)$ are generated by $X - aI$ and by $Y - bI$, which are therefore m-dissipative. Since

$$Z - (a+b)I = X - aI + Y - bI \ \ \text{on} \mathcal{D} \tag{5.33}$$

and $\mathcal{D}$ is a core of Z, then $Z - (a+b)I$ is dissipative. Since it generates the strongly continuous semigroup

$$t \mapsto e^{-(a+b)t}R(t), \tag{5.34}$$

it is m-dissipative, and the semigroup (5.34) is a contraction semigroup. By Theorem 5.4.3 and by (5.33),

$$e^{-(a+b)t}R(t)x = \lim_{n \to +\infty} \left(\tilde{T}\left(\frac{t}{n}\right) \tilde{S}\left(\frac{t}{n}\right) \right)^n x$$

$$= \lim_{n \to +\infty} \left(e^{-a\frac{t}{n}} e^{-b\frac{t}{n}} T\left(\frac{t}{n}\right) S\left(\frac{t}{n}\right) \right)^n x$$

$$= e^{-(a+b)t} \lim_{n \to +\infty} \left(T\left(\frac{t}{n}\right) S\left(\frac{t}{n}\right) \right)^n x,$$

showing that (5.31) holds. ∎

As a consequence of this corollary (or of Theorem 4.5 of [10]), the following proposition holds.

Proposition 5.4.6 *Let X and Y be bounded perturbations of two densely defined, m-dissipative operators X_o and Y_o. Let T and S be* the strongly continuous semigroups generated by X and Y, and let Z be the infinitesimal generator of a strongly continuous semigroup R. If there is a core $\mathcal{D}$ of Z such that $\mathcal{D} \subset \mathcal{D}(X) \cap \mathcal{D}(Y)$, and which satisfies (5.30), then (5.31) holds.

The proof follows from Corollary 5.4.5 and from the fact that, by Theorem 5.1.1, T and S satisfy (5.32), where a and b are the norms of the bounded perturbations of X_o and Y_o.

Theorem 5.4.7 *[10]Let* $F : \mathbf{R}_+ \rightarrow \mathcal{L}(\mathcal{E})$ *be such that* $F(0) = I$ *and* $\|F(t)\| \leq 1$ *for all* $t \geq 0$. *Let* $T : \mathbf{R}_+ \rightarrow \mathcal{L}(\mathcal{E})$ *be a strongly continuous semigroup of contractions such that, for all* $x \in \mathcal{E}$,

$$\lim_{n \rightarrow +\infty} F\left(\frac{t}{n}\right)^n x = T(t)x \qquad (5.35)$$

uniformly on all compact subsets in $\mathbf{R}_+$. *Then, the infinitesimal generator* X *of* T *is an extension of the derivative* $F'(0)$ *of* F *at* 0.

The domain $\mathcal{D}(F'(0))$ of $F'(0)$ is not assumed to be dense in $\mathcal{E}$.

We begin the proof by establishing the following lemma.

Lemma 5.4.8 *If* $\{k_n\}$ *is a sequence of positive integers such that the sequence* $\{\frac{k_n}{n}\}$ *converges to some* $k > 0$ *when* $n \rightarrow +\infty$, *then*

$$\lim_{n \rightarrow +\infty} F\left(\frac{t}{n}\right)^{k_n} x = T(kt)x \qquad (5.36)$$

for all $x \in \mathcal{E}$, *uniformly on all compact subsets of* $\mathbf{R}_+$.

Proof Setting $t_n = \frac{k_n}{n}t$, then $\lim_{n \rightarrow +\infty} t_n = kt$ and

$$F\left(\frac{t}{n}\right)^{k_n} = F\left(\frac{t_n}{k_n}\right)^{k_n}.$$

Since (5.36) holds uniformly on all compact subsets of $\mathbf{R}_+$, given any $s \in \mathbf{R}_+$, any $x \in \mathcal{E}$ and any $\epsilon > 0$, there exist $\delta > 0$ and a positive integer n_0 such that

$$\left\|F\left(\frac{r}{n}\right)^n x - T(r)x\right\| < \epsilon$$

whenever $r \in (s - \delta, s + \delta) \cap \mathbf{R}_+$ and $n \geq n_0$. Because $\{k_n\}$ diverges to $+\infty$, there is some $n_1 \geq n_0$ such that

$$\left\| F\left(\frac{r}{k_n}\right)^{k_n} x - T(r)x \right\| < \epsilon$$

whenever $r \in (s - \delta, s + \delta)) \cap \mathbf{R}_+$ and $n \geq n_1$.

Choose $s = kt$, and let $n_2 \geq n_1$ be such that, if $n \geq n_2$, then $t_n \in (kt - \delta, kt + \delta) \cap \mathbf{R}_+$. Thus, for $n \geq n_2$,

$$\left\| F\left(\frac{t_n}{k_n}\right)^{k_n} x - T(t_n)x \right\| < \epsilon$$

i. e.

$$\left\| F\left(\frac{t}{n}\right)^{k_n} x - T(t_n)x \right\| < \epsilon.$$

Since $\lim_{n \to +\infty} T(t_n)x = T(kt)x$, there is $n_3 \geq n_2$ such that

$$\left\| F\left(\frac{t}{n}\right)^{k_n} x - T(kt)x \right\| < 2\epsilon$$

whenever $n \geq n_3$. ∎

To complete the proof of the theorem, it will be shown now that, if $x \in \mathcal{D}(F'(0))$, then $x \in \mathcal{D}(X)$, i.e.,

$$\lim_{t \downarrow 0} \frac{1}{t}(T(t)x - x) = F'(0)x.$$

For $x \in \mathcal{D}(F'(0))$ and $t > 0$,

$$\frac{1}{t}(T(t) - I)x = \lim_{n \to +\infty} \frac{1}{t}\left(F\left(\frac{t}{n}\right)^n - I\right)x$$

$$= \lim_{n \to +\infty} \frac{1}{t} \sum_{j=0}^{n-1} F\left(\frac{t}{n}\right)^j \left(F\left(\frac{t}{n}\right)^n - I\right)x.$$

Let $[s]$ denote the integral part of $s \in \mathbf{R}$. Then

$$\int_0^1 F\left(\frac{t}{n}\right)^{[ns]} xds = \int_0^{\frac{1}{n}} ds\, x + \int_{\frac{1}{n}}^{\frac{2}{n}} F\left(\frac{t}{n}\right) xds +$$

$$\int_{\frac{2}{n}}^{\frac{3}{n}} F\left(\frac{t}{n}\right)^2 xds + \cdots + \int_{1-\frac{1}{n}}^1 F\left(\frac{t}{n}\right)^{n-1} xds$$

$$= \frac{1}{n}\sum_{j=0}^{n-1} F\left(\frac{t}{n}\right)^j x,$$

and therefore

$$\frac{1}{t}(T(t) - I)x = \lim_{n\to+\infty} \int_0^1 F\left(\frac{t}{n}\right)^{[ns]} ds\, \frac{n}{t}\left(F\left(\frac{t}{n}\right) - I\right)x.$$

By (5.36)

$$\lim_{n\to+\infty} \int_0^1 F\left(\frac{t}{n}\right)^{[ns]} xds = \int_0^1 T(st)xds.$$

Since

$$\lim_{n\to+\infty} \frac{n}{t}\left(F\left(\frac{t}{n}\right) - I\right)x = \lim_{t\downarrow 0} \frac{1}{t}(F(t) - I)x = F'(0)x,$$

then

$$\frac{1}{t}(T(t) - I)x = \frac{1}{t}\int_0^1 T(st)F'(0)xds$$

$$= \int_0^t T(s)F'(0)xds.$$

Passing to the limit as $t \downarrow 0$, we obtain $x \in \mathcal{D}(X)$, and furthermore

$$Xx = \lim_{t\downarrow 0} \frac{1}{t}(T(t) - I)x = F'(0)x.$$

■

If A and B are linear operators defined on $\mathcal{E}$, a direct computation yields

$$(A+B)^n - A^n = \sum_{j=1}^{n} (A+B)^{n-j} B A^{j-1} \quad \forall n = 1, 2, \ldots . \qquad (5.37)$$

Lemma 5.4.9 *[10]Let $\{A_n\}$ be a sequence of operators $A_n \in \mathcal{L}(\mathcal{E})$ such that $\|A_n\| \leq 1 + \frac{\alpha}{n}$ for some constant $\alpha \geq 0$ and all $n = 1, 2, \ldots$. If $\{A_n{}^n\}$ converges strongly to $C \in \mathcal{L}(\mathcal{E})$ and if $\{B_n\}$ is a sequence of operators $B_n \in \mathcal{L}(\mathcal{E})$ with $\|B_n\| = o(\frac{1}{n})$, then $\{A_n + B_n\}$ converges to C.*

Proof It follows from (5.37) that

$$\|(A_n + B_n)^n - A_n{}^n\| \leq \sum_{k=1}^{n} (\|A_n\| + \|B_n\|)^{n-k} \|B_n\| \, \|A_n\|^{k-1}$$

$$\leq \sum_{k=1}^{n} \left(1 + \frac{\alpha}{n} + \|B_n\|\right)^{n-k} \|B_n\| \left(1 + \frac{\alpha}{n}\right)^{k-1}$$

$$\leq \sum_{k=1}^{n} \left(1 + \frac{\alpha}{n} + \|B_n\|\right)^{n-k} \|B_n\| \times$$

$$\left(1 + \frac{\alpha}{n} + \|B_n\|\right)^{k-1}$$

$$= n \left(1 + \frac{\alpha}{n} + \|B_n\|\right)^{n-1} \|B_n\|, \qquad (5.38)$$

which tends to zero when $n \to +\infty$, because $\lim_{n \to +\infty} n\|B_n\| = 0$.

Hence, for any $x \in \mathcal{E}$,

$$\|(A_n + B_n)^n x - Cx\| \leq \|(A_n + B_n)^n x - A_n{}^n x\| + \|A_n{}^n x - Cx\|,$$

which tends to zero when $n \to +\infty$, because of (5.38) and of the hypothesis whereby $\lim_{n \to +\infty} A_n{}^n x = Cx$. ∎

Theorem 5.4.10 *[10] Let* $T : \mathbf{R}_+ \to \mathcal{L}(\mathcal{E})$, $S : \mathbf{R}_+ \to \mathcal{L}(\mathcal{E})$ *and* $R : \mathbf{R}_+ \to \mathcal{L}(\mathcal{E})$ *be strongly continuous semigroups of contractions. Let* $\tilde{T} : \mathbf{R}_+ \to \mathcal{L}(\mathcal{E})$ *and* $\tilde{R} : \mathbf{R}_+ \to \mathcal{L}(\mathcal{E})$ *be the strongly continuous semigroups generated by the bounded perturbations of the infinitesimal generators of* T *and* R *by an operator* $A \in \mathcal{L}(\mathcal{E})$.
If, for all $t \in \mathbf{R}_+$,

$$\lim_{n \to +\infty} \left(T\left(\frac{t}{n}\right) S\left(\frac{t}{n}\right) \right)^n = R(t), \qquad (5.39)$$

for the strong topology, uniformly when t *varies on bounded, closed intervals in* $\mathbf{R}_+$*, then*

$$\lim_{n \to +\infty} \left(\tilde{T}\left(\frac{t}{n}\right) S\left(\frac{t}{n}\right) \right)^n = \tilde{R}(t)$$

for the strong topology, for all $t \in \mathbf{R}_+$.

Sketch of Proof The perturbation series yields

$$\tilde{T}(t)x = T(t)x + \int_0^t T(s)AT(t-s)x\,ds + O(t^2),$$

for all $t \in \mathcal{E}$ and all $t \in \mathbf{R}_+$. In view of Lemma 5.4.9, it suffices to show that

$$\lim_{n \to +\infty} \left(\left(T\left(\frac{t}{n}\right) + \int_0^{\frac{t}{n}} T(s)AT\left(\frac{t}{n} - s\right) ds \right) S\left(\frac{t}{n}\right) \right)^n x = \tilde{R}(t)x,$$

{i. e.}, setting $F(t) = T(t)S(t)$ and

$$K(t) = \int_0^t T(s)AT(t-s)S(t)ds,$$

then

$$\lim_{n \to +\infty} \left(F\left(\frac{t}{n}\right) + K\left(\frac{t}{n}\right) \right)^n x = \tilde{R}(t)x.$$

First of all,

$$(F + K)^n(t) = F^n(t)+$$

$$\sum_{0 \le k_1 \le \cdots \le k_r \le n-1} F^{k_1} K F^{k_2-k_1} K F^{k_3-k_2} K \cdots K F^{n-k_r-r}(t) + \cdots$$

$$+ K^n(t).$$

The r − th term $(r = 1, 2, \ldots n)$ in this expansion is estimated by the r − th term in the expansion of $(\|F(t)\| + \|K(t)\|)^n$, which is

$$\binom{n}{r} \|F(t)\|^{n-r} \|K(t)\|^r \le \binom{n}{r} \|K(t)\|^r.$$

Since

$$\left\| K\left(\frac{t}{n}\right) \right\| = \left\| \int_0^{\frac{t}{n}} T(s) A T\left(\frac{t}{n} - s\right) S\left(\frac{t}{n}\right) ds \right\|$$

$$\le \|A\| \int_0^{\frac{t}{n}} ds = \frac{t}{n} \|A\|,$$

then

$$\binom{n}{r} \left\| F\left(\frac{t}{n}\right) \right\|^{n-r} \left\| K\left(\frac{t}{n}\right) \right\|^r \le$$

$$\le \frac{n(n-1)\cdots(n-r+1)}{r!} \frac{t^r}{n^r} \|A\|^r$$

$$\le \frac{\overbrace{n.n.\cdots .n}^{r}}{r!} \frac{t^r}{n^r} \|A\|^r = \frac{t^r}{r!} \|A\|^r.$$

Hence

$$\left\| \left(F\left(\frac{t}{n}\right) + K\left(\frac{t}{n}\right) \right)^n \right\| \le \sum_{r=0}^{+\infty} \frac{t^r}{r!} \|A\|^r = e^{t\|A\|},$$

and, as $n \to +\infty$, the limit can be computed term by term.

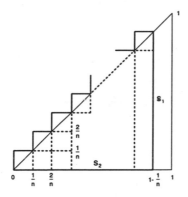

Figure 5.2:

Furthermore, denoting by $[l]$ the integral part of l,

$$
t \int_0^1 F\left(\frac{t}{n}\right)^{[ns]} \frac{n}{t} K\left(\frac{t}{n}\right) F\left(\frac{t}{n}\right)^{[n(1-s)]} ds =
$$

$$
n \left\{ \int_0^{\frac{1}{n}} K\left(\frac{t}{n}\right) F\left(\frac{t}{n}\right)^{n-1} ds + \right.
$$

$$
\int_{\frac{1}{n}}^{\frac{2}{n}} F\left(\frac{t}{n}\right)^{\frac{1}{n}} K\left(\frac{t}{n}\right) F\left(\frac{t}{n}\right)^{n-2} ds +
$$

$$
\left. + \cdots + \int_{1-\frac{1}{n}}^1 F\left(\frac{t}{n}\right)^{n-1} K\left(\frac{t}{n}\right) ds \right\}
$$

$$
= K\left(\frac{t}{n}\right) F\left(\frac{t}{n}\right)^{n-1} + F\left(\frac{t}{n}\right) K\left(\frac{t}{n}\right) F\left(\frac{t}{n}\right)^{n-2} + \cdots +
$$

$$
F\left(\frac{t}{n}\right)^{n-1} K\left(\frac{t}{n}\right).
$$

Choose now $r = 2$, and let Δ_2 be the two-simplex

$$
\Delta_2 = \{(s_1, s_2) \in \mathbf{R}^2 : 0 \le s_1 \le s_2 \le 1\}.
$$

For $n > 1$, let $\Delta_2(n)$ be the bounded set described in the figure.

Note that $\lim_{n \to +\infty} \Delta_2(n) = \Delta_2$. Then

$$\int\int_{\Delta_2(t)} F\left(\frac{t}{n}\right)^{[ns_1]} K\left(\frac{t}{n}\right) F\left(\frac{t}{n}\right)^{[ns_2]-[ns_1]} \times$$

$$K\left(\frac{t}{n}\right) F\left(\frac{t}{n}\right)^{[n-1-ns_2]} ds_1 ds_2$$

$$= \int_0^{\frac{1}{n}} \left(\int_0^{\frac{1}{n}} K\left(\frac{t}{n}\right)^2 ds_1 \right) F\left(\frac{t}{n}\right)^{n-2} ds_2 +$$

$$+ \int_{\frac{1}{n}}^{\frac{2}{n}} \left(\int_0^{\frac{1}{n}} K\left(\frac{t}{n}\right) F\left(\frac{t}{n}\right) K\left(\frac{t}{n}\right) ds_1 \right) F\left(\frac{t}{n}\right)^{n-3} ds_2 +$$

$$+ \int_{\frac{1}{n}}^{\frac{2}{n}} \left(\int_{\frac{1}{n}}^{\frac{2}{n}} F\left(\frac{t}{n}\right) K\left(\frac{t}{n}\right)^2 ds_1 \right) F\left(\frac{t}{n}\right)^{n-3} ds_2 + \cdots +$$

$$+ \int_{1-\frac{2}{n}}^{1-\frac{1}{n}} \left(\int_0^{\frac{1}{n}} K\left(\frac{t}{n}\right) F\left(\frac{t}{n}\right)^{n-2} K\left(\frac{t}{n}\right) ds_1 \right) ds_2 +$$

$$+ \int_{1-\frac{2}{n}}^{1-\frac{1}{n}} \left(\int_{\frac{1}{n}}^{\frac{2}{n}} F\left(\frac{t}{n}\right) K\left(\frac{t}{n}\right) F\left(\frac{t}{n}\right)^{n-3} K\left(\frac{t}{n}\right) ds_1 \right) ds_2 + \cdots +$$

$$+ \int_{1-\frac{2}{n}}^{1-\frac{1}{n}} \left(\int_{1-\frac{2}{n}}^{1-\frac{1}{n}} F\left(\frac{t}{n}\right)^{n-2} K\left(\frac{t}{n}\right)^2 ds_1 \right) ds_2$$

$$= \frac{1}{n^2} \left\{ K\left(\frac{t}{n}\right)^2 F\left(\frac{t}{n}\right)^{n-2} + \right.$$

$$+ \left(K\left(\frac{t}{n}\right) F\left(\frac{t}{n}\right) K\left(\frac{t}{n}\right) F\left(\frac{t}{n}\right)^{n-3} + \right.$$

$$+ F\left(\frac{t}{n}\right) K\left(\frac{t}{n}\right)^2 F\left(\frac{t}{n}\right)^{n-3} \bigg) +$$

$$+ \cdots + \left(K\left(\frac{t}{n}\right) F\left(\frac{t}{n}\right)^{n-2} K\left(\frac{t}{n}\right) + \right.$$

$$+ F\left(\frac{t}{n}\right) K\left(\frac{t}{n}\right) F\left(\frac{t}{n}\right)^{n-3} K\left(\frac{t}{n}\right) +$$

$$\left. + \cdots + F\left(\frac{t}{n}\right)^{n-2} K\left(\frac{t}{n}\right)^2 \right) \bigg\},$$

{i. e.},

$$t^2 \int\int_{\Delta_2(n)} F\left(\frac{t}{n}\right)^{[ns_1]} \frac{n}{t} K\left(\frac{t}{n}\right) F\left(\frac{t}{n}\right)^{[ns_2]-[ns_1]} \times$$

$$\frac{n}{t} K\left(\frac{t}{n}\right) F\left(\frac{t}{n}\right)^{[n-1-ns_2]} ds_1 ds_2 =$$

$$((K^2 F^{n-2} + (KFKF^{n-3} + FK^2 F^{n-3}) + \cdots +$$

$$(KF^{n-2}K + FKF^{n-3}K + \cdots + F^{n-2}K^2))(t).$$

By Lemma 5.4.9,

$$\lim_{n\to+\infty} F\left(\frac{t}{n}\right)^{[ns]} = R(st),$$

$$\lim_{n\to+\infty} F\left(\frac{t}{n}\right)^{[ns_2]-[ns_1]} = R(((s_2 - s_1)t),$$

$$\lim_{n\to+\infty} F\left(\frac{t}{n}\right)^{[n-1-ns]} = \lim_{n\to+\infty} F\left(\frac{t}{n}\right)^{[n(1-s)-\frac{1}{n}]}$$

$$= \lim_{n\to+\infty} F\left(\frac{t}{n}\right)^{[n(1-s)]}$$

$$= R((1-s)t),$$

for the strong topology. Since, moreover,

$$\lim_{n\to+\infty} \frac{n}{t} K\left(\frac{t}{n}\right) x = \lim_{n\to+\infty} \frac{n}{t} \int_0^{\frac{t}{n}} T(s)AT\left(\frac{t}{n} - s\right) S\left(\frac{t}{n}\right) x ds$$

$$= \lim_{t\downarrow 0} \int_0^t T(s)AT(t-s)S(s)x ds = Ax$$

for all $x \in \mathcal{E}$, then, for the strong topology,

$$\lim_{n\to+\infty} t \int_0^1 F\left(\frac{t}{n}\right)^{[ns]} \frac{n}{t} K\left(\frac{t}{n}\right) F\left(\frac{t}{n}\right)^{[n(1-s)]} ds =$$

$$t \int_0^1 R(st)AR((1-s)t)ds$$

$$= \int_0^t R(s)AR(t-s)ds,$$

which is the linear term in the perturbation series of R:

$$\tilde{R}(t) = R(t) + \int_0^t R(s)AR(t-s)ds +$$

$$\int\int_{t\Delta(2)} R(s_1)AR(s_2-s_1)AR(t-s_2)ds_1ds_2 + \cdots,$$

converging on $t\Delta_2 = \{(s_1, s_2) \in \mathbf{R} : 0 \leq s_1 \leq s_2 \leq t\}$ for the strong topology.

Similarly,

$$\lim_{n\to+\infty} t^2 \int\int_{\Delta_2(n)} F\left(\frac{t}{n}\right)^{[ns_1]} \frac{n}{t} K\left(\frac{t}{n}\right) F\left(\frac{t}{n}\right)^{[ns_2]-[ns_1]} \times$$

$$\frac{n}{t} K\left(\frac{t}{n}\right) F\left(\frac{t}{n}\right)^{[n-1-ns_2]} ds_1 ds_2 =$$

$$= t^2 \int\int_{\Delta_2} R(s_1 t)AR((s_2-s_1)t)AR((1-s_2)t)ds_1 ds_2$$

$$= \int\int_{t\Delta_2} R(s_1)AR(s_2-s_1)AR(t-s_2)ds_1 ds_2,$$

which is the quadratic term in the perturbation series of R.

Iteration of this procedure to higher dimensional simplices yields the conclusion. ∎

Corollary 5.4.11 *Let T, S and R satisfy the hypotheses of Theorem 5.4.10. Let $\tilde{T}$, $\tilde{S}$ and $\tilde{R}$ be the strongly continuous semigroups generated by the bounded perturbations of the infinitesimal generators of T, S and R by the operators $A \in \mathcal{L}(\mathcal{E})$, $B \in \mathcal{L}(\mathcal{E})$ and $A + B$. Then*

$$\lim_{n\to+\infty} \left(\tilde{T}\left(\frac{t}{n}\right) \tilde{S}\left(\frac{t}{n}\right)\right)^n = \tilde{R}(t)$$

for the strong topology, for all $t \in \mathbf{R}_+$.

The limit (5.39) may exist and define a strongly continuous semigroup without $X + Y$ being an infinitesimal generator. Examples exhibited by E.Nelson [60] and by P.R.Chernoff [10] (but see also [34], [22], [65]) show that, even if (5.39) holds for all $t \in \mathbf{R}_+$, $\mathcal{D}(X) \cap \mathcal{D}(Y)$ is not necessarily dense, and it can even happen that $\mathcal{D}(X) \cap \mathcal{D}(Y) = \{0\}$. If the limit (5.39) exists, the infinitesimal generator Z of R has been called by Chernoff the *generalized sum* (or *Lie sum*) of X and Y; in symbols: $Z = X +_L Y$. In the case of semigroups acting on a Hilbert space, Chernoff has shown in [10]) that, if $X +_L Y$ exists for every self-adjoint operator Y, then X is bounded.

Going back to the Lie group G considered at the beginning of this section, the one-parameter groups generated by X, Y and by $[X, Y]$ are related by the *perturbation formula*

$$\exp(t[X, Y]) =$$

$$= \lim_{n \to +\infty} \left[\exp\left(\frac{t}{n}X\right) \cdot \exp\left(\frac{t}{n}Y\right) \cdot \exp\left(-\frac{t}{n}X\right) \cdot \exp\left(-\frac{t}{n}Y\right) \right]^{n^2}.$$

This formula has been extended by J.A.Goldstein [34] to the case in which T and S are strongly continuous groups of isometries, and X and Y satisfy suitable hypotheses concerning the domain of $[X, Y]$.

Bibliography

[1] W. Arendt, A. Grabosch, G. Greiner, U. Groh, H.P. Lotz, U. Moustakas, R. Nagel (ed.), F. Neubrander, U. Schlotterbeck, One-parameter Semigroups of Positive Operators, Lecture Notes in Mathematics, n. 1184, Springer-Verlag, Berlin/Heidelberg/New York/Tokyo, 1986.

[2] W.Arveson, *On groups of automorphisms od operator algebras*, J.Functional Analysis, 15 (1974), 217-243.

[3] R.Beals, Topics in Operator Theory, University of Chicago Press, Chicago and London, 1971.

[4] A.Belleni-Morante, Applied Semigroups and Evolution Equations, Clarendon Press, Oxford, 1979.

[5] F.F.Bonsall and J.Duncan, Numerical Ranges, London Math. Lecture Note Series, I and II.

[6] Soo Bong Chae, Holomorphy and Calculus in Normed Spaces, Marcel Dekker, New York/Basel, 1985.

[7] H.Brézis, Opérateurs maximaux monotones et semi-groupes de contraction dans les espaces de Hilbert, Math. Studies, n. 5, North Holland, Amsterdam, 1973.

[8] H.Brézis, Analyse fonctionnelle. Théorie et applications, Masson, Paris, 1983.

[9] A.Browder, Introduction to Function Algebras, Benjamin, New York, 1969.

[10] P.R.Chernoff,, *Product Formulas, Nonlinear Semigroups and Addition of Unbounded Operators*, Mem. Amer.Math. Soc. n. 140 (1974).

[11] P.R.Chernoff, *Two Counterexamples in Semigroup Theory in Hilbert Space*, Proc.Amer.Math.Soc., 56 (1976), 253-255.

[12] Ph.Clement, H.J.A.M.Heijmans, S.Angenent, C.J.vanDuijn, B.deBagter, One Parameter Semigroups, North Holland, Amsterdam/New York/Oxford/Tokyo, 1987.

[13] M.G.Crandall, *Nonlinear semigroups and evolution governed by accretive operators*, Proc. Symp. Pure Math., 45, Part I, American Mathematical Society, Providence, R.I., 1986, 305-337.

[14] G.Da Prato, Applications croissantes et équations d'évolution dans les espaces de Banach, Institutiones Mathematicae, Academic Press, London/New York, 1976.

[15] E.B. Davies, One-Parameter Semigroups, Academic Press, London-New York, 1980.

[16] E.B. Davies, Spectral Theory and Differential Operators, Cambridge University Press, Cambridge, 1995.

[17] G.de Rham, Variétés différentiables. Hermann, Paris, III Ed., 1973.

[18] J.Diestel, Geometry of Banach spaces - Selected topics, Lecture Notes in Mathematics, n. 485, Springer-Verlag, Berlin/Heidelberg/New York, 1975.

[19] N.Dunford and J.T.Schwartz, Linear operators, Part I, Interscience, New York, 1958.

[20] F.J.Dyson, *The radiation theories of Tomonaga, Schwinger and Feynman*, Phys.Rev., 75 (1949), 486-502.

[21] K-J.Engel and R.Nagel, One-Parameter Semigroups for Linear Evolution Equations, Springer-Verlag, New York, 2001.

[22] W.G.Faris, *The Product Formula for Semigroups Defined by Friedrichs Extensions*, Pac.J.Math., 22 (1967), 47-70.

[23] D.E.Evans, *On the spectrum of a one-parameter strongly continuous representation*, Math.Scand., 39 (1976), 80-82.

[24] H.O.Fattorini, The Cauchy Problem, Addison-Wesley, Reading, Mass., 1983.

[25] H.O.Fattorini, Second Order Linear Differential Equations in Banach Spaces, North-Holland, Amsterdam, 1985.

[26] T.Franzoni and E.Vesentini, Holomorphic Maps and Invariant Distances, North-Holland, Amsterdam, 1980.

[27] T.W.Gamelin, Uniform algebras, Prentice-Hall, Englewood Cliffs, N.J., 1969.

[28] L. Gårding, *Note on Continuous Representations of Lie Groups*, Proc. Nat. Acad. USA, 33 (1947), 331-332.

[29] L. Gårding, *Vecteurs Analytiques dans les Représentations des Groupes de Lie*, Bull.Soc.Math.France, 88 (1960), 73-93.

[30] I.M. Gelfand, *On One Parametrical Groups of Operators in a Normed Space*, Dokl.Akad.Nauk.SSSR, N.S. 25 (1959), 713-718.

[31] G. Gentili, F. Podestà, E.Vesentini, Lezioni di geometria differenziale, Bollati Boringhieri, Torino, 1995.

[32] A.Gleason, *A Characterization of Maximal Ideals*, J.Analyse Math., 19 (1967), 171-172.

[33] S.L.Goldberg, Curvature and Homology, Academic Press, New York, 1962; Dover, New York, 1982.

[34] J.A. Goldstein, *A Lie Product Formula for One Parameter Groups of Isometries on a Banach Space*, Math. Ann., 186 (1970), 299-306.

[35] J.A. Goldstein, Semigroups of Operators and Applications, Oxford University Press, 1985.

[36] J.A. Goldstein, C.Radin and R.E.Showalter, *Convergence Rates of Ergodic Limits for Semigroups and Cosine Functions*, Semigroup Forum, 16 (1978), 89-95.

[37] T.Hawkins, *The Birth of Lie's Theory of Groups*, The Mathematical Intelligencier, 16, 2 (1994), 6-17.

[38] P.R.Halmos, A Hilbert Space Problem Book, Van Nostrand, Princeton N.J., 1967.

[39] R.Hermann, Lie Groups for Physicists, Benjamin, New York/ Amsterdam, 1966.

[40] E.Hille and R.S.Phillips, Functional Analysis and Semigroups, Amer.Math.Soc.Coll.Publ., Vol. 31, Providence R.I.,1957.

[41] V.E.Kacnelson, *A Conservative Operator has Norm Equal to its Spectral Radius*, Mat.Issled. 5, 3 (17) (1970), 186-189.

[42] J.-P.Kahane, Séries de Fourier Absolument Convergentes. Springer-Verlag, Berlin/Heidelberg/New York, 1970.

[43] J.-P.Kahane and W.Zelazko, *A Characterization of Maximal Ideals in Commutative Banach Algebras*, Studia Math. 29 (1968), 339-343.

[44] T.Kato, *On the Product of Semigroups of Operators*, Proc. Amer. Math. Soc., 10 (1959), 545-551.

[45] T.Kato, Perturbation Theory for Linear operators, Springer-Verlag, Berlin/Heidelberg/New York, 1966.

[46] I.Katznelson, An introduction to harmonic analysis, 2nd ed., Dover, New York, 1976.

[47] A.Kirillov, Eléments de la Théorie des Représentations, Mir, Moscou, 1974.

[48] M.G.Krein, Linear Differential Equations in Banach Spaces, Amer.Math.Soc.Translations, Vol. 29, Providence, R.I., 1971.

[49] S.G.Krein and M.I.Khazan, *Differential Equations in a Banach Space*, Itogi Nauki i Tekhniki Seriya Matematicheskii Analiz, 21 (1980), 130-264.

[50] G.Lumer, *Semi-inner-product Spaces*, Trans.Amer.Math.Soc., 100 (1961), 29-43.

[51] G.Lumer and R.S.Phillips, *Dissipative Operators in a Banach Space*, Pacific J. Math., 11 (1959), 679-698.

[52] A.Lunardi, Analytic Semigroups and Optimal Regularity in Parabolic Problems, Birkhäuser Verlag, Basel, 1995.

[53] P.Masani, *Ergodic Theorems for Locally Integrable Semigroups of Continuous Linear Operators on a Banach Space*, Advances in Mth., 21 (1976), 202-228.

[54] L.Nachbin, Topology in Spaces of Holomorphic Functions, Springer-Verlag, Berlin/Heidelberg/New York, 1969.

[55] B.Sz.Nagy and C.Foias, Harmonic Analysis of Operators in Hilbert Space, North Holland, Amsterdam, 1970.

[56] B.Sz.Nagy, *Spectral mapping theorems for semigroups of operators*, Acta Sci. Math. Szeged, 38 (1976), 343-351.

[57] M.Naimark and A.Stern, Théorie des Représentations des Groupes, Mir, Moscou,1979.

[58] J.van Neerven, The Asymptotic Behaviour of Semigroups of Linear Operators, Birkhäuser Verlag, Basel, 1996.

[59] E.Nelson, *Analytic Vectors*, Ann.of Math., 70(1959), 572-615.

[60] E.Nelson, Topics in Dynamics, I: Flows, Princeton University Press, Princeton, 1970.

[61] J.W.Neuberger, *Existence of a spectrum for nonlinear transformations*, Pacific J. of Math., 31 (1969), 157-159.

[62] J.Neveu, *Théorie des Semi-groupes de Markov*, Univ. of California Publ. Statistics, 2 (1958), 319-394.

[63] A. Pazy, Semigroups of Linear Operators and Applications to Partial Differential Equations, Springer-Verlag, New York/Berlin/Heidelberg/Tokyo, 1983.

[64] L. Perko, Differential Equations and Dynamical Systems, Springer-Verlag, New York, 1991.

[65] R.S.Phillips, *Perturbation theory for semi-groups of linear operators*, Trans. Amer.Math.Soc., 74 (1953), 199-221.

[66] L.Pitt, *Products of Markovian Semigroups of Operators*, Z. Warsch. Geb., 12 (1969) 241-254.

[67] F.Riesz et B.Sz.Nagy, Lecons d'Analyse Fonctionnelle, Académie des Sciences de Hongrie, 1955.

[68] H.L.Royden, Real Analysis, Macmillan, London, 1968.

[69] W.Rudin, Real and Complex Analysis, Mc-Graw Hill, New York, 1966.

[70] W.Rudin, Functional Analysis, Mc-Graw Hill, New York, 1973.

[71] Siddiqi, Can. Math. Bull., 13 (1970), 219-220.

[72] M.H.Stone, Linear Transformations in Hilbert Space, Amer. Math. Soc. Colloquium Publications, Vol. 15, Providence, R.I., 1932.

[73] H.Trotter, *Approximations of Semigroups of Operators*, Pacific J. Math., 8 (1958), 887-919.

[74] H. Trotter, *On the Product of Semigroups of Operators*, Proc. Amer. Math. Soc., 10 (1959), 545-551.

[75] E. Vesentini, *Semigruppi Fortemente Continui in Algebre di Banach ed in Sistemi di Spin*, Rend. Sem. Mat. Fis. Milano, 60 (1990), 157-165.

[76] E. Vesentini, Funzioni Olomorfe in Spazi di Banach e Applicazioni alla Teoria Spettrale, SISSA, Trieste, 1991.

[77] E. Vesentini, *Semigroups of Holomorphic Isometries*, in : P.M. Gauthier and G. Sabidussi (ed.), Complex potential theory, Kluwer Academic Publishers, Dordrecht/Boston/London, 1994, 475-548.

[78] E. Vesentini, *Conservative Operators*, in : P. Marcellini, G. Talenti and E. Vesentini (ed.) "Topics in Partial Differential Equations and Applications", Marcel Dekker, New York/Basel/Hong Kong, 1996.

[79] E. Vesentini, *Conservative perturbations of conservative operators*, Rend.Circ. Mat.Palermo, (2) 47 (1998), 353-362.

[80] I.M.Yaglom, Felix Klein and Sophus Lie. Evolution of the Idea of Symmetry in the Nineteenth Century, Birkhäuser, Boston/Basel, 1988.

[81] K.Yosida, Functional Analysis, Second Edition, Springer-Verlag, Berlin/Heidelberg/New York, 1968.

[82] J.Zabczyk, *A Note on C_o Semigroups*, Bull.Acad.Polon.Sci., 23 (1975), 895-898.

[83] E.H.Zarantonello, *Conical Spectral Theory*, in: F.Strocchi, E.Zarantonello, E.De Giorgi, G.Dal Maso-L.Modica, Topics in Functional Analysis 1980-81, Scuola Normale Superiore, Pisa, 1981; 37-116.

[84] W.Zelazko, *A Characterization of Multiplicative Linear Functionals in Complex Banach Algebras*, Studia Math., 30 (1968), 83-85.

Elenco dei volumi della collana
"Appunti"
pubblicati dall'Anno Accademico 1994/95

SAURO SUCCI, *An Introduction to Computational Physics. Part I: Grid Methods,* 2002

DORIN BUCUR, GIUSEPPE BUTTAZZO, *Variational Methods in Some Shape Optimization Problems,* 2002

EDOARDO VESENTINI, *Introduction to continuous semigroups,* 2002.

ANNA MINGUZZI, MARIO TOSI, *Introduction to the Theory of Many-Body Systems,* 2002.

Fotocomposizione "CompoMat" Loc. Braccone, 02040 Configni (RI), Italy
Finito di stampare per conto della "CompoMat" dalla Nuova Grafica 86 nel gennaio 2003